R.M. Mikulin

ELIZABETH PEET gained a degree in Mathematics from the University of Essex and subsequently a Postgraduate Certificate of Education at the University of London. She went on to teach at an outer London comprehensive where she was in charge of the Mathematics Department. She now lectures in Mathematics at a College of Further Education: Kilburn Polytechnic.

key facts

GCE A-Level Passbooks

BIOLOGY, H. Rapson, B.Sc.

CHEMISTRY, J. E. Chandler, B.Sc. and R. J. Wilkinson, Ph.D.

ECONOMICS, R. Maile, B.A.

GEOGRAPHY, R. Bryant, B.A., R. Knowles, M.A. and J. Wareing, B.A., M.Sc.

PHYSICS, J. R. Garrood, M.A., Ph.D.

PURE MATHEMATICS, R. A. Parsons, B.Sc. and A. G. Dawson, B.Sc.

PURE AND APPLIED MATHEMATICS, R. A. Parsons, B.Sc. and A. G. Dawson, B.Sc.

GCE A-Level Passbook

Applied Mathematics

E. M. Peet, B.A.

key facts

Published by
Charles Letts and Company Limited
London, Edinburgh and New York

First published 1980 by Intercontinental Book Productions

Published 1982 by Charles Letts & Co Ltd
Diary House, Borough Road, London SE1 1DW

Reprinted 1983
1st edition 4th impression

Made and printed by Charles Letts (Scotland) Ltd
ISBN 0 85097 394 5

Contents

Introduction

The contents of this book cover the majority of topics listed by the syllabuses of the various 'A' level Examination Boards. Each chapter contains the theory of the topic under consideration followed by worked examples of a typical 'A' level standard.

An adequate preparation for the 'A' level examination will include the working of many past examination questions and students are advised to obtain and work through the past papers of the relevant Board.

Where questions have been reproduced verbatim from the past papers of 'A' level Examination Boards, they have been acknowledged by the appropriate initials of the Board:

A.E.B. .... The Associated Examining Board;
S.U.J.B. .... Southern Universities' Joint Board;
W.J.E.C. .... Welsh Joint Education Committee.

Chapter 1
Mechanics: an Introduction

Newton's laws of motion are the basis of Theoretical Mechanics at this level. They are hypotheses and have been successfully confirmed at a terrestrial level. We use them to form a mathematical model of a physical situation and then apply Pure Mathematics to this model.

Conventions

In order to simplify these models, we adopt certain conventions concerning certain quantities whose effect on a situation is so minimal that they may be ignored. These are as follows: a **light** string or wire is considered to be **weightless**; a **smooth** surface is **without friction**; **air resistance is ignored** unless specifically stated otherwise; a **particle** is without size or shape and has its **mass concentrated at a point**; a **lamina** is without thickness having the **dimensions of area only**; a **hollow shell** is **without thickness**; a **thin** wire has **no thickness**.

Units

We use the **SI system of units** throughout this book. The **three basic quantities** are **mass**, measured in kilograms, **length**, measured in metres and **time**, measured in seconds. Every other quantity is derived from these basic ones and will be defined as needed.

Scalars and vectors

We need to distinguish between two different kinds of quantities: **scalars** and **vectors**. A **scalar** has **magnitude only**; a **vector** has **magnitude and direction**. **Speed** is a **scalar**, whereas **velocity** is a **vector**. The speed of a car tells us how fast it is travelling; the velocity tells us not only how fast it is going, but also the direction in which it is travelling. When, in the text, we wish to distinguish between vectors and scalars, we shall denote **vectors** in **bold roman type**, and scalars in italic type, e.g., velocity **v**, speed v (or ***v***).

Newton's laws of motion and some necessary definitions

Mechanics is about the effect that forces have on bodies and the analysis of the motion produced.

Newton's first law states that a body continues in a state of rest or of uniform motion in a straight line unless acted on by an external force.

A **force**, then, is that quantity which changes or tends to change the existing state of rest or of uniform velocity of a body. The **unit of force** is the **Newton** (N) and is defined to be that magnitude of force that is needed to give a mass of 1 kg an acceleration of 1 ms^{-2}.

A **system** is a collection of particles and/or bodies and the forces acting between them. An **external force acts from outside this system**.

The **displacement**, **s**, of a body is its distance from a fixed point in a certain direction and thus is a **vector**. The **distance**, *s*, of a body from this point is merely how far it is away from it, with no account being taken of direction, and thus is a **scalar**. The **velocity**, **v**, of a body is its rate of change of displacement, and thus is a **vector**, its size being measured in ms^{-1}. The **acceleration**, **a**, of a body is the rate of change of velocity, also a **vector**, with its size measured in ms^{-2}. Thus,

$$\mathbf{v} = \frac{d\mathbf{s}}{dt} \text{ and } \mathbf{a} = \frac{d\mathbf{v}}{dt}$$

The **momentum** of a body is defined to be the **product of its mass and its velocity** and is measured in kg ms^{-1}. It is a **vector**; **momentum** = $m\mathbf{v}$ kg ms^{-1}, in the **direction of the velocity**.

Newton's second law states that the rate of change of momentum of a body is proportional to the force acting on it and takes place in the direction of that force, i.e.,

$$\frac{d}{dt}(m\mathbf{v}) = k\mathbf{f} \text{ giving } m\frac{d\mathbf{v}}{dt} = k\mathbf{f}$$

for constant mass where k is the constant of proportionality.

This constant of proportionality, k, is 1 in the SI system of units and hence we can write Newton's second law as $\mathbf{f} = m\mathbf{a}$ which is a fundamental equation of mechanics.

It follows that unless an external force is acting on a system then the momentum of that system will remain constant. This is known as the **Law of Conservation of Linear Momentum**.

If a body is dropped it will fall to the ground. From this we can deduce that a force is acting on it, otherwise it would remain in mid-air (Newton's first law). We call this force its **gravitational force**, or its **weight**, $\mathbf{w}$. If we define the acceleration due to gravity to be $\mathbf{g}$, then from Newton's second law, $\mathbf{w} = m\mathbf{g}$.

Newton's third law states that to every action there is an equal and opposite reaction.

Action and **reaction** are the names given to those forces which act in pairs within a system, balancing each other out, e.g., if a book is resting on a table and not falling to the ground under the **action** of its own weight, then it must be due to the **reaction** of the table to it, balancing it out.

Friction is the name given to that **force** which comes into play **between surfaces in contact** which acts in such a way as to **oppose potential and actual motion**.

In general, each of the forces, action and reaction, between two surfaces in contact may be considered to be made up of two perpendicular parts (called components): one is **normal (perpendicular)** to the reacting surface and called the **normal reaction**; the other is **parallel** to the reacting surface and is called **friction**. If the contact is **smooth**, then there is no frictional force and the only reaction is the **normal reaction**.

Chapter 2
Vectors

A **vector** has both **magnitude** and **direction**. It can be represented completely by a **line segment** where the direction is that of the vector and the length is proportional to the magnitude of the vector. The **magnitude**, or **size** of a vector, **v**, is written as $|\mathbf{v}|$. It is also called the **modulus** of the vector, and is of course a **scalar**. A velocity vector, **v**, 20 ms^{-1} due south, could be represented by:

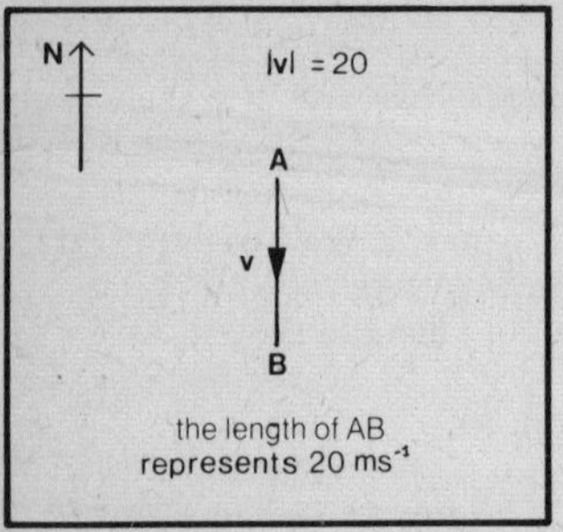

Figure 1.

Alternative notation for **v** is **AB** where the order of the letters indicates the direction.

Tied vectors

A **tied vector** is such that the magnitude and direction of the vector is not sufficient to specify it completely: more information about its particular location(s) in space must also be given. For example, in order to predict the result of applying a force to a body, we need to know its line of action. **Force** is a **tied, line-localised vector**. Displacement, velocity, acceleration, are **free vectors**: they are completely described by their magnitude and direction.

Equal vectors

Two (free) vectors of **equal magnitude** and the **same direction** are said to be **equal**. Hence a (free) vector may be represented by any one of an **infinite number of parallel line segments, equal in length**.

A **tied, line-localised** vector may be represented by any one of an infinite number of line segments equal in length as long as they are along the same line and in the same sense.

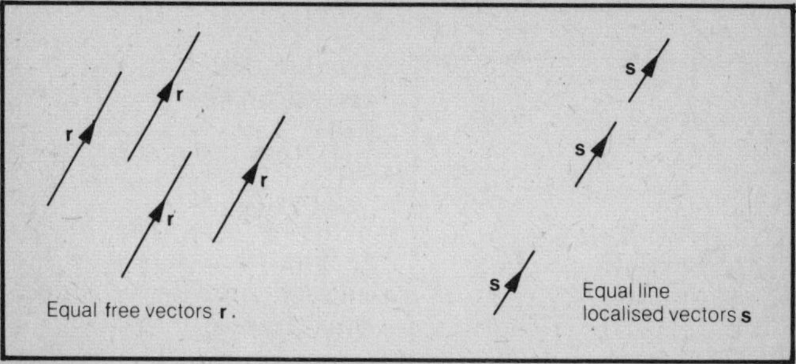

Figure 2.

Unless a vector, other than a force which is always line-localised, is specifically stated to be **tied**, we assume it to be a **free** vector.

Addition of vectors

The **addition of vectors** is defined by the **Triangle Law** which states that if two vectors **a** and **b** are **represented completely** by the sides **AB** and **BC** respectively of a triangle *ABC*, then their **resultant vector c** is **represented completely** by the side **AC**. We say that **AC** is the same as, or **equivalent** to, (**AB** + **BC**).

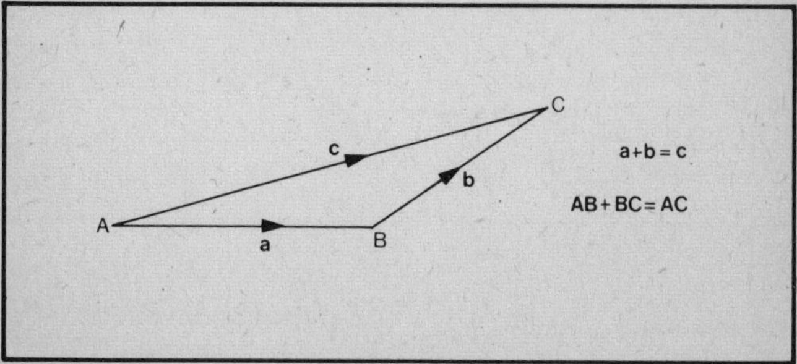

Figure 3.

The Parallelogram Law also defines vector addition. It states that if two vectors **a** and **b** are **represented completely** by the sides **AB**, **AD** respectively of a parallelogram then they are **equivalent** to, or their **resultant** is, **c**, the vector **represented completely** by the diagonal **AC**.

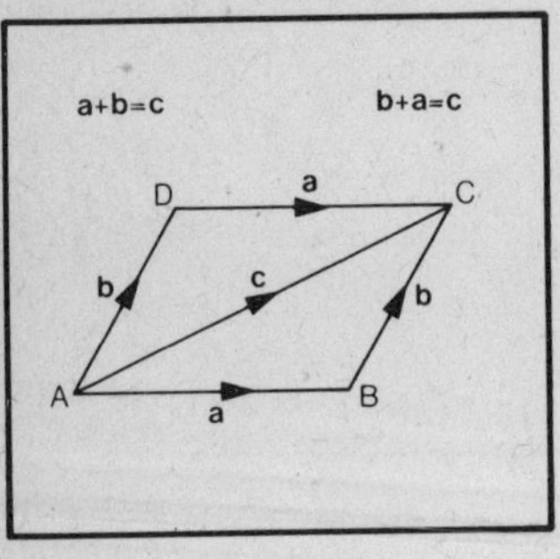

Figure 4.

We can see that this is in agreement with the **Triangle Rule** since **AD** = **BC**, then **AB** + **AD** = **AB** + **BC** = **AC**.

Both of these rules for the addition of vectors show how **vector addition** is **commutative** i.e.,

a + **b** = **b** + **a**.

These rules generalise to the addition of any number of vectors by the **Vector Polygon Rule**. This states that if vectors **a**, **b**, **c**, **d**, **e**, ... are represented completely by the sides **AB**, **BC**, **CD**, **DE**, **EF**, ... of a polygon *ABCDEF*... then their **resultant vector** is **represented completely** by the **side of the Polygon needed to close it**.

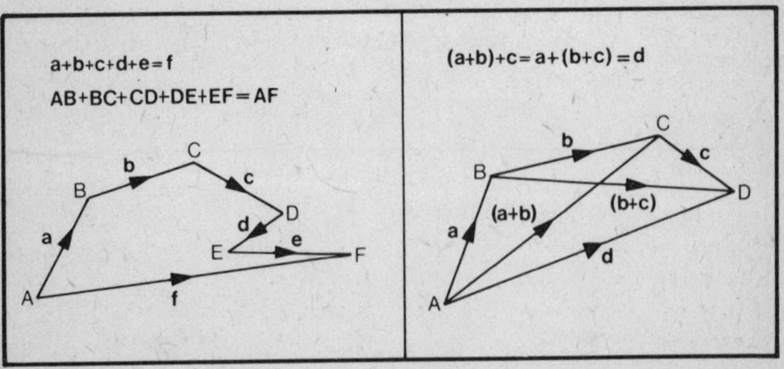

Figure 5.

We say that **a**, **b**, **c**, **d**, **e**, are **components** of vector **f**. It is clear from figure 5 that **vector addition** is **associative** i.e.,

(**a** + **b**) + **c** = **a** + (**b** + **c**).

Perpendicular Components of a Vector

It is often very useful to consider a vector as being the **resultant of two perpendicular components**. If we consider a vector **u** inclined at an angle θ to the horizontal then we can see that it is also the **resultant** of vectors **u cos** θ in the **horizontal** direction and **u sin** θ in the **vertical** direction.

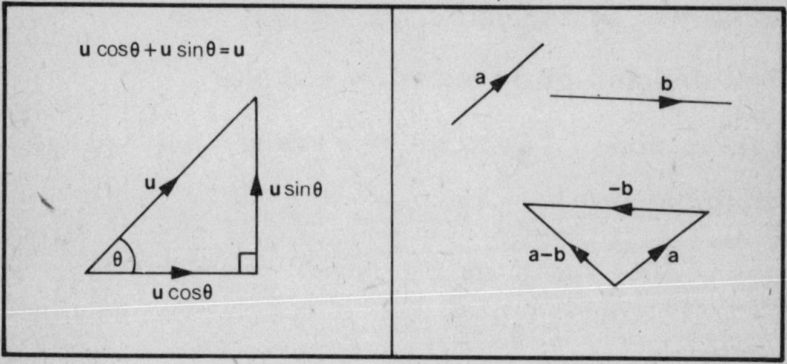

Figure 6.

The null vector

A **null vector** has **zero magnitude** and is written as **0**. When it is added to any other vector the vector is unchanged. **a** + **0** = **a**.

Negative vectors

A **negative** sign in vector work indicates a **reversal of sense**; e.g., **AB** = −**BA**.

Subtraction of vectors

The **subtraction of vectors** is defined as the **addition of the negative vector** i.e., **AB** − **BC** = **AB** + (−**BC**) = **AB** + **CB**. In particular, **AB** − **AB** = **AB** + (−**BA**) = **0**.

Unit vectors

Unit vectors have **magnitude one**, or **unity**. They are often used to specify direction, as we shall see later. The **unit vector** in the **same direction** as **a** is written as **â**.

Multiplication of a vector by a scalar

$\mathbf{a} + \mathbf{a}$ will give us a vector in the **same direction** as $\mathbf{a}$, but with **twice the magnitude**. We can write this as $2\mathbf{a}$. The vector $\lambda\mathbf{a}$, where λ is a scalar, is in the **same direction** as $\mathbf{a}$ if λ is **positive**, and in the **opposite direction** to $\mathbf{a}$ if λ is **negative**. $\lambda\mathbf{a}$ has magnitude $|\lambda\mathbf{a}| = |\lambda||a|$ (i.e., $\lambda|\mathbf{a}|$ if $\lambda > 0$ and $-\lambda|\mathbf{a}|$ if $\lambda < 0$). The distributive law holds for **multiplication of vectors by a scalar** i.e., $\lambda(\mathbf{a} + \mathbf{b}) = \lambda\mathbf{a} + \lambda\mathbf{b}$.

The division of a vector by a scalar

The **division of a vector a by a scalar** λ is defined as the **multiplication** by $\frac{1}{\lambda}$ i.e., $\frac{\mathbf{a}}{\lambda} = \frac{1}{\lambda}\mathbf{a}$.

If two vectors $\mathbf{a}$ and $\mathbf{b}$ are in the same direction then

$$\hat{\mathbf{a}} = \hat{\mathbf{b}} = \frac{\mathbf{a}}{|\mathbf{a}|} = \frac{\mathbf{b}}{|\mathbf{b}|}.$$

If two vectors $\mathbf{a}$ and $\mathbf{b}$ are parallel (but not necessarily in the same sense), then $\mathbf{a} = \lambda\mathbf{b}$.

Resultant vector theorem

If two vectors may be represented by $\lambda\mathbf{OA}$ and $\mu\mathbf{OB}$, their resultant is represented by the vector $(\lambda + \mu)\mathbf{OC}$ where the point C on AB is such that $AC : CB = \mu : \lambda$.

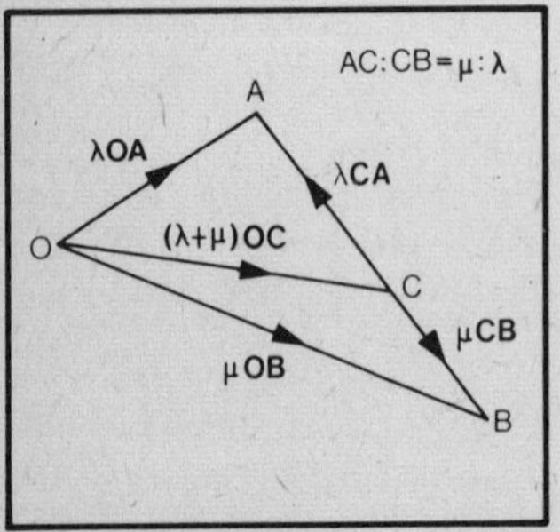

Figure 7.

$\lambda\mathbf{OA} = \lambda\mathbf{OC} + \lambda\mathbf{CA}$, and
$\mu\mathbf{OB} = \mu\mathbf{OC} + \mu\mathbf{CB}$

Adding these gives:

$$\lambda\mathbf{OA} + \mu\mathbf{OB} = (\lambda + \mu)\mathbf{OC} + \lambda\mathbf{CA} + \mu\mathbf{CB},$$

but $\lambda\mathbf{CA} + \mu\mathbf{CB} = \mathbf{0}$ since $AC : CB = \mu : \lambda$ and $\mathbf{CA}, \mathbf{CB}$ are in opposite directions.

Hence

$$\lambda\mathbf{OA} + \mu\mathbf{OB} = (\lambda + \mu)\mathbf{OC}.$$

Coplanar vectors

Coplanar vectors are all **parallel to the same plane. Coplanar forces** are in the **same plane** since **force is a tied line-localized vector.**

Collinear vectors

Collinear vectors are all **parallel to the same line. Collinear forces** have the **same line of action** since **force is a tied line-localized vector.**

Cartesian components of a vector

It is often convenient to consider a vector as being the **resultant of its components** in the ***Ox* and *Oy* directions** in the case of a vector in **two dimensions**, and in the case of a vector in **three dimensions**, its components in the ***Ox*, *Oy*, *Oz*, directions**. In general, a vector can be expressed as the **resultant of scalar multiples** of any **three non-coplanar vectors**. However we shall limit this discussion to **Cartesian components. Particular letters** are given to the **unit vectors** in each of the three directions: **i** is the **unit vector** in the ***Ox* direction**; **j** is the **unit vector** in the ***Oy* direction**; **k** is the **unit vector** in the ***Oz* direction**.

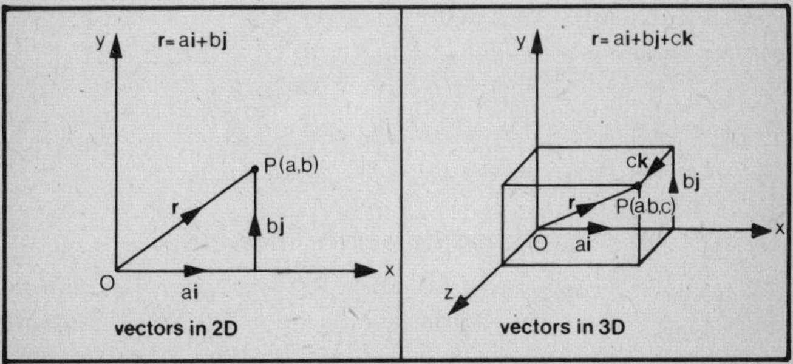

Figure 8.

As we can see from figure 8, a vector which may be represented by the line **OP** where P has coordinates (a, b, c) in the Cartesian system of reference, may be expressed as the resultant of vectors $a\mathbf{i}$, $b\mathbf{j}$, and $c\mathbf{k}$; i.e., $\mathbf{r} = a\mathbf{i} + b\mathbf{j} + c\mathbf{k}$. The length OP is easily calculated from the coordinates, i.e., $OP = |\mathbf{r}| = \sqrt{a^2 + b^2 + c^2}$. In two dimensions $c = 0$, thus $|\mathbf{r}| = \sqrt{a^2 + b^2}$.

To avoid confusion we use the **right-handed system of reference directions** i.e., the **i** and **j** directions are drawn first, perpendicular to each other, the **j** direction usually being upwards. The **k** direction is then taken to be perpendicular to the plane containing the **i** and **j** vectors, pointing in the direction that a corkscrew would move when the handle is wound from the positive **i** direction to the positive **j** direction in one 90° turn.

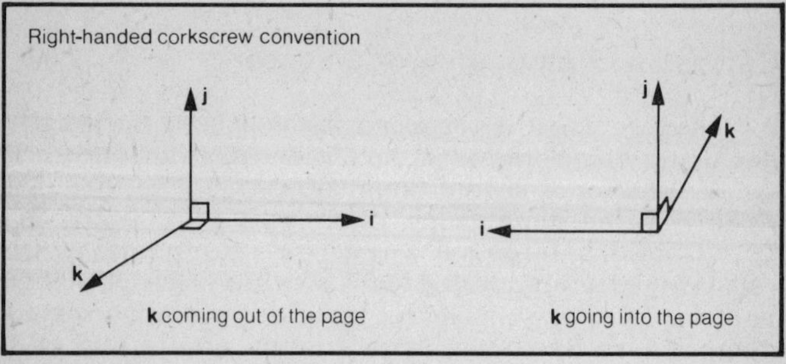

Figure 9.

When vectors are represented in their cartesian form, their **addition** becomes very straightforward.

If $\mathbf{r}_1 = a_1\,\mathbf{i} + b_1\,\mathbf{j} + c_1\,\mathbf{k}$ and $\mathbf{r}_2 = a_2\,\mathbf{i} + b_2\,\mathbf{j} + c_2\,\mathbf{k}$ then

$$\begin{aligned}\mathbf{r}_1 + \mathbf{r}_2 &= a_1\,\mathbf{i} + b_1\,\mathbf{j} + c_1\,\mathbf{k} + a_2\,\mathbf{i} + b_2\,\mathbf{j} + c_2\,\mathbf{k}\\ &= (a_1 + a_2)\mathbf{i} + (b_1 + b_2)\mathbf{j} + (c_1 + c_2)\mathbf{k}.\end{aligned}$$

In general, if $\mathbf{r}_n = a_n\,\mathbf{i} + b_n\,\mathbf{j} + c_n\,\mathbf{k}$, then

$$\sum \mathbf{r}_n = (\sum a_n)\mathbf{i} + (\sum b_n)\mathbf{j} + (\sum c_n)\mathbf{k}.$$

Direction ratios and direction cosines

If a vector **r** is represented by the line **OP** where P has coordinates (a, b, c) then the **ratios** $a : b : c$ are called the **direction ratios** of **r**. However $P(-a, \ -b, \ -c)$ would give the same direction ratios for OP, i.e., **vectors with equal direction ratios are parallel**, but we do not know their sense.

The direction cosines are

$$\cos\alpha = \frac{a}{|\mathbf{r}|}, \ \cos\beta = \frac{b}{|\mathbf{r}|}, \ \cos\gamma = \frac{c}{|\mathbf{r}|},$$

and are **unique for like parallel vectors. Unlike parallel vectors** have **direction cosines equal in magnitude**, but of the **opposite sign**. Negative direction cosines indicate that the angle made by OP with the positive direction of the appropriate axis is greater than 90°. The **direction cosines** are often given the letters l, m, n, where $l = \cos\alpha$, $m = \cos\beta$, $n = \cos\gamma$.

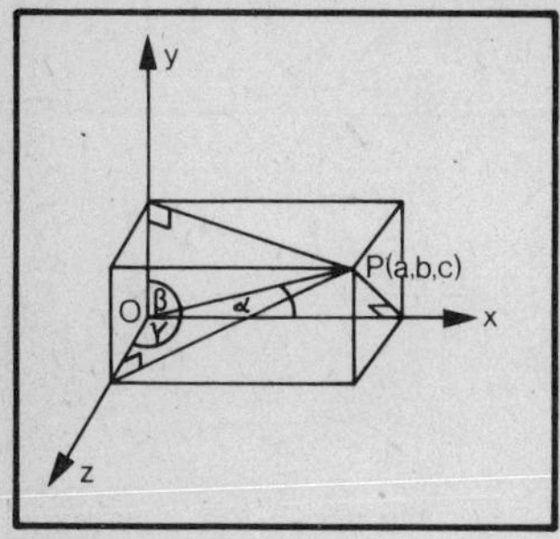

Figure 10.

We can see that:

$$l^2 + m^2 + n^2 = \frac{a^2}{|\mathbf{r}|^2} + \frac{b^2}{|\mathbf{r}|^2} + \frac{c^2}{|\mathbf{r}|^2}$$

$$= \frac{a^2 + b^2 + c^2}{|\mathbf{r}|^2} = \frac{|\mathbf{r}|^2}{|\mathbf{r}|^2} = 1.$$

i.e., **the sum of the squares of the direction cosines of any vector is always unity.**

Direction vectors

A vector which is used to **specify the direction of another vector** is called a **direction vector**. An obvious and useful direction vector is the **unit vector** in the direction which needs to be specified. **Any vector may be expressed as the product of its magnitude and the unit vector in its direction.**

If $\mathbf{r} = a\mathbf{i} + b\mathbf{j} + c\mathbf{k}$, then $\hat{\mathbf{r}} = \dfrac{\mathbf{r}}{|\mathbf{r}|} = \dfrac{a\mathbf{i} + b\mathbf{j} + c\mathbf{k}}{\sqrt{a^2 + b^2 + c^2}}$

i.e., $\hat{\mathbf{r}} = l\mathbf{i} + m\mathbf{j} + n\mathbf{k}$, and $\mathbf{r} = |\mathbf{r}|\hat{\mathbf{r}}$.

Multiplication of vectors

The only multiplication of vectors we have considered so far is the multiplication of a vector by a scalar. We shall now consider ways of 'multiplying' two vectors together. Clearly any idea of multiplying vectors in a similar way to numbers would be meaningless; what physical quantity would be the result of multiplying 20 ms^{-1} due east by 30 ms^{-1} due north? We must look for alternative ideas of the product of two vectors.

Scalar product (or 'dot' product) of two vectors

The **scalar product** of two vectors **a** and **b** is written as **a . b** and is defined to be $|\mathbf{a}||\mathbf{b}|\cos\theta$ where θ is the angle between the vectors. It is a scalar quantity, hence the name. The scalar product is **commutative**, i.e., **a . b** = **b . a** since $\mathbf{a} \cdot \mathbf{b} = |\mathbf{a}||\mathbf{b}| \cos\theta = |\mathbf{b}||\mathbf{a}| \cos\theta = \mathbf{b} \cdot \mathbf{a}$.

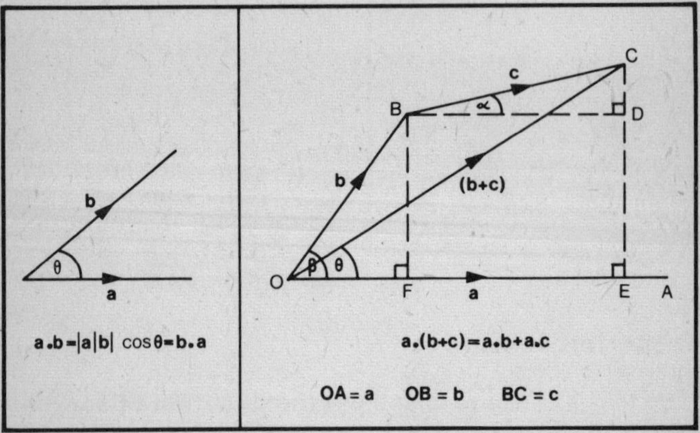

Figure 11.

Another property of the scalar product is that it is **distributive over addition** i.e., $\mathbf{a} \cdot (\mathbf{b} + \mathbf{c}) = \mathbf{a} \cdot \mathbf{b} + \mathbf{a} \cdot \mathbf{c}$. From figure 11, we see that:

$$\mathbf{a} \cdot (\mathbf{b} + \mathbf{c}) = (OA)(OC)\cos\theta = (OA)(OE)$$

Also, $\mathbf{a} \cdot \mathbf{b} = (OA)(OB)\cos\beta = (OA)(OF)$ and

$$\mathbf{a} \cdot \mathbf{c} = (OA)(BC)\cos\alpha = (OA)(BD) = (OA)(FE),$$

so that

$$\mathbf{a} \cdot \mathbf{b} + \mathbf{a} \cdot \mathbf{c} = (OA)(OF) + (OA)(FE) = (OA)(OF + FE) = (OA)(OE)$$

Hence $\mathbf{a} \cdot (\mathbf{b} + \mathbf{c}) = \mathbf{a} \cdot \mathbf{b} + \mathbf{a} \cdot \mathbf{c}$

If two vectors are **parallel** the angle between them is zero, thus $\mathbf{a} \cdot \mathbf{b} = |\mathbf{a}||\mathbf{b}|$. If two vectors are perpendicular the angle between them is 90°, thus $\mathbf{a} \cdot \mathbf{b} = 0$. The scalar product of a vector, **a**, with itself gives $|\mathbf{a}|^2$.
N.B., $\mathbf{i} \cdot \mathbf{i} = \mathbf{j} \cdot \mathbf{j} = \mathbf{k} \cdot \mathbf{k} = 1$ and $\mathbf{i} \cdot \mathbf{j} = \mathbf{j} \cdot \mathbf{i} = \mathbf{i} \cdot \mathbf{k} = \mathbf{k} \cdot \mathbf{i} = \mathbf{j} \cdot \mathbf{k} = \mathbf{k} \cdot \mathbf{j} = 0$.

Calculation of the scalar product

Let $\mathbf{r}_1 = a_1\,\mathbf{i} + b_1\,\mathbf{j} + c_1\,\mathbf{k}$ and $\mathbf{r}_2 = a_2\,\mathbf{i} + b_2\,\mathbf{j} + c_2\,\mathbf{k}$ so that

$$\begin{aligned}\mathbf{r}_1 \cdot \mathbf{r}_2 &= (a_1\mathbf{i} + b_1\mathbf{j} + c_1\mathbf{k}) \cdot (a_2\,\mathbf{i} + b_2\,\mathbf{j} + c_2\,\mathbf{k}) \\ &= a_1\mathbf{i} \cdot a_2\,\mathbf{i} + a_1\mathbf{i} \cdot b_2\,\mathbf{j} + a_1\mathbf{i} \cdot c_2\,\mathbf{k} \\ &\quad + b_1\mathbf{j} \cdot a_2\,\mathbf{i} + b_1\mathbf{j} \cdot b_2\,\mathbf{j} + b_1\mathbf{j} + c_2\,\mathbf{k} \\ &\quad + c_1\mathbf{k} \cdot a_2\,\mathbf{i} + c_1\mathbf{k} \cdot b_2\,\mathbf{j} + c_1\mathbf{k} \cdot c_2\,\mathbf{k} \\ &= a_1a_2\,\mathbf{i} \cdot \mathbf{i} + a_1b_2\,\mathbf{i} \cdot \mathbf{j} + a_1c_2\,\mathbf{i} \cdot \mathbf{k} \\ &\quad + b_1a_2\,\mathbf{j} \cdot \mathbf{i} + b_1b_2\,\mathbf{j} \cdot \mathbf{j} + b_1c_2\,\mathbf{j} \cdot \mathbf{k} \\ &\quad + c_1a_2\,\mathbf{k} \cdot \mathbf{i} + c_1b_2\,\mathbf{k} \cdot \mathbf{j} + c_1 \cdot c_2\,\mathbf{k} \cdot \mathbf{k}\end{aligned}$$

but $\mathbf{i} \cdot \mathbf{i} = \mathbf{j} \cdot \mathbf{j} = \mathbf{k} \cdot \mathbf{k}$ and $\mathbf{i} \cdot \mathbf{j} = \mathbf{i} \cdot \mathbf{k} = \mathbf{j} \cdot \mathbf{k} = \mathbf{k} \cdot \mathbf{j} = \mathbf{k} \cdot \mathbf{i} = \mathbf{j} \cdot \mathbf{i} = 0$. Therefore, $\mathbf{r}_1 \cdot \mathbf{r}_2 = a_1a_2 + b_1b_2 + c_1c_2$.

Calculation of the angle θ between two vectors $\mathbf{r}_1$ and $\mathbf{r}_2$

We know that $\mathbf{r}_1 \cdot \mathbf{r}_2 = |\mathbf{r}_1||\mathbf{r}_2| \cos\theta$, therefore

$$\cos\theta = \frac{\mathbf{r}_1 \cdot \mathbf{r}_2}{|\mathbf{r}_1||\mathbf{r}_2|} = \frac{a_1a_2}{|\mathbf{r}_1||\mathbf{r}_2|} + \frac{b_1b_2}{|\mathbf{r}_1||\mathbf{r}_2|} + \frac{c_1c_2}{|\mathbf{r}_1||\mathbf{r}_2|},$$

but $\dfrac{a_1}{|\mathbf{r}_1|}, \dfrac{b_1}{|\mathbf{r}_1|}, \dfrac{c_1}{|\mathbf{r}_1|}$ are the **direction cosines** l_1, m_1, n_1, of $\mathbf{r}_1$

and $\dfrac{a_2}{|\mathbf{r}_2|}, \dfrac{b_2}{|\mathbf{r}_2|}, \dfrac{c_2}{|\mathbf{r}_2|}$ are the **direction cosines** l_2, m_2, n_2 of $\mathbf{r}_2$

i.e., $\boldsymbol{\cos\theta = l_1l_2 + m_1m_2 + n_1n_2}$. This is the scalar product of the unit vectors in directions $\mathbf{r}_1$ and $\mathbf{r}_2$ i.e., $\hat{\mathbf{r}}_1 \cdot \hat{\mathbf{r}}_2 = \cos\theta = l_1l_2 + m_1m_2 + n_1n_2$.

Vector product (or 'cross' product) of two vectors

The **vector product** of two vectors **a** and **b** which are inclined at an angle θ is written $\mathbf{a} \times \mathbf{b}$ (or $\mathbf{a} \wedge \mathbf{b}$) and is defined as $\mathbf{a} \times \mathbf{b} = |\mathbf{a}||\mathbf{b}| \sin\theta\,\hat{\mathbf{n}}$, where $\hat{\mathbf{n}}$ is the unit vector perpendicular to the plane containing **a** and **b** in the direction of a **right-handed corkscrew** turned from **a** to **b** (see page 18). $\mathbf{a} \times \mathbf{b}$ is a **vector**, hence the name.

As we can see from figure 12, the **vector product** is **not commutative** i.e. $\mathbf{a} \times \mathbf{b}$ is **not equal** to $\mathbf{b} \times \mathbf{a}$ but $\mathbf{a} \times \mathbf{b} = -\mathbf{b} \times \mathbf{a}$.

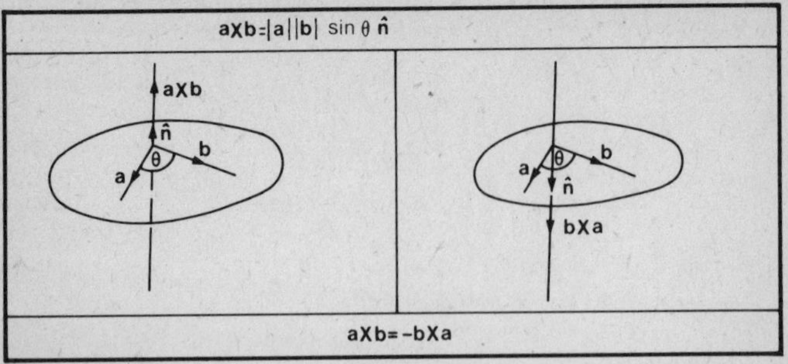

Figure 12.

If vectors **a** and **b** are **parallel** the angle between them is 0°, thus $\mathbf{a} \times \mathbf{b} = \mathbf{0}$. If the vectors are **perpendicular** to each other the angle between them is 90°, thus $\mathbf{a} \times \mathbf{b} = |\mathbf{a}||\mathbf{b}|\hat{\mathbf{n}}$, where $\hat{\mathbf{n}}$ is the unit vector in the direction of a right-handed screw from **a** to **b**; i.e., if **a** and **b** are perpendicular to each other then **a**, **b**, and $\mathbf{a} \times \mathbf{b}$ form a **right-handed set of three mutually perpendicular vectors**. This result is particularly important in the case of the unit vectors **i**, **j**, **k**: $\mathbf{i} \times \mathbf{j} = \mathbf{k}$, $\mathbf{j} \times \mathbf{k} = \mathbf{i}$, $\mathbf{k} \times \mathbf{i} = \mathbf{j}$; $\mathbf{k} \times \mathbf{j} = -\mathbf{i}$, $\mathbf{j} \times \mathbf{i} = -\mathbf{k}$, $\mathbf{i} \times \mathbf{k} = -\mathbf{j}$; $\mathbf{i} \times \mathbf{i} = \mathbf{j} \times \mathbf{j} = \mathbf{k} \times \mathbf{k} = \mathbf{0}$.

The scalar triple product

Consider a parallelepiped with vectors **a**, **b**, **c** as sides as shown in the diagram below. θ is the angle between **c** and the perpendicular height **h** of the parallelepiped.

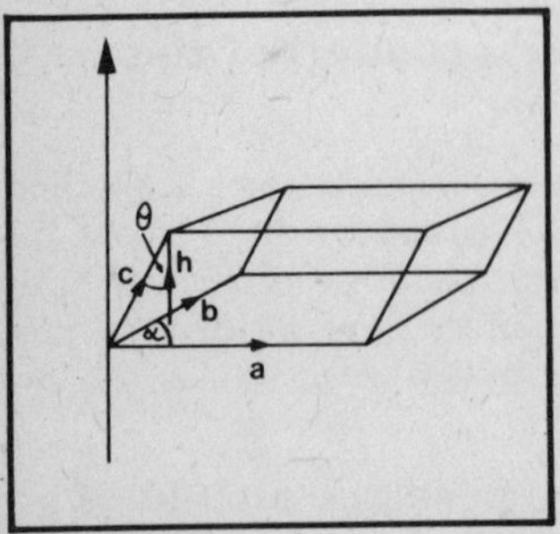

Figure 13.

The volume of the parallelepiped

$= \text{base area} \times \text{height}$
$= |\mathbf{a}||\mathbf{b}| \sin \alpha |\mathbf{h}|$
$= |\mathbf{a} \times \mathbf{b}||\mathbf{h}|$
$= |\mathbf{a} \times \mathbf{b}||\mathbf{c}| \cos \theta$

but since **h** is perpendicular to the plane containing **a** and **b**, θ is also the angle between $\mathbf{a} \times \mathbf{b}$ and **c**, i.e., the volume may be written as $\mathbf{a} \times \mathbf{b} \cdot \mathbf{c}$.

There is no confusion in this form as to which operation to do first since $\mathbf{a} \times (\mathbf{b} \cdot \mathbf{c})$ would involve the vector product of a vector with a scalar which is meaningless. The order is obviously $(\mathbf{a} \times \mathbf{b}) \cdot \mathbf{c}$ and the brackets are unnecessary.

The volume of the parallelepiped could also have been evaluated in a similar manner by $\mathbf{a} \cdot \mathbf{b} \times \mathbf{c}$. This demonstrates a useful property, called the **triple scalar product property**. This property is, that when three vectors are to be combined with a **vector product** and a **scalar product**, the **order of the products may be interchanged**, i.e., $\mathbf{a} \times \mathbf{b} \cdot \mathbf{c} = \mathbf{a} \cdot \mathbf{b} \times \mathbf{c}$, and the product is unaffected by the **cyclic** interchange of the letters.

i.e., $(\mathbf{a} \times \mathbf{b} \cdot \mathbf{c}) = (\mathbf{c} \times \mathbf{a} \cdot \mathbf{b}) = (\mathbf{b} \times \mathbf{c} \cdot \mathbf{a})$.

The **vector product** is **distributive over addition** i.e. $\mathbf{a} \times (\mathbf{b} + \mathbf{c}) = \mathbf{a} \times \mathbf{b} + \mathbf{a} \times \mathbf{c}$. To show this we shall need to use the **triple scalar product property**. Consider

$\mathbf{p} \cdot (\mathbf{a} \times \mathbf{b} + \mathbf{a} \times \mathbf{c}) = \mathbf{p} \cdot \mathbf{a} \times \mathbf{b} + \mathbf{p} \cdot \mathbf{a} \times \mathbf{c}$ (since the scalar product is distributive over addition)

$= \mathbf{p} \times \mathbf{a} \cdot \mathbf{b} + \mathbf{p} \times \mathbf{a} \cdot \mathbf{c}$ (triple scalar product property)

$= \mathbf{p} \times \mathbf{a} \cdot (\mathbf{b} + \mathbf{c})$ (scalar product is distributive)

$= \mathbf{p} \cdot \mathbf{a} \times (\mathbf{b} + \mathbf{c})$ (triple scalar product property)

i.e., $\mathbf{p} \cdot (\mathbf{a} \times \mathbf{b} + \mathbf{a} \times \mathbf{c}) = \mathbf{p} \cdot \mathbf{a} \times (\mathbf{b} + \mathbf{c})$ thus,

$$\mathbf{a} \times (\mathbf{b} + \mathbf{c}) = \mathbf{a} \times \mathbf{b} + \mathbf{a} \times \mathbf{c}.$$

Calculation of the vector product of vectors in cartesian form

Consider $\mathbf{r}_1 = a_1\,\mathbf{i} + b_1\,\mathbf{j} + c_1\,\mathbf{k}$ and $\mathbf{r}_2 = a_2\,\mathbf{i} + b_2\,\mathbf{j} + c_2\,\mathbf{k}$

$$\begin{aligned}\mathbf{r}_1 \times \mathbf{r}_2 &= (a_1\mathbf{i} + b_1\mathbf{j} + c_1\mathbf{k}) \times (a_2\,\mathbf{i} + b_2\,\mathbf{j} + c_2\,\mathbf{k}) \\ &= a_1a_2\,\mathbf{i} \times \mathbf{i} + a_1b_2\,\mathbf{i} \times \mathbf{j} + a_1c_2\,\mathbf{i} \times \mathbf{k} + b_1a_2\,\mathbf{j} \times \mathbf{i} \\ &\quad + b_1b_2\,\mathbf{j} \times \mathbf{j} + b_1c_2\,\mathbf{j} \times \mathbf{k} + c_1a_2\,\mathbf{k} \times \mathbf{i} + c_1b_2\,\mathbf{k} \times \mathbf{j} \\ &\quad + c_1c_2\,\mathbf{k} \times \mathbf{k}\end{aligned}$$

since $\mathbf{i} \times \mathbf{i} = \mathbf{j} \times \mathbf{j} = \mathbf{k} \times \mathbf{k} = 0$ and $\mathbf{i} \times \mathbf{j} = \mathbf{k}$, $\mathbf{j} \times \mathbf{k} = \mathbf{i}$, $\mathbf{k} \times \mathbf{i} = \mathbf{j}$, $\mathbf{j} \times \mathbf{i} = -\mathbf{k}$, $\mathbf{k} \times \mathbf{j} = -\mathbf{i}$, $\mathbf{i} \times \mathbf{k} = -\mathbf{j}$, then

$$\begin{aligned}\mathbf{r}_1 \times \mathbf{r}_2 &= a_1b_2\,\mathbf{k} - a_1c_2\,\mathbf{j} - b_1a_2\,\mathbf{k} + b_1c_2\,\mathbf{i} + c_1a_2\,\mathbf{j} - c_1b_2\,\mathbf{i} \\ &= (b_1c_2 - c_1b_2)\mathbf{i} - (a_1c_2 - c_1a_2)\mathbf{j} + (a_1b_2 - b_1a_2)\mathbf{k}.\end{aligned}$$

This result is the expansion of the determinant $\begin{vmatrix} \mathbf{i} & \mathbf{j} & \mathbf{k} \\ a_1 & b_1 & c_1 \\ a_2 & b_2 & c_2 \end{vmatrix}$

The vector equation of a straight line

The position vector with respect to a **fixed** origin, (**position** vectors are **tied** to a fixed origin, **displacement** vectors are **free**), of any general point on a particular line, may be specified by giving the position vector of a particular known point on the line, and then adding on to it a scalar multiple of a direction vector for that line; i.e., if $\mathbf{a}$ is the position vector of a point on the line and $\mathbf{b}$ is a direction vector for the line, then $\mathbf{r} = \mathbf{a} + \lambda\mathbf{b}$, where λ is a scalar, will be the position vector of a general point on the line: each particular value of λ will give the position vector of a particular point on the line.

$\mathbf{r} = \mathbf{a} + \lambda\mathbf{b}$ is called a **vector equation** of the straight line.

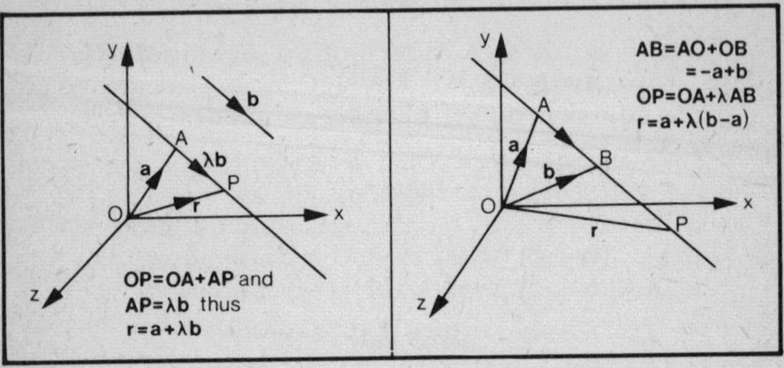

Figure 14.

If we know two points on the line, A and B, with position vectors $\mathbf{a}$ and $\mathbf{b}$ then $\mathbf{AB} = (\mathbf{b} - \mathbf{a})$ is a **direction vector** for that line and $\mathbf{r} = \mathbf{a} + \lambda(\mathbf{b} - \mathbf{a})$ is another form of the **vector equation** of the line.

Consider the vector equation of a line $\mathbf{r} = \mathbf{a} + \lambda\mathbf{b}$; writing

$$\mathbf{a} = x_1\,\mathbf{i} + y_1\,\mathbf{j} + z_1\,\mathbf{k} \text{ and } \mathbf{b} = x_2\,\mathbf{i} + y_2\,\mathbf{j} + z_2\,\mathbf{k}$$

we see that

$$\mathbf{r} = x_1\,\mathbf{i} + y_1\,\mathbf{j} + z_1\,\mathbf{k} + \lambda(x_2\,\mathbf{i} + y_2\,\mathbf{j} + z_2\,\mathbf{k})$$

Collecting components in the $\mathbf{i}$, $\mathbf{j}$, $\mathbf{k}$, directions,

$$\mathbf{r} = (x_1 + \lambda x_2)\mathbf{i} + (y_1 + \lambda y_2)\mathbf{j} + (z_1 + \lambda z_2)\mathbf{k}$$

Thus the cartesian coordinates of any point $P(x, y, z)$ on the line are: $x = x_1 + \lambda x_2$, $y = y_1 + \lambda y_2$ and $z = z_1 + \lambda z_2$. These are the **parametric equations of the line**. Rearranging these gives

$$\frac{x - x_1}{x_2} = \frac{y - y_1}{y_2} = \frac{z - z_1}{z_2} = \lambda$$

which are the **cartesian equations of the line**. We notice that the **direction ratios** $x_2 : y_2 : z_2$ are the **coefficients of** λ **in the parametric equations** and are the **denominators in the cartesian equations**. This is often a useful property for solving problems.

Two lines may be **parallel** to each other, may **intersect**, or they may do neither of these things, i.e. they are non-parallel and non-intersecting, in which case they are said to be **skew**. If they are **parallel** then we have seen already that they have **equal direction ratios**. If they **intersect**, then, letting their vector equations be $\mathbf{r}_1 = \mathbf{a}_1 + \lambda\mathbf{b}_1$ and $\mathbf{r}_2 = \mathbf{a}_2 + \mu\mathbf{b}_2$, there will be unique values of λ and μ for which $\mathbf{a}_1 + \lambda\mathbf{b}_1 = \mathbf{a}_2 + \mu\mathbf{b}_2$. If no such values may be found then the lines are **skew**.

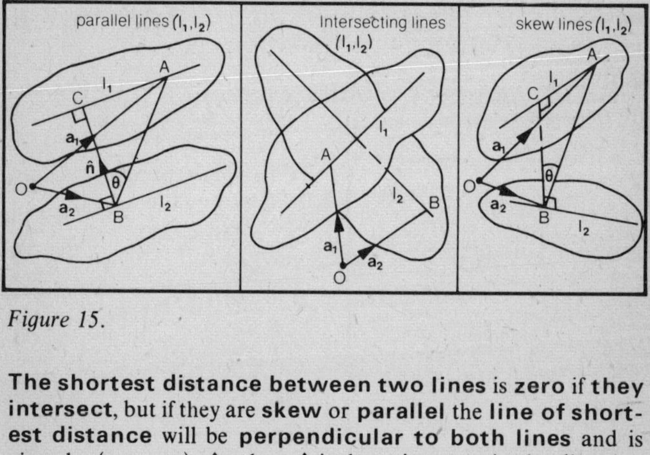

Figure 15.

The shortest distance between two lines is **zero** if **they intersect**, but if they are **skew** or **parallel** the **line of shortest distance** will be **perpendicular to both lines** and is given by $(\mathbf{a}_1 - \mathbf{a}_2) \cdot \hat{\mathbf{n}}$, where $\hat{\mathbf{n}}$ is the unit vector in the direction perpendicular to both lines.

If $\mathbf{OA} = \mathbf{a}_1$, $\mathbf{OB} = \mathbf{a}_2$ and $|\mathbf{BC}|$ is the shortest distance between the lines, then $\mathbf{BA} = \mathbf{a}_1 - \mathbf{a}_2$ and $BC = BA \cos\theta$ (see figure 15).

Therefore $BC = (\mathbf{a}_1 - \mathbf{a}_2) \cdot \hat{\mathbf{n}}$.

Using the scalar product as we did just now to find $AB \cos\theta$ illustrates a useful application of the scalar product: that of **resolving a vector into components** in given directions. If we wish to know the component of vector $\mathbf{f}$ in the direction

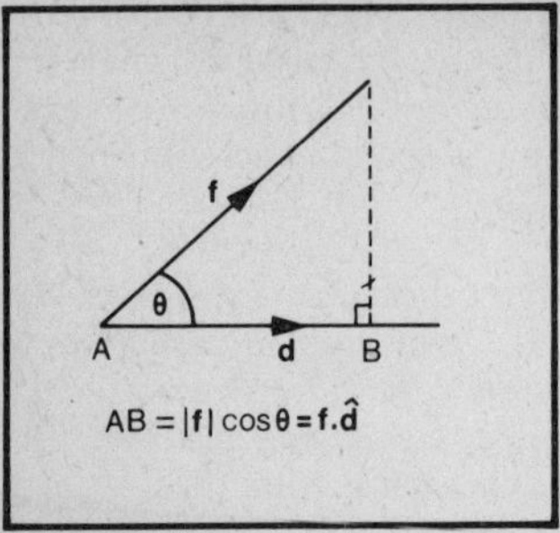

Figure 16.

of **d** where **d** and **f** are inclined at angle θ then we see that $\mathbf{f} \cdot \mathbf{d} = |\mathbf{f}||\mathbf{d}| \cos \theta$.

Thus $|\mathbf{f}| \cos \theta$, which is the value we require, is

$$\frac{\mathbf{f} \cdot \mathbf{d}}{|\mathbf{d}|} = \mathbf{f} \cdot \hat{\mathbf{d}}$$

We often need to resolve a vector into two perpendicular components and this method is commonly used.

It is useful to realize that vectors $x\mathbf{i} + y\mathbf{j}$ and $y\mathbf{i} - x\mathbf{j}$ are **perpendicular** to each other, as are $x\mathbf{i} + y\mathbf{j}$ and $-y\mathbf{i} + x\mathbf{j}$ since their **scalar products are zero.**

Position vector of a point dividing a line into a given ratio

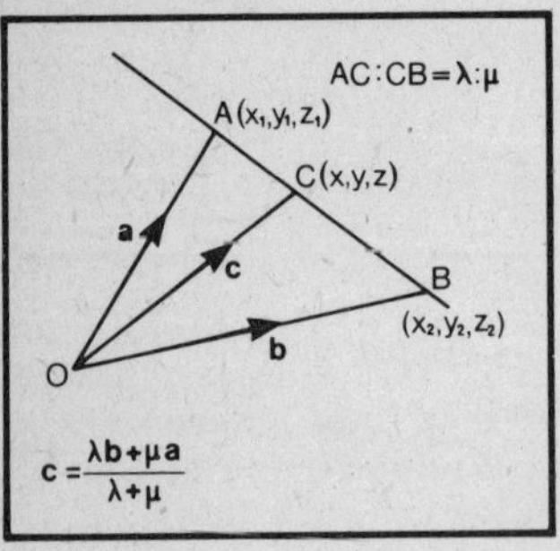

Figure 17.

If $C(x, y, z)$ is the point dividing AB into the ratio $\lambda : \mu$ then

$$\mathbf{OC} = \mathbf{OA} + \mathbf{AC}$$

$$= \mathbf{OA} + \frac{\lambda}{\lambda + \mu} \mathbf{AB}, \text{ thus}$$

$$\mathbf{c} = \mathbf{a} + \frac{\lambda}{\lambda + \mu} (\mathbf{b} - \mathbf{a}), \text{ i.e.}$$

$$\mathbf{c} = \frac{\lambda \mathbf{b} + \mu \mathbf{a}}{\lambda + \mu}.$$

For external division, the ratio is used in the form $-\lambda : \mu$ and the same result applies.

If we write the expression for **c** in its **equivalent cartesian** form, we see that,

$$\mathbf{c} = \frac{\lambda(x_2 \mathbf{i} + y_2 \mathbf{j} + z_2 \mathbf{k}) + \mu(x_1 \mathbf{i} + y_1 \mathbf{j} + z_1 \mathbf{k})}{\lambda + \mu}$$

Thus, $$x = \frac{\lambda x_2 + \mu x_1}{\lambda + \mu}, \quad y = \frac{\lambda y_2 + \mu y_1}{\lambda + \mu}, \quad z = \frac{\lambda z_2 + \mu z_1}{\lambda + \mu}.$$

Vector equation of a plane

There are many ways of defining a plane, but the most usual way is to give the **normal to the plane**, and the **perpendicular distance of the plane from the origin**. If the unit vector normal to the plane is $\hat{\mathbf{n}}$ and d is the distance from the origin then we see from the diagram that the position vector, $\mathbf{r}$, of a general point P on the plane, is given by $\mathbf{r} \cdot \hat{\mathbf{n}} = d$. This is a **vector equation of the plane**.

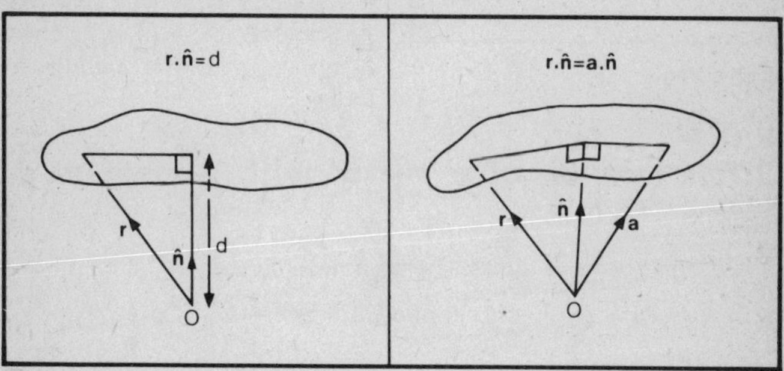

Figure 18.

If instead of being given the distance from the origin, d, we were given the position vector, $\mathbf{a}$, of a point on the plane we could write the **vector equation of the plane** as $\mathbf{r} \cdot \hat{\mathbf{n}} = \mathbf{a} \cdot \hat{\mathbf{n}}$ since $\mathbf{a} \cdot \hat{\mathbf{n}}$ will give us the distance of the plane from the origin. The more general form of the vector equation of a plane is $\mathbf{r} \cdot \mathbf{n} = D$ where $\mathbf{n}$ is normal to the plane but not necessarily the unit vector.

We see that this easily converts to $\mathbf{r} \cdot \hat{\mathbf{n}} = \dfrac{D}{|\mathbf{n}|}$ where $\dfrac{D}{|\mathbf{n}|}$ will be the distance from the origin of the plane.

The Equivalent Cartesian Form is

$$(x\mathbf{i} + y\mathbf{j} + z\mathbf{k}) \cdot (l\mathbf{i} + m\mathbf{j} + n\mathbf{k}) = d, \text{ i.e., } xl + ym + zn = d$$

where l, m, n, are the **direction cosines of the normal to the plane**.

More generally, the cartesian form of the vector equation of a plane is $xL + yM + zN = D$ where $L : M : N$ are the **direction ratios of the normal to the plane**.

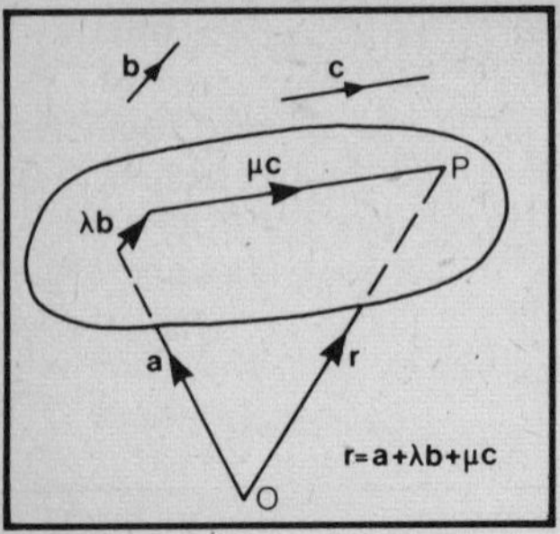

Figure 19.

The **parametric form** of the vector equation of a plane is

$$\mathbf{r} = \mathbf{a} + \lambda\mathbf{b} + \mu\mathbf{c}$$

where **a** is the **position vector of a particular point on the plane**, **b** and **c** are two **non-parallel vectors parallel to the plane.** As we can see, from figure 19, $\mathbf{r} = \mathbf{a} + \lambda\mathbf{b} + \mu\mathbf{c}$ will give the position vector, **r**, of any point, P, on the plane for suitable λ and μ.

Differentiation of a vector with respect to a scalar variable

If **a** is a function of the scalar variable λ, so that

$\mathbf{a}(\lambda) = f(\lambda)\mathbf{i} + g(\lambda)\mathbf{j} + h(\lambda)\mathbf{k}$ and $\delta\mathbf{a} = \mathbf{a}(\lambda + \delta\lambda) - \mathbf{a}(\lambda)$

then $\dfrac{\mathrm{d}\mathbf{a}}{\mathrm{d}\lambda} = \lim\limits_{\delta\lambda \to 0} \dfrac{\delta\mathbf{a}}{\delta\lambda}$

$$= \lim_{\delta\lambda \to 0} \frac{f(\lambda+\delta\lambda)\mathbf{i}+g(\lambda+\delta\lambda)\mathbf{j}+h(\lambda+\delta\lambda)\mathbf{k}-f(\lambda)\mathbf{i}-g(\lambda)\mathbf{j}-h(\lambda)\mathbf{k}}{\delta\lambda}$$

$$= \left(\lim_{\delta\lambda \to 0} \frac{f(\lambda + \delta\lambda) - f(\lambda)}{\delta\lambda}\right)\mathbf{i} + \left(\lim_{\delta\lambda \to 0} \frac{g(\lambda + \delta\lambda) - g(\lambda)}{\delta\lambda}\right)\mathbf{j}$$

$$+ \left(\lim_{\delta\lambda \to 0} \frac{h(\lambda + \delta\lambda) - h(\lambda)}{\delta\lambda}\right)\mathbf{k}$$

Hence $\dfrac{\mathrm{d}a}{\mathrm{d}\lambda} = \dfrac{\mathrm{d}f}{\mathrm{d}\lambda}\mathbf{i} + \dfrac{\mathrm{d}g}{\mathrm{d}\lambda}\mathbf{j} + \dfrac{\mathrm{d}h}{\mathrm{d}\lambda}\mathbf{k}.$

In general, **a** and $\dfrac{\mathrm{d}\mathbf{a}}{\mathrm{d}\lambda}$ will have different directions.

Integration of a vector with respect to a scalar variable

As we may expect from the previous section, if **a** is a vector which is a function of a scalar variable, λ, i.e., $\mathbf{a}(\lambda) = f(\lambda)\mathbf{i} + g(\lambda)\mathbf{j} + h(\lambda)\mathbf{k}$ then $\int \mathbf{a}\,\mathrm{d}\lambda = (\int f\,\mathrm{d}\lambda)\mathbf{i} + (\int g\,\mathrm{d}\lambda)\mathbf{j} + (\int h\,\mathrm{d}\lambda)\mathbf{k}$. In general, $\int \mathbf{a}\,\mathrm{d}\lambda$ and **a** will have different directions.

These results are particularly useful when we are considering the **displacement**, **velocity** and **acceleration** of a body. **Velocity** is defined as the rate of change of displacement, and **acceleration** as the rate of change of velocity, i.e., if $\mathbf{r}$ is the **position vector** of a body, $\mathbf{v}$ its **velocity vector** and $\mathbf{a}$ its **acceleration vector** then

$$\mathbf{v} = \frac{d\mathbf{r}}{dt}, \quad \mathbf{a} = \frac{d\mathbf{v}}{dt}, \quad \mathbf{v} = \int \mathbf{a}\, dt \text{ and } \mathbf{r} = \int \mathbf{v}\, dt.$$

Worked examples

Example 1 The position vectors of the points A, B, C, are $\mathbf{a} = \mathbf{i} - \mathbf{j} + 2\mathbf{k}$, $\mathbf{b} = 2\mathbf{i} + \mathbf{j} + 4\mathbf{k}$, $\mathbf{c} = 3\mathbf{i} + 4\mathbf{k}$, respectively. Find (i) the angle CAB to the nearest degree; (ii) a vector equation of the plane containing A, B, and C; (iii) the area of triangle ABC; (iv) the distance from the plane containing A, B, C, to the point with position vector $\mathbf{i} + \mathbf{j} + 2\mathbf{k}$.

(i) Since $(\mathbf{b} - \mathbf{a}) \cdot (\mathbf{c} - \mathbf{a}) = |\mathbf{b} - \mathbf{a}|\,|\mathbf{c} - \mathbf{a}| \cos\theta$, where $C\hat{A}B = \theta$,

then $\cos\theta = \dfrac{(\mathbf{b} - \mathbf{a}) \cdot (\mathbf{c} - \mathbf{a})}{|\mathbf{b} - \mathbf{a}|\,|\mathbf{c} - \mathbf{a}|}$. We know that:

$$(\mathbf{b} - \mathbf{a}) = (2\mathbf{i} + \mathbf{j} + 4\mathbf{k} - (\mathbf{i} - \mathbf{j} + 2\mathbf{k})) = \mathbf{i} + 2\mathbf{j} + 2\mathbf{k}$$

Thus $|\mathbf{b} - \mathbf{a}| = \sqrt{1 + 4 + 4} = 3$.

Also, $(\mathbf{c} - \mathbf{a}) = (3\mathbf{i} + 4\mathbf{k} - (\mathbf{i} - \mathbf{j} + 2\mathbf{k})) = 2\mathbf{i} + \mathbf{j} + 2\mathbf{k}$

Thus $|\mathbf{c} - \mathbf{a}| = \sqrt{4 + 1 + 4} = 3$.

Hence, $\cos\theta = \dfrac{(\mathbf{i} + 2\mathbf{j} + 2\mathbf{k}) \cdot (2\mathbf{i} + \mathbf{j} + 2\mathbf{k})}{9} = \dfrac{2 + 2 + 4}{9} = \dfrac{8}{9}$

and **angle CAB = 27°** (to the nearest degree).

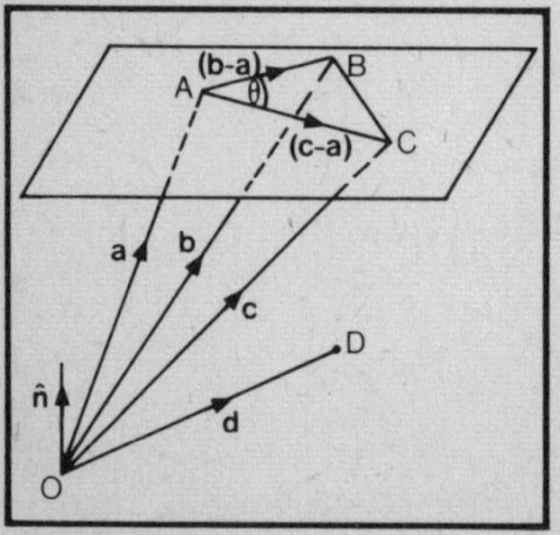

Figure 20.

(ii) $(\mathbf{b} - \mathbf{a})$ and $(\mathbf{c} - \mathbf{a})$ are two non-parallel vectors (the angle between them is 27°) which are parallel to the plane and A is a point in the plane. Thus a **vector equation of the plane** is

$$\mathbf{r} = \mathbf{a} + \lambda(\mathbf{b} - \mathbf{a}) + \mu(\mathbf{c} - \mathbf{a})$$

where λ and μ are scalars, thus,

$$\mathbf{r} = \mathbf{i} - \mathbf{j} + 2\mathbf{k} + \lambda(\mathbf{i} + 2\mathbf{j} + 2\mathbf{k}) + \mu(2\mathbf{i} + \mathbf{j} + 2\mathbf{k})$$

(iii) The area of triangle ABC is $\frac{1}{2}(AC)(AB \sin \theta)$

$$= \tfrac{1}{2}|\mathbf{AC} \times \mathbf{AB}|.$$

Since $\mathbf{AC} \times \mathbf{AB} = (\mathbf{c} - \mathbf{a}) \times (\mathbf{b} - \mathbf{a})$

$$= (2\mathbf{i} + \mathbf{j} + 2\mathbf{k}) \times (\mathbf{i} + 2\mathbf{j} + 2\mathbf{k})$$

$$= \begin{vmatrix} \mathbf{i} & \mathbf{j} & \mathbf{k} \\ 2 & 1 & 2 \\ 1 & 2 & 2 \end{vmatrix} = -2\mathbf{i} - 2\mathbf{j} + 3\mathbf{k},$$

then $|\mathbf{AC} \times \mathbf{AB}| = \sqrt{4 + 4 + 9} = \sqrt{17}$.

Hence **area of triangle *ABC*** $= \dfrac{\sqrt{17}}{2}$.

(We could also have found this result by $\frac{1}{2}(AC)(AB \sin \theta) = \frac{1}{2}(3)(3) \sin 27°$ from part (i).)

(iv) The distance from the plane ABC to the point D with position vector $\mathbf{i} + \mathbf{j} + 2\mathbf{k}$ is given by:

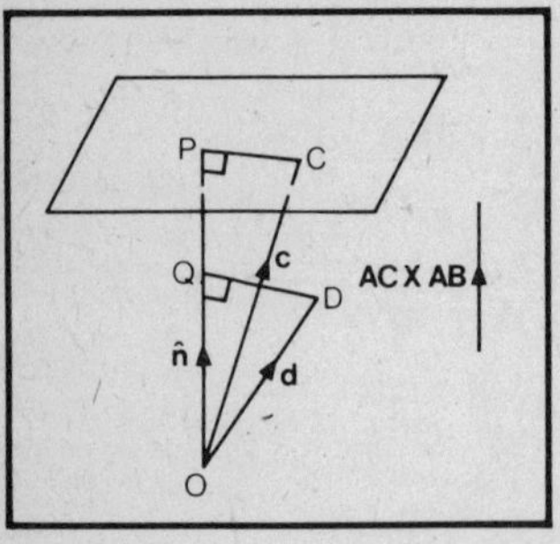

Figure 21.

$\mathbf{OP} - \mathbf{OQ} = \mathbf{c} \cdot \hat{\mathbf{n}} - \mathbf{d} \cdot \hat{\mathbf{n}}$ where $\hat{\mathbf{n}}$ is the unit vector normal to the plane (see figure 21).

We have just found a vector normal to the plane

$$(\mathbf{AC} \times \mathbf{AB}) = -2\mathbf{i} - 2\mathbf{j} + 3\mathbf{k}$$

Thus

$$\hat{\mathbf{n}} = \frac{1}{\sqrt{17}}(-2\mathbf{i} - 2\mathbf{j} + 3\mathbf{k})$$

and $\mathbf{PQ} = (3\mathbf{i} + 4\mathbf{k}) \cdot \dfrac{1}{\sqrt{17}}(-2\mathbf{i} - 2\mathbf{j} + 3\mathbf{k})$

$$- (\mathbf{i} + \mathbf{j} + 2\mathbf{k}) \cdot \frac{1}{\sqrt{17}}(-2\mathbf{i} - 2\mathbf{j} + 3\mathbf{k})$$

$$= \frac{1}{\sqrt{17}}(-6 + 12) - \frac{1}{\sqrt{17}}(-2 - 2 + 6) = \frac{4}{\sqrt{17}}$$

(If the result had been **negative** we would have known that D was **above** the plane.)

Example 2 Show that vectors **u**, **v**, **w**, where $\mathbf{u} = \mathbf{i} + \mathbf{j} + \mathbf{k}$, $\mathbf{v} = \mathbf{i} - \frac{1}{2}\mathbf{j} - \frac{1}{2}\mathbf{k}$, $\mathbf{w} = \mathbf{j} - \mathbf{k}$, are mutually perpendicular and find the unit vector in the direction of the vector **u**. Find constants α, β and γ such that $\mathbf{i} = \alpha\mathbf{u} + \beta\mathbf{v} + \gamma\mathbf{w}$. If P, Q, and R have position vectors $5\mathbf{i} - \mathbf{j} - \mathbf{k}$, **u** and **v** respectively with respect to the origin O, show that O, P, Q and R are coplanar. Find the cosine of the angle POQ. If the vector **c** lies in the xy plane, find **c** such that $\mathbf{c} \times \mathbf{w} = \mathbf{u}$. Hence find the value of $\mathbf{w} \cdot (\mathbf{c} \times \mathbf{w})$.

If $\mathbf{u} \cdot \mathbf{v} = \mathbf{v} \cdot \mathbf{w} = \mathbf{w} \cdot \mathbf{u} = 0$ then the vectors **u**, **v**, **w** are mutually perpendicular.

$$\mathbf{u} \cdot \mathbf{v} = (\mathbf{i} + \mathbf{j} + \mathbf{k}) \cdot (\mathbf{i} - \tfrac{1}{2}\mathbf{j} - \tfrac{1}{2}\mathbf{k}) = 1 - \tfrac{1}{2} - \tfrac{1}{2} = 0$$
$$\mathbf{v} \cdot \mathbf{w} = (\mathbf{i} - \tfrac{1}{2}\mathbf{j} - \tfrac{1}{2}\mathbf{k}) \cdot (\mathbf{j} - \mathbf{k}) = -\tfrac{1}{2} + \tfrac{1}{2} = 0$$
$$\mathbf{w} \cdot \mathbf{u} = (\mathbf{j} - \mathbf{k}) \cdot (\mathbf{i} + \mathbf{j} + \mathbf{k}) = 1 - 1 = 0$$

thus **u**, **v**, **w** are **mutually perpendicular**.

$$\hat{\mathbf{u}} = \frac{\mathbf{u}}{|\mathbf{u}|} = \frac{\mathbf{i} + \mathbf{j} + \mathbf{k}}{\sqrt{3}} = \frac{\mathbf{i}}{\sqrt{3}} + \frac{\mathbf{j}}{\sqrt{3}} + \frac{\mathbf{k}}{\sqrt{3}}.$$

If $\mathbf{i} = \alpha\mathbf{u} + \beta\mathbf{v} + \gamma\mathbf{w}$
$= \alpha(\mathbf{i} + \mathbf{j} + \mathbf{k}) + \beta(\mathbf{i} - \frac{1}{2}\mathbf{j} - \frac{1}{2}\mathbf{k}) + \gamma(\mathbf{j} - \mathbf{k})$,

then by comparing coefficients of vectors **i**, **j**, and **k**

$1 = \alpha + \beta$ (i), $0 = \alpha = \dfrac{\beta}{2} + \gamma$ (ii), and $0 = \alpha - \dfrac{\beta}{2} - \gamma$ (iii),

(ii) and (iii) give $\gamma = 0$

(i) and (ii) give $\beta = \dfrac{2}{3}$, thus $\alpha = \dfrac{1}{3}$ from (i).

Hence $\mathbf{i} = \dfrac{1}{3}\mathbf{u} + \dfrac{2}{3}\mathbf{v}$.

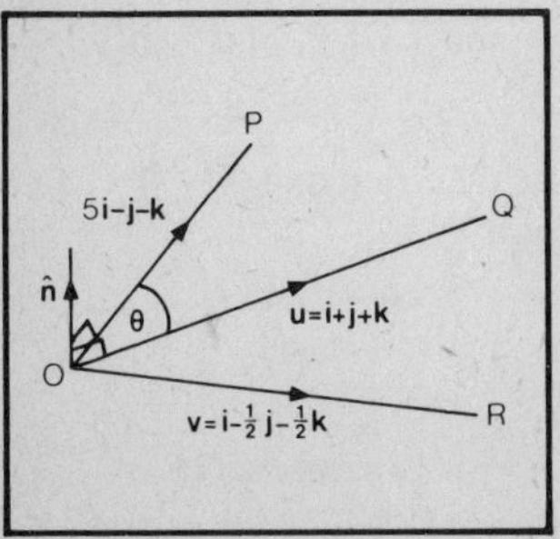

Figure 22.

To find out if O, P, Q, R are coplanar we shall first find the equation of the plane OPQ and then test to see if R lies in it. A normal **n** to the plane OPQ is given by

$$\mathbf{OQ} \times \mathbf{OP} = (\mathbf{i} + \mathbf{j} + \mathbf{k}) \times (5\mathbf{i} - \mathbf{j} - \mathbf{k})$$
$$= \begin{vmatrix} \mathbf{i} & \mathbf{j} & \mathbf{k} \\ 1 & 1 & 1 \\ 5 & -1 & -1 \end{vmatrix} = 6\mathbf{j} - 6\mathbf{k}$$

Thus $\hat{\mathbf{n}} = \dfrac{\mathbf{n}}{|\mathbf{n}|} = \dfrac{\mathbf{j}}{\sqrt{2}} - \dfrac{\mathbf{k}}{\sqrt{2}}$.

Hence the equation of the plane is $\mathbf{r} \cdot \hat{\mathbf{n}} = 0$. (The distance from the origin is 0 since the origin is in the plane.) We must now see if **OR** satisfies this equation.

$$\mathbf{OR} \cdot \hat{\mathbf{n}} = (\mathbf{i} - \tfrac{1}{2}\mathbf{j} - \tfrac{1}{2}\mathbf{k}) \cdot \left(\frac{\mathbf{j}}{\sqrt{2}} - \frac{\mathbf{k}}{\sqrt{2}}\right) = 0$$

thus ***OPQR* are coplanar**. From figure 22,

$$\cos\theta = \frac{\mathbf{OQ} \cdot \mathbf{OP}}{|\mathbf{OQ}|\,|\mathbf{OP}|} = \frac{(\mathbf{i} + \mathbf{j} + \mathbf{k}) \cdot (5\mathbf{i} - \mathbf{j} + \mathbf{k})}{\sqrt{3}\sqrt{27}}$$

$$= \frac{5 - 1 - 1}{9} = \frac{1}{3}, \qquad \cos P\hat{O}Q = \frac{1}{3}$$

If vector $\mathbf{c}$ lies in the xy plane then $\mathbf{c}$ may be written as $\mathbf{c} = a\mathbf{i} + b\mathbf{j}$ where a and b are constants. If $\mathbf{c} \times \mathbf{w} = \mathbf{u}$ then $(a\mathbf{i} + b\mathbf{j}) \times (\mathbf{j} - \mathbf{k}) = \mathbf{i} + \mathbf{j} + \mathbf{k}$. Since

$$(a\mathbf{i} + b\mathbf{j}) \times (\mathbf{j} - \mathbf{k}) = \begin{vmatrix} \mathbf{i} & \mathbf{j} & \mathbf{k} \\ a & b & 0 \\ 0 & 1 & -1 \end{vmatrix} = -b\mathbf{i} + a\mathbf{j} + a\mathbf{k}$$

then this must equal $\mathbf{i} + \mathbf{j} + \mathbf{k}$. Thus $\boldsymbol{b} = \mathbf{-1}$ and $\boldsymbol{a} = \mathbf{1}$, giving $\mathbf{c} = \mathbf{i} - \mathbf{j}$. Since $\mathbf{c} \times \mathbf{w} = \mathbf{u}$, then $\mathbf{w} \cdot (\mathbf{c} \times \mathbf{w}) = \mathbf{w} \cdot \mathbf{u} = 0$ ($\mathbf{w}$ and $\mathbf{u}$ are perpendicular).

Example 3 *ABCDEFGH* is a cube with horizontal faces *ABCD* and *EFGH* and vertical edges *AE*, *BF*, *CG*, *DH*. Taking $\mathbf{AB} = \mathbf{i}$, $\mathbf{AD} = \mathbf{j}$ and $\mathbf{AE} = \mathbf{k}$, obtain vectors in terms of $\mathbf{i}, \mathbf{j}, \mathbf{k}$, perpendicular to the planes *ACGE* and *CDEF*, and hence find the angle between these two planes.

Two vectors parallel to the plane *ACGE* are $\mathbf{k}$ and $(\mathbf{i} + \mathbf{j})$. Hence a vector perpendicular to this plane is

$$\mathbf{k} \times (\mathbf{i} + \mathbf{j}) = \begin{vmatrix} \mathbf{i} & \mathbf{j} & \mathbf{k} \\ 0 & 0 & 1 \\ 1 & 1 & 0 \end{vmatrix} = -\mathbf{i} + \mathbf{j}.$$

Thus the **unit vector normal to the plane *ACGE*** is

$$\hat{\mathbf{n}}_1 = -\frac{\mathbf{i}}{\sqrt{2}} + \frac{\mathbf{j}}{\sqrt{2}}$$

Two vectors parallel to the plane *CDEF* are **i** and $(-\mathbf{j}+\mathbf{k})$. Hence a vector perpendicular to this plane is

$$\mathbf{i} \times (-\mathbf{j}+\mathbf{k}) = \begin{vmatrix} \mathbf{i} & \mathbf{j} & \mathbf{k} \\ 1 & 0 & 0 \\ 0 & -1 & 1 \end{vmatrix} = -\mathbf{j}-\mathbf{k}$$

and the unit vector perpendicular to plane *CDEF* is

$$\hat{\mathbf{n}}_2 = -\frac{\mathbf{j}}{\sqrt{2}} - \frac{\mathbf{k}}{\sqrt{2}}$$

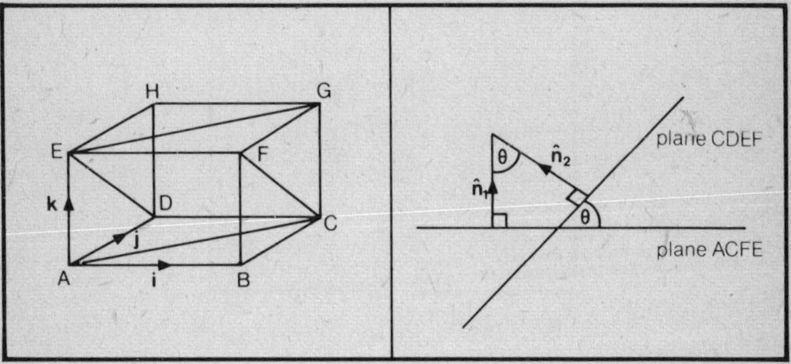

Figure 23.

As we can see from the diagram above, **the angle between the planes, θ, is the angle between the two unit normal vectors** i.e.,

$$\cos\theta = \hat{\mathbf{n}}_1 \cdot \hat{\mathbf{n}}_2 = \left(-\frac{\mathbf{i}}{\sqrt{2}} + \frac{\mathbf{j}}{\sqrt{2}}\right) \cdot \left(-\frac{\mathbf{j}}{\sqrt{2}} - \frac{\mathbf{k}}{\sqrt{2}}\right) = -\frac{1}{2}$$

Therefore, **the acute angle between the planes is 60°**.

Example 4 Given that $\mathbf{OA} = 3\mathbf{i} + 4\mathbf{j} + 5\mathbf{k}$, $\mathbf{OB} = 4\mathbf{i} + 6\mathbf{j} + 7\mathbf{k}$, $\mathbf{OC} = \mathbf{i} + 5\mathbf{j} + 3\mathbf{k}$, find (i) the angle *BAC* (ii) the area of triangle *ABC* (iii) the direction cosines of the normal to the plane *ABC*, (iv) the volume of tetrahedron *OABC*.

(i) The angle *BAC* is the angle between vectors **AC** and **AB**, i.e.,

$$\cos B\hat{A}C = \frac{\mathbf{AC} \cdot \mathbf{AB}}{|\mathbf{AC}|\,|\mathbf{AB}|}$$

$$\begin{aligned}\mathbf{AC} = \mathbf{AO} + \mathbf{OC} &= -(3\mathbf{i} + 4\mathbf{j} + 5\mathbf{k}) + \mathbf{i} + 5\mathbf{j} + 3\mathbf{k} \\ &= -2\mathbf{i} + \mathbf{j} - 2\mathbf{k}\end{aligned}$$

and $\mathbf{AB} = \mathbf{AO} + \mathbf{OB} = -(3\mathbf{i} + 4\mathbf{j} + 5\mathbf{k}) + 4\mathbf{i} + 6\mathbf{j} + 7\mathbf{k}$
$= \mathbf{i} + 2\mathbf{j} + 2\mathbf{k}$

Thus, $|\mathbf{AC}| = 3$ and $|\mathbf{AB}| = 3.$

$$\text{Hence, } \cos C\hat{A}B = \frac{(-2\mathbf{i} + \mathbf{j} - 2\mathbf{k}) \cdot (\mathbf{i} + 2\mathbf{j} + 2\mathbf{k})}{9} = -\frac{4}{9}$$

and $C\hat{A}B = 116{\cdot}4°.$

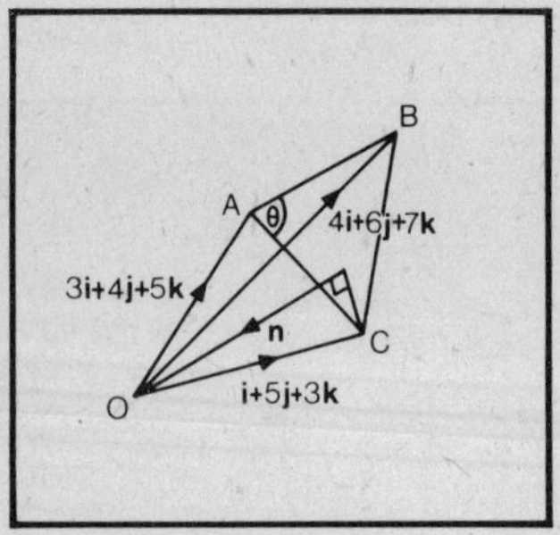

Figure 24.

(ii) **The area of triangle *ABC*** is
$\frac{1}{2}(AC)(AB \sin\theta) = \frac{1}{2}|\mathbf{AC} \times \mathbf{AB}|$
$= \frac{1}{2}|(-2\mathbf{i} + \mathbf{j} - 2\mathbf{k}) \times (\mathbf{i} + 2\mathbf{j} + 2\mathbf{k})|.$

Since $(-2\mathbf{i} + \mathbf{j} - 2\mathbf{k}) \times (\mathbf{i} + 2\mathbf{j} + 2\mathbf{k})$

$$= \begin{vmatrix} \mathbf{i} & \mathbf{j} & \mathbf{k} \\ -2 & 1 & -2 \\ 1 & 2 & 2 \end{vmatrix} = 6\mathbf{i} + 2\mathbf{j} - 5\mathbf{k},$$

then $\frac{1}{2}|\mathbf{AC} \times \mathbf{AB}| = \frac{1}{2}\sqrt{65}.$

(iii) A normal to the plane, $\mathbf{n}$, is $\mathbf{AC} \times \mathbf{AB} = 6\mathbf{i} + 2\mathbf{j} - 5\mathbf{k}$. Therefore, the **unit** vector normal to the plane, $\hat{\mathbf{n}}$, is

$$\frac{6}{\sqrt{65}}\mathbf{i} + \frac{2}{\sqrt{65}}\mathbf{j} - \frac{5}{\sqrt{65}}\mathbf{k}.$$

Hence the **direction cosines of the normal to the plane** are

$$\frac{6}{\sqrt{65}}, \quad \frac{2}{\sqrt{65}}, \quad -\frac{5}{\sqrt{65}}.$$

(iv) **The volume of tetrahedron *OABC***

$= \frac{1}{3}$(base area)(perp. ht.)

$= \frac{1}{3}$(Area of triangle ABC)(perp. ht.) $= \frac{1}{3}(\frac{1}{2}\sqrt{65})(\mathbf{OC}\,.\,\hat{\mathbf{n}})$

$$= \frac{1}{6}\sqrt{65}\left((\mathbf{i} + 5\mathbf{j} + 3\mathbf{k})\,.\,\left(\frac{6}{\sqrt{65}}\mathbf{i} + \frac{2}{\sqrt{65}}\mathbf{j} - \frac{5}{\sqrt{65}}\mathbf{k}\right)\right) = \frac{1}{6}$$

Example 5 The lines l_1 and l_2 are given in the parametric forms $\mathbf{r}_1 = (3\mathbf{i} + 2\mathbf{j} + \mathbf{k}) + \lambda(\mathbf{i} + 2\mathbf{j} + 2\mathbf{k})$ and $\mathbf{r}_2 = (2\mathbf{i} + 3\mathbf{j} + 2\mathbf{k}) + \mu(2\mathbf{i} + \mathbf{j} - 2\mathbf{k})$. Show that l_1 and l_2 are mutually perpendicular. Find values of λ and μ if the vector $(\mathbf{r}_2 - \mathbf{r}_1)$ is perpendicular to each of the given lines. Hence find the shortest distance between the lines.

A **direction** vector for l_1 is $\mathbf{i} + 2\mathbf{j} + 2\mathbf{k}$, and for l_2 is $2\mathbf{i} + \mathbf{j} - 2\mathbf{k}$. Since $(\mathbf{i} + 2\mathbf{j} + 2\mathbf{k})\,.\,(2\mathbf{i} + \mathbf{j} - 2\mathbf{k}) = 2 + 2 - 4 = 0$ then l_1 and l_2 are **perpendicular**.

$$(\mathbf{r}_1 - \mathbf{r}_2) = (-1 + 2\mu - \lambda)\mathbf{i} + (1 + \mu - 2\lambda)\mathbf{j} + (1 - 2\mu - 2\lambda)\mathbf{k}$$

and if $(\mathbf{r}_2 - \mathbf{r}_1)$ is perpendicular to l_1 and l_2 then it will have the **same direction ratios** as $(\mathbf{i} + 2\mathbf{j} + 2\mathbf{k}) \times (2\mathbf{i} + \mathbf{j} - 2\mathbf{k})$.

Since $(\mathbf{i} + 2\mathbf{j} + 2\mathbf{k}) \times (2\mathbf{i} + \mathbf{j} - 2\mathbf{k}) = \begin{vmatrix} \mathbf{i} & \mathbf{j} & \mathbf{k} \\ 1 & 2 & 2 \\ 2 & 1 & -2 \end{vmatrix}$

$= -6\mathbf{i} + 6\mathbf{j} - 3\mathbf{k}$, which has direction ratios $2 : -2 : 1$, then

$$(-1 + 2\mu - \lambda) : (1 + \mu - 2\lambda) = -1.$$

Thus, $\mu = \lambda$. Also, $(1 + \mu - 2\lambda) : (1 - 2\mu - 2\lambda) = -2$. Thus, $\mu = \lambda = \frac{1}{3}$.

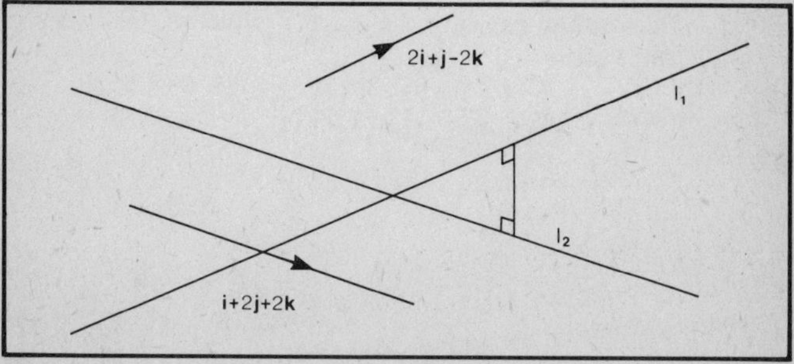

Figure 25.

The shortest distance between the lines l_1 and l_2 is given by $|(r_2 - r_1)|$ when $\lambda = \mu = \frac{1}{3}$, since the shortest distance will be the length of the common perpendicular from l_1 to l_2. Hence **the shortest distance** is

$$\left| -\frac{2i}{3} + \frac{2j}{3} - \frac{1k}{3} \right| = \sqrt{\frac{4}{9} + \frac{4}{9} + \frac{1}{9}} = \mathbf{1\ unit}$$

Example 6 The position vector $\mathbf{r}$ of a particle at time t is given by $\mathbf{r} = 2t^2\mathbf{i} + (t^2 - 4t)\mathbf{j} + (3t - 5)\mathbf{k}$. Find the velocity and acceleration of the particle at time t. Show that when $t = \frac{2}{5}$ the velocity and acceleration are perpendicular to each other. Find the velocity and acceleration components parallel to the vector $\mathbf{i} - 3\mathbf{j} + 2\mathbf{k}$ and find the value of these components when $t = \frac{2}{5}$.

Since $\mathbf{r} = 2t^2\mathbf{i} + (t^2 - 4t)\mathbf{j} + (3t - 5)\mathbf{k}$,

then $$\mathbf{v} = \frac{d\mathbf{r}}{dt} = 4t\mathbf{i} + (2t - 4)\mathbf{j} + 3\mathbf{k},$$

and $$\mathbf{a} = \frac{d\mathbf{v}}{dt} = 4\mathbf{i} + 2\mathbf{j}.$$

When $t = \frac{2}{5}$, $\mathbf{v} = \frac{8\mathbf{i}}{5} - \frac{16\mathbf{j}}{15} + 3\mathbf{k}$, and $\mathbf{a} = 4\mathbf{i} + 2\mathbf{j}$.

Since $$\left(\frac{8\mathbf{i}}{5} - \frac{16\mathbf{j}}{5} + 3\mathbf{k}\right) . (4\mathbf{i} + 2\mathbf{j}) = 0,$$

then **the velocity and acceleration vectors are perpendicular when $t = \frac{2}{5}$.**

To resolve the velocity and acceleration vectors parallel to $\mathbf{i} - 3\mathbf{j} + 2\mathbf{k}$, we shall consider the scalar product of these vectors with the unit vector

$$\left(\frac{\mathbf{i}}{\sqrt{14}} - \frac{3\mathbf{j}}{\sqrt{14}} + \frac{2\mathbf{k}}{\sqrt{14}}\right).$$

We have

$$\mathbf{v} . \left(\frac{\mathbf{i}}{\sqrt{14}} - \frac{3\mathbf{j}}{\sqrt{14}} + \frac{2\mathbf{k}}{\sqrt{14}}\right)$$

$$= (4t\mathbf{i} + (2t - 4)\mathbf{j} + 3\mathbf{k}) . \left(\frac{\mathbf{i}}{\sqrt{14}} - \frac{3\mathbf{j}}{\sqrt{14}} + \frac{2\mathbf{k}}{\sqrt{14}}\right)$$

$$= \frac{4t}{\sqrt{14}} - \frac{3(2t - 4)}{\sqrt{14}} + \frac{6}{\sqrt{14}},$$

which is the magnitude of the component of velocity in the required direction and

$$\mathbf{a} \cdot \left(\frac{\mathbf{i}}{\sqrt{14}} - \frac{3\mathbf{j}}{\sqrt{14}} + \frac{2\mathbf{k}}{\sqrt{14}}\right) = (4\mathbf{i} + 2\mathbf{j}) \cdot \left(\frac{\mathbf{i}}{\sqrt{14}} - \frac{3\mathbf{j}}{\sqrt{14}} + \frac{2\mathbf{k}}{\sqrt{14}}\right)$$

$$= \frac{4}{\sqrt{14}} - \frac{6}{\sqrt{14}} = \frac{-2}{\sqrt{14}}$$

which is the **magnitude** of the component of acceleration in the required direction.

Hence, when $t = \frac{2}{5}$ the **component of velocity parallel to** $\mathbf{i} - 3\mathbf{j} + 2\mathbf{k}$ is

The component of **v** in the direction (i−3j+2k)

$$= v \cos\theta = \frac{\mathbf{v}.(\mathbf{i}-3\mathbf{j}+2\mathbf{k})}{|(\mathbf{i}-3\mathbf{j}+2\mathbf{k})|}$$

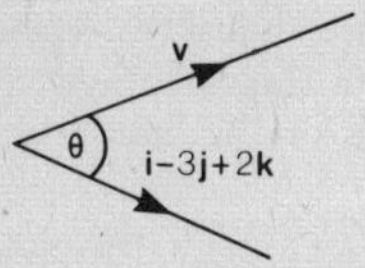

$$\frac{1}{\sqrt{14}}\left(\frac{8}{5} + \frac{48}{5} + 6\right) \times \left(\frac{\mathbf{i}}{\sqrt{14}} - \frac{3\mathbf{j}}{\sqrt{14}} + \frac{2\mathbf{k}}{\sqrt{14}}\right)$$

$$= \frac{43}{35}(\mathbf{i} - 3\mathbf{j} + 2\mathbf{k})$$

Figure 26.

The component of acceleration (which is constant) **in this direction is**

$$-\frac{2}{\sqrt{14}}\left(\frac{\mathbf{i}}{\sqrt{14}} - \frac{3\mathbf{j}}{\sqrt{14}} + \frac{2\mathbf{k}}{\sqrt{14}}\right) = -\frac{1}{7}\mathbf{i} + \frac{3}{7}\mathbf{j} - \frac{2}{7}\mathbf{k}$$

Key terms

A **vector** has magnitude and direction, a **scalar** has magnitude only.

Vectors are either free or tied. A **free** vector has no particular location in space and may be represented by an infinite number of equal, like, parallel line segments. A **tied** vector is not completely described by its magnitude and direction. Further information is needed about its location in space. Unless stated otherwise, a vector, other than a **force** or a **position vector** is free. **Force** is a **line-localised** vector and a position vector is **tied** to a fixed origin.

The **vector polygon law of addition** of vectors is an extension of the triangle law and of the parallelogram law of addition. It states that when vectors are represented completely by the sides of a polygon taken in order, then their resultant is represented completely by that side of the polygon needed to close it. $\mathbf{AB} + \mathbf{BC} + \mathbf{CD} + \mathbf{DE} + \mathbf{EF} + \mathbf{FG} + \mathbf{GH} = \mathbf{AH}$.

$(\mathbf{a} + \mathbf{b}) + \mathbf{c} = \mathbf{a} + (\mathbf{b} + \mathbf{c})$ and $\mathbf{a} + \mathbf{b} = \mathbf{b} + \mathbf{a}$. If λ is a scalar then $\lambda(\mathbf{a} + \mathbf{b}) = \lambda\mathbf{a} + \lambda\mathbf{b}$.

The **subtraction of vectors** is defined as the **addition of negative vectors** i.e.,

$$\mathbf{AB} - \mathbf{BC} = \mathbf{AB} + (-\mathbf{BC}) = \mathbf{AB} + \mathbf{CB}.$$

A **unit** vector has **magnitude one**. Any vector may be expressed as the product of its unit direction vector and its modulus, $\mathbf{v} = |\mathbf{v}|\,\hat{\mathbf{v}}$.

The **resultant vector theorem** states that if two vectors may be represented by $\lambda\mathbf{OA}$ and $\mu\mathbf{OB}$, then their resultant is represented by $(\lambda + \mu)\mathbf{OC}$ where C is a point on the line AB such that $AC : CB = \mu : \lambda$.

The vectors **i**, **j**, **k** are **mutually perpendicular** (i.e., orthogonal) **unit vectors** obeying the **right-handed corkscrew rule**. A vector written in terms of its components in the **i**, **j**, **k** directions is in its **Cartesian form**.

If $\mathbf{v} = x\mathbf{i} + y\mathbf{j} + z\mathbf{k}$ then its **direction ratios** are $x : y : z$, and its **direction cosines** are

$$\frac{x}{|\mathbf{v}|}, \frac{y}{|\mathbf{v}|}, \frac{z}{|\mathbf{v}|}.$$

In general if the **direction cosines** of a vector are l, m, n, respectively then $l^2 + m^2 + n^2 = 1$. **Unlike parallel vectors** have **equal direction ratios** but have **direction cosines equal in magnitude but of opposite sign.**

The **scalar product** of two vectors **a** and **b** inclined at angle θ is written as $\mathbf{a} \cdot \mathbf{b}$ and defined by $\mathbf{a} \cdot \mathbf{b} = |\mathbf{a}||\mathbf{b}| \cos \theta$, which is a scalar. If $\mathbf{a} = x_1\mathbf{i} + y_1\mathbf{j} + z_1\mathbf{k}$, and $\mathbf{b} = x_2\mathbf{i} + y_2\mathbf{j} + z_2\mathbf{k}$, then $\mathbf{a} \cdot \mathbf{b} = x_1x_2 + y_1y_2 + z_1z_2$.

$\mathbf{a} \cdot \mathbf{b} = \mathbf{b} \cdot \mathbf{a}$ and $\mathbf{a} \cdot (\mathbf{b} + \mathbf{c}) = (\mathbf{a} \cdot \mathbf{b}) + (\mathbf{a} \cdot \mathbf{c})$.

The angle θ between two vectors $\mathbf{a}$ and $\mathbf{b}$ is given by

$$\cos\theta = \frac{\mathbf{a}\,.\,\mathbf{b}}{|\mathbf{a}|\,|\mathbf{b}|}$$

The **vector product** of two vectors $\mathbf{a}$ and $\mathbf{b}$ inclined at angle θ is written as $\mathbf{a} \times \mathbf{b}$ and defined by $\mathbf{a} \times \mathbf{b} = |\mathbf{a}||\mathbf{b}|\sin\theta\,\hat{\mathbf{n}}$ where $\hat{\mathbf{n}}$ is the **unit vector in the direction perpendicular to the plane containing a and b in the sense of a right-handed screw from a to b**. The vector product gives a vector.

$\mathbf{a} \times \mathbf{b} = -\mathbf{b} \times \mathbf{a}$, and $\mathbf{a} \times (\mathbf{b} + \mathbf{c}) = \mathbf{a} \times \mathbf{b} + \mathbf{a} \times \mathbf{c}$.

The **scalar triple product property** states that when three vectors are to be combined with a **scalar product** and a **vector product**, provided the **order of the vectors is unchanged**, or the vectors are **cyclically interchanged**, then the **order of operations is irrelevant**.

$$\mathbf{a} \times \mathbf{b}\,.\,\mathbf{c} = \mathbf{a}\,.\,\mathbf{b} \times \mathbf{c}.$$

If two vectors $\mathbf{a}$ and $\mathbf{b}$ are **parallel** then $\mathbf{a}\,.\,\mathbf{b} = |\mathbf{a}||\mathbf{b}|$ and $|\mathbf{a} \times \mathbf{b}|$.

If two vectors $\mathbf{a}$ and $\mathbf{b}$ are **perpendicular** then $\mathbf{a}\,.\,\mathbf{b} = 0$ and $|\mathbf{a} \times \mathbf{b}| = |\mathbf{a}||\mathbf{b}|$.

The **vector equation of a line** may be given by $\mathbf{r} = \mathbf{a} + \lambda\mathbf{b}$ where $\mathbf{r}$ is the **general position vector of a point on the line**, $\mathbf{a}$ is a **particular point on the line** and $\mathbf{b}$ is a **direction vector for the line.**

The **length of line of shortest distance** between two lines $\mathbf{r}_1 = \mathbf{a}_1 + \lambda\mathbf{b}_1$ and $\mathbf{r}_2 = \mathbf{a}_2 + \mu\mathbf{b}_2$ is perpendicular to both of the lines i.e., in the direction of $\mathbf{b}_1 \times \mathbf{b}_2$ and is given by $(\mathbf{a}_1 - \mathbf{a}_2)\,.\,\hat{\mathbf{n}}$, where $\hat{\mathbf{n}}$ is the unit vector in the direction perpendicular to both lines.

The **position vector of a point dividing a line** AB in the ratio $\lambda : \mu$ is given by

$$\frac{\lambda\mathbf{OB} + \mu\mathbf{OA}}{\lambda + \mu}.$$

A vector, **a**, may be **resolved into its component** in the direction of another vector, **b**, where the vectors are inclined at angle θ, by

$$\frac{\mathbf{a}\,.\,\mathbf{b}}{|\mathbf{b}|} = |\mathbf{a}| \cos \theta.$$

The **vector equation of a plane** may be given as $\mathbf{r}\,.\,\mathbf{n} = D$, where **r** is the **general position vector of a point on the plane**, **n** is a vector **normal to the plane**. The **distance of the plane from the origin** is given by

$$\frac{D}{|\mathbf{n}|}$$

If a vector **a** is a function of a scalar variable λ i.e., $\mathbf{a}(\lambda) = f(\lambda)\mathbf{i} + g(\lambda)\mathbf{j} + h(\lambda)\mathbf{k}$ then its **derivative with respect to the scalar variable** is:

$$\frac{d\mathbf{a}}{d\lambda} = f'(\lambda)\mathbf{i} + g'(\lambda)\mathbf{j} + h'(\lambda)\mathbf{k}$$

The **integral of the vector a** with **respect to the scalar variable** λ is:

$$\int \mathbf{a}\, d\lambda = \int f(\lambda)\, d\lambda\, \mathbf{i} + \int g(\lambda)\, d\lambda\, \mathbf{j} + \int h(\lambda)\, d\lambda\, \mathbf{k}.$$

In particular if **r** is the position vector of a body, **v** its velocity vector, and **a** its acceleration vector then,

$$\frac{d\mathbf{r}}{dt} = \mathbf{v}, \quad \frac{d\mathbf{v}}{dt} = \mathbf{a} \quad \text{and} \quad \int \mathbf{v}\, dt = \mathbf{r}, \quad \int \mathbf{a}\, dt = \mathbf{v}.$$

Chapter 3
Motion of Particles Moving in Straight Lines

If a particle is moving in a **straight line** it has only **two directions** in which to travel and these are distinguished by giving them **positive** and **negative** signs.

Velocity

Velocity, **v**, is the **rate of change of the displacement**, **s**, i.e., $\mathbf{v} = \dfrac{d\mathbf{s}}{dt}$ and is a vector. It follows that $\mathbf{s} = \int \mathbf{v}\, dt$.

Speed is the magnitude of the velocity, i.e. speed $= |\mathbf{v}|$, and is a scalar.

If a particle has **uniform velocity** then it has **constant speed in a fixed direction**.

The **average velocity** is defined as $\dfrac{\text{increase in displacement}}{\text{time taken}}$

whereas the **average speed** is $\dfrac{\text{total distance travelled}}{\text{time taken}}$.

Consider the **displacement-time** graphs shown below:

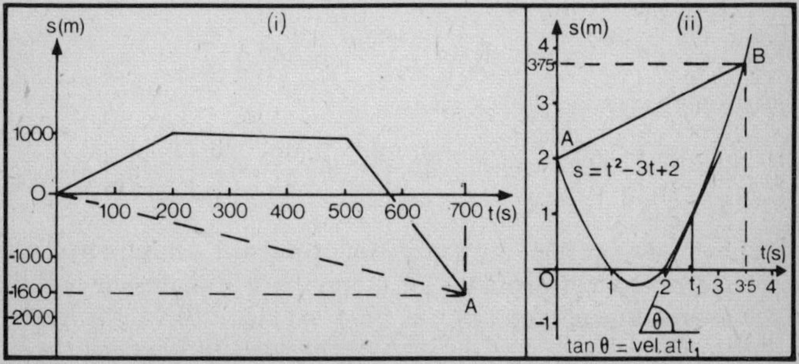

Figure 27

Figure 27 (i) illustrates the journey of a particle moving on a straight line starting from a point O and travelling away from O in the positive direction at a **constant speed** of 5 ms^{-1} for a distance of 1000 m, resting for 300 s, and then travelling back in the negative direction for 2600 m at a **constant speed** of 13 ms^{-1}. We can see that while the particle is moving with **constant velocity** the **displacement-time graph is a straight line**, the **gradient** of which gives the **velocity**.

The **average velocity** is the $\dfrac{\textbf{increase in displacement}}{\textbf{time taken}}$

which is of course the **gradient of the chord *OA*** and is **$-16/7$ ms^{-1}**. The total distance travelled is 3600 m, and the time taken is 700 s. Hence the **average speed** is **36/7 ms^{-1}**.

Figure 27(ii) illustrates the journey of a particle moving on a straight line with its displacement, s, from a fixed point, O, given by $s = t^2 - 3t + 2$. Clearly the velocity is not constant, therefore we can only consider the **velocity at an instant**, which is given by the **gradient of the tangent to the displacement-time curve at that instant**. The **average velocity** over the first 3·5 s is given by the **gradient of the chord *AB***, and is

$$\frac{3{\cdot}75 - 2}{3{\cdot}5} = \mathbf{0{\cdot}5\ ms^{-1}}.$$

Acceleration

The **acceleration, a**, is the **rate of change of velocity** i.e. $\mathbf{a} = \mathrm{d}\mathbf{v}/\mathrm{d}t$ and is a vector. It follows that $\mathbf{v} = \int \mathbf{a}\,\mathrm{d}t$.

Since $a = \dfrac{\mathrm{d}v}{\mathrm{d}t}$ and $\dfrac{\mathrm{d}v}{\mathrm{d}t} = \dfrac{\mathrm{d}s}{\mathrm{d}t}\dfrac{\mathrm{d}v}{\mathrm{d}s}$, then $a = v\dfrac{\mathrm{d}v}{\mathrm{d}s}$

which may also be written

$$\frac{\mathrm{d}}{\mathrm{d}s}\left(\frac{1}{2}v^2\right) \quad \text{thus} \quad a = \frac{\mathrm{d}v}{\mathrm{d}t} = v\frac{\mathrm{d}v}{\mathrm{d}s} = \frac{\mathrm{d}}{\mathrm{d}s}\left(\frac{1}{2}v^2\right).$$

We shall now consider motion under **constant acceleration**. The **velocity-time curve** for motion in a straight line under **constant acceleration** will be a **straight line**, the **gradient** of which will give the **acceleration**. Consider a particle moving in a straight line under constant acceleration a, taking it from initial velocity u, to final velocity v, in time t, over displacement s:

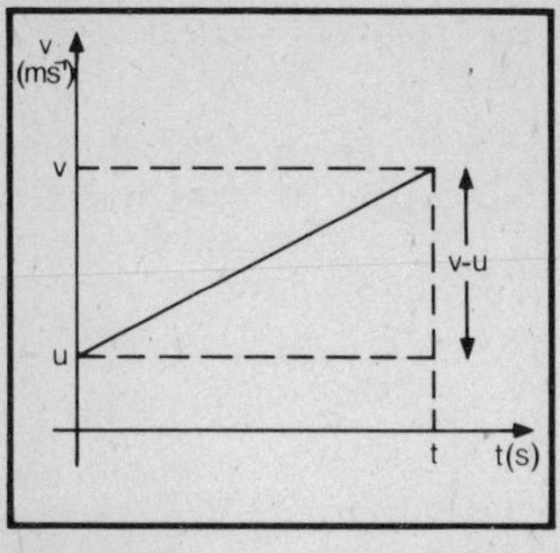

Figure 28.

$a = \frac{v-u}{t}$, giving $\boldsymbol{v = u + at}$,

and since, as we have seen earlier, $s = \int v\,dt$, the **displacement** is the **area under the velocity-time graph, i.e.,**

$$s = ut + \tfrac{1}{2}(v-u)t,$$

thus

$$\boldsymbol{s = ut + \tfrac{1}{2}at^2}.$$

Substituting $t = \frac{v-u}{a}$ into this equation, $\boldsymbol{v^2 = u^2 + 2as}$.

These equations are the **basic equations for motion in a straight line under constant acceleration** and may be quoted,

i.e., $\boldsymbol{v = u + at}$, $\boldsymbol{v^2 = u^2 + 2as}$, $\boldsymbol{s = ut + \tfrac{1}{2}at^2}$.

Also $\boldsymbol{s = vt - \tfrac{1}{2}at^2}$ and $\boldsymbol{s = \tfrac{1}{2}(u + v)t}$.

If the straight line graph were to cross the t-axis then the total **distance** travelled would be the **sum of the areas of the two triangles** formed, but the **nett displacement** would be the **sum of the areas taking account of sign**, i.e., (the area above the t-axis) – (the area below the t-axis).

If we are given **experimental data** about the motion of a particle in a straight line under **variable acceleration**, then depending on the nature of the data, we can draw appropriate graphs to gather more information about the motion.

We know that $a = dv/dt$, and so if we are given corresponding **velocities** and **times**, we can draw the **velocity-time curve**.

The **acceleration at any instant** will be given by the **gradient of the tangent** to the curve at that instant.

Since $s = \int v\,dt$, we know that the **displacement** from t_1 to t_2 will be the **area under the curve** for that interval.

If we are given corresponding **velocities and displacements**, since

$$a = v\frac{dv}{ds} = \frac{d}{ds}\left(\frac{1}{2}v^2\right),$$

then the **gradient of the tangent** to the graph of $\frac{1}{2}v^2$ **against** s at any instant, will give the **acceleration at that instant.**

Also from this data since

$$v = \frac{ds}{dt}, \qquad t = \int \frac{1}{v}\,ds,$$

then the area under the graph of

$$\frac{1}{v} \textbf{ against } s$$

from one displacement s_1 to another s_2 will give the time taken for that journey.

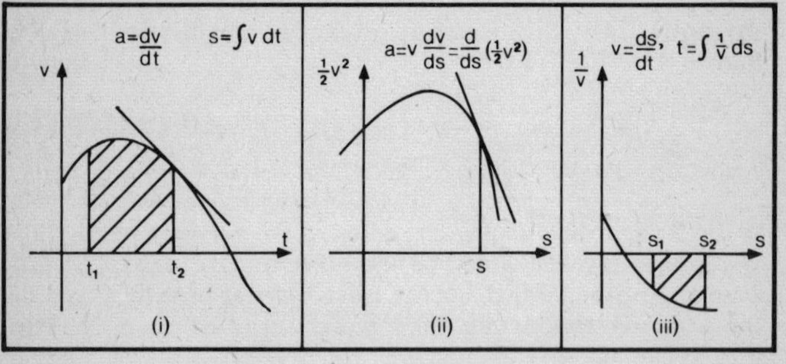

Figure 29

If instead of being given experimental data we are given the variable **acceleration** as a **function of time, velocity, or displacement**, then we can use **calculus** to give us more information about the motion.

If $a = f(t)$, then $v = \int f(t)\,dt$, and $s = \int v(t)\,dt$.

If $a = f(v)$, then since

$$a = \frac{dv}{dt}, \quad t = \int \frac{1}{f(v)}\,dv$$

and since $\quad a = v\dfrac{dv}{ds} \quad$ then $\quad s = \displaystyle\int \frac{v}{f(v)}\, dv.$

If $a = f(s)$, since $a = v\dfrac{dv}{ds}$, then $\displaystyle\int f(s)\, ds = \int v\, dv.$

Motion in a straight line of connected particles

We shall now consider the motion of particles connected by a **light inextensible string** passing over **smooth pulleys of negligible mass**. Their motion is in a **straight line**, the direction being either **upwards** or **downwards**. **Tension** is assumed to be **transmitted through the string** since it is **light** and the pulley is **smooth**.

We write down the **equations of motion** for each particle using **Newton's second law**, eliminate the tensions, and hence find the acceleration for each of the masses.

Consider masses m_1 and m_2 $(m_1 > m_2)$ connected by a light inextensible string passing over a light smooth fixed pulley. Let us find the acceleration for each of the masses and the tension in the string.

Let T be the tension in the string which will be transmitted throughout its length.

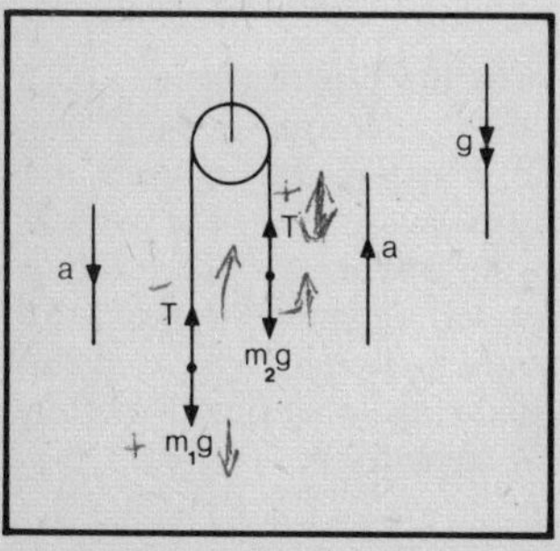

Figure 30.

If a is the acceleration of mass m_1 **downwards**, then m_2 will have acceleration a **upwards** since the length of the string is constant. Applying Newton's second law to each of the masses we have:

$$f = ma$$

$$m_1 g - T = m_1 a \quad \text{and}$$
$$T - m_2 g = m_2 a$$

which on eliminating T gives $\boldsymbol{a} = \dfrac{\boldsymbol{m_1 - m_2}}{\boldsymbol{m_1 + m_2}}\,\boldsymbol{g}$

Substituting this value for a into either of the equations of motion gives

$$\boldsymbol{T} = \frac{2m_1 m_2}{m_1 + m_2} g.$$

When dealing with problems of this kind it is important to be clear about which forces are **external** and which forces are **internal** to the system under consideration: only the **external forces** will affect the motion.

Consider a light scalepan with a particle A of mass m_1 on it attached by a light inextensible string, passing over a smooth fixed pulley to another particle, B of mass m_2 ($m_2 > m_1$).

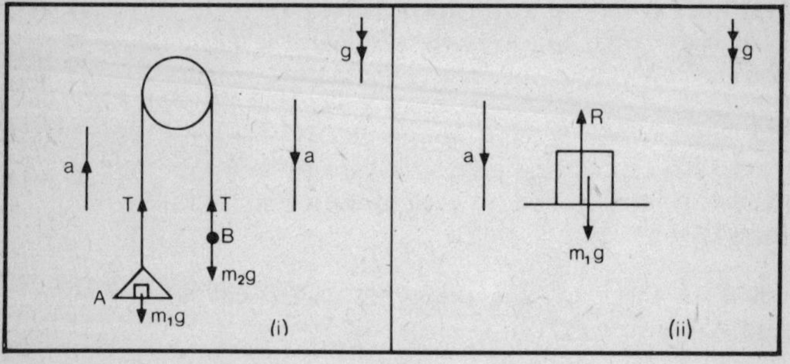

Figure 31.

Taking the scalepan and its contents as the system (figure 31(i)), the **external forces** acting on it are the tension in the string, T, and the weight of A, m_1g. The reaction, R, between the scalepan and A is an **internal** force. However, taking the particle A as the system (figure 31(ii)), then the **external** forces acting on it are its weight, m_1g, and the reaction, R, between it and the scalepan. For B the **external** forces are its weight, $m_2 g$, and the tension, T, in the string. Hence the following equations of motion may be written down from Newton's second law, '$f = ma$':

$T - m_1 g = m_1 a$ for the scalepan and contents;

$R - m_1g = m_1a$ for A; $m_2 g - T = m_2 a$ for B.

We shall consider more complicated examples of connected particles, in Chapter 4.

Worked examples

Example 1 A particle A is travelling in a straight line in the positive direction, with acceleration proportional to the square of its speed v. At time $t = 0$ it passes through a fixed point O with speed gT and with acceleration g. Show that the acceleration is v^2/gT^2, and hence obtain an expression for v in terms of x, g and T.

Prove that, at time t, $x = gT^2 \ln\left(\dfrac{T}{T-t}\right)$

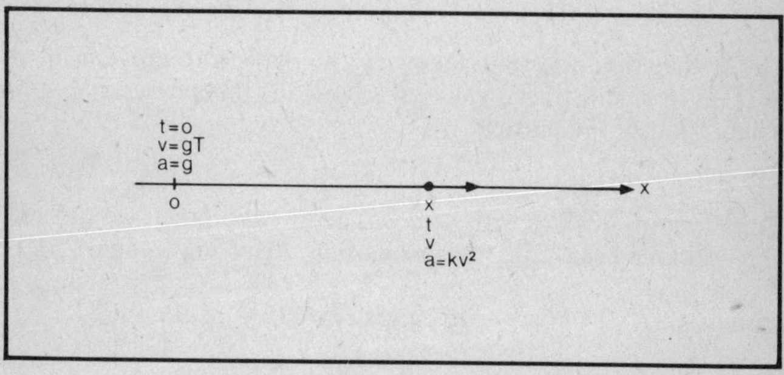

Figure 32.

Let the direction of motion be Ox as shown. The acceleration, a, is proportional to the square of the speed, thus $a \propto v^2$ and $a = kv^2$ where k is the constant of proportionality. We know that when $v = gT$, then $a = g$. Thus

$$k = \frac{1}{gT^2}, \quad \text{and} \quad \boldsymbol{a = \frac{v^2}{gT^2}}. \quad a = v\frac{dv}{dx} = \frac{v^2}{gT^2}$$

and since, when $x = 0$, $v = gT$, then

$$\int_{gT}^{v} \frac{1}{v}\,dv = \frac{1}{gT^2}\int_0^x dx. \quad \text{Integrating,} \quad \Big[\ln v\Big]_{gT}^{v} = \frac{x}{gT^2}.$$

$$\text{Thus } \ln\frac{v}{gT} = \frac{x}{gT^2} \quad \text{and} \quad \boldsymbol{v = gTe^{x/gT^2}}.$$

To find x we shall write v as $\dfrac{dx}{dt}$ to give $\dfrac{dx}{dt} = gTe^{x/gT^2}$,

and since when $x = 0$, $t = 0$ we have $\displaystyle\int_0^x e^{-x/gT^2}\,dx = gT\int_0^t dt.$

Integrating, $-gT^2\left[e^{-x/gT^2}\right]_0^x = gTt.$

Thus $e^{-x/gT^2} = \dfrac{T-t}{T}$, and $\boldsymbol{x = gT^2 \ln\left(\dfrac{T}{T-t}\right)}$.

Example 2 A particle is uniformly accelerated from A to B, a distance of 96 m and is then uniformly retarded from B to C, a distance of 30 m. The speeds of the particle at A and B are 6 ms^{-1} and u ms^{-1} respectively, and the particle comes to rest at C. Express in terms of u only, the times taken to move from A to B and from B to C.

Given that the total time taken by the particle to move from A to C is 18 s, find (i) the value of u, and (ii) the acceleration and retardation of the particle.

(A.E.B. part)

Since we are dealing with uniform acceleration then we can use the quotable basic equations of motion. Referring to figure 33,

from A to B: '$v^2 = u^2 + 2as$' gives $u^2 = 36 + 2a_1 96$, thus

$$a_1 = \frac{u^2 - 36}{192}.$$

$$\text{'}v = u + at\text{' gives } v = 6 + \left(\frac{u^2 - 36}{192}\right)t_1.$$

Thus $t_1 = \dfrac{192(u-6)}{u^2 - 36} = \boldsymbol{\dfrac{192}{u+6}}$.

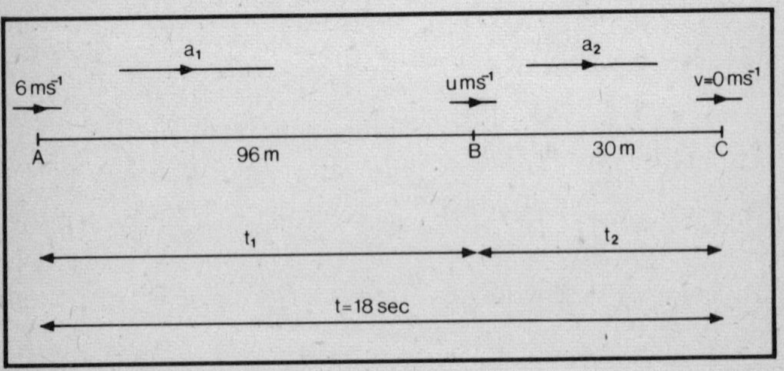

Figure 33.

From B to C, '$v^2 = u^2 + 2as$' gives $0 = u^2 + 2a_2 30$.

Thus $a_2 = -\dfrac{u^2}{60}$.

'$v = u + at$' gives $0 = u - \dfrac{v^2}{60} t_2$. Thus $\boldsymbol{t_2} = \dfrac{\mathbf{60}}{\boldsymbol{n}}$.

If the total time for the journey is 18 s, then $t_1 + t_2 = 18$, i.e.,

$$\frac{192}{u+6} + \frac{60}{u} = 18,$$

$192u + 60(u + 6) = 18u(u + 6)$, and $(u - 10)(u + 2) = 0$.

Rejecting the negative value for u, we have $\boldsymbol{u} = \mathbf{10\ ms^{-1}}$.
We know that the acceleration, a, of the particle is

$$\frac{u^2 - 36}{192}, \text{ therefore } \boldsymbol{a_1} = \frac{\mathbf{1}}{\mathbf{3}}\,\mathbf{ms^{-2}}$$

and $\boldsymbol{a_2} = -\dfrac{u^2}{60} = -\dfrac{\mathbf{5}}{\mathbf{3}}\,\mathbf{ms^{-2}}$.

Example 3 A particle of mass m is projected vertically upwards against gravity in a medium which offers a resistance of $mg(v/k)^2$, where v is the speed of the particle and k is a constant. If u is the speed of projection, find the greatest height of the particle above the point of projection.

Show that $\dfrac{1}{V^2} - \dfrac{1}{u^2} = \dfrac{1}{k^2}$.

where V is the speed of the particle when it returns to the point of projection.

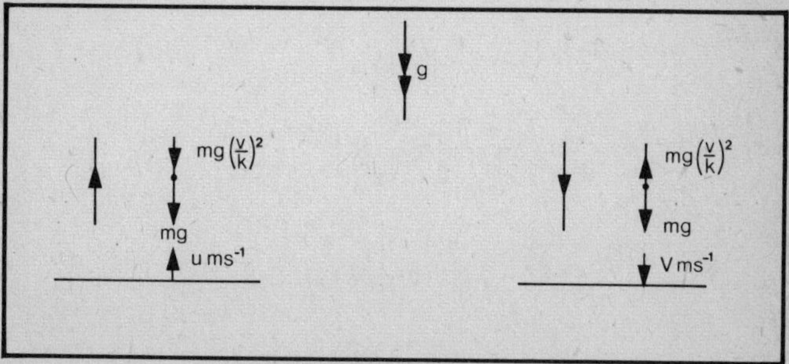

Figure 34.

From Newton's second law '$f = ma$', for the upward part of the journey,

$-mg\left(\frac{v}{k}\right)^2 - mg = ma$, where a is the acceleration of the particle,

thus

$$a = -\left(\frac{gv^2}{k^2} + g\right).$$

Since $a = v\frac{dv}{dx}$, then $v\frac{dv}{dx} = -\left(\frac{gv^2 + gk^2}{k^2}\right)$.

The greatest height, h, of the particle will occur when $v = 0$ (the particle will have just stopped travelling upward and be on the point of travelling downwards).

Thus $\frac{k^2}{g}\int_u^0 \frac{v}{v^2 + k^2}\,dv = -\int_0^h dx$.

Integrating, $h = \frac{k^2}{2g}\ln\left(\frac{u^2 + k^2}{k^2}\right)$

For the return, downward journey, from Newton's second law,

$mg - mg\left(\frac{v}{k}\right)^2 = ma$ where a is the acceleration of the particle.

The distance it will travel downwards to the point of projection will be the value of the greatest height. Thus

$$a = v\frac{dv}{dx} = g\left(\frac{k^2 - v^2}{k^2}\right)$$

and $$\frac{k^2}{g}\int_0^V \frac{v}{k^2 - v^2}\,dv = \int_0^{\frac{k^2}{2g}\ln\left(\frac{v^2 + k^2}{k^2}\right)} dx$$

Integrating, $-\frac{k^2}{2g}\ln\frac{k^2 - V^2}{k^2} = \frac{k^2}{2g}\ln\left(\frac{u^2 + k^2}{k^2}\right)$,

therefore

$$\frac{k^2}{k^2 - V^2} = \frac{u^2 + k^2}{k^2} \quad \text{and} \quad k^4 = k^2u^2 - V^2k^2 - V^2u^2 + k^4.$$

Thus $\frac{1}{V^2} - \frac{1}{u^2} = \frac{1}{k^2}$

Example 4 The table below gives corresponding values of displacement and velocity for a particle moving on a straight line. By drawing a suitable graph, find the acceleration of the particle when its displacements are 2 m and 10 m. By drawing another graph find the time to the nearest second taken to increase its displacement from 2 m to 12 m.

s m	0	2	4	6	8	10	12
v ms^{-1}	2	3	4	5·1	6·3	7·5	9

To find the acceleration of the particle from a graph, we use

$$a = v\frac{dv}{ds} = \frac{d}{ds}\left(\frac{1}{2}v^2\right),$$

so that if we plot $\frac{1}{2}v^2$ against s, then the slope of the tangent to the curve at any point will give the acceleration at that instant.

From the graph of $\frac{1}{2}v^2$ against s, figure 35, when $\mathbf{s = 2}$, and $\mathbf{s = 10}$, the gradients of the tangents to the curve, and hence the accelerations for these displacements, are **1·6 ms^{-1}** and **5 ms^{-2}** respectively.

In order to find the time taken for the particle to increase its displacement from 2 m to 12 m we use

$$v = \frac{ds}{dt}, \text{ so that } \int \frac{1}{v}\,ds = \int dt.$$

Hence, if we plot a graph of $1/v$ against s, the area under the curve between $s = 2$ and $s = 12$ will give the time taken for that journey, see figure 36.

From this graph, we see that the area under the curve, between ordinates $s = 2$ and $s = 12$, and hence the time taken for that journey, is 1·9 s, i.e. **2 s** correct to the nearest second.

Example 5 A train, when braking, is subject to a retardation of

$$\left(1 + \frac{v}{100}\right) \text{ms}^{-2},$$

where v is its speed at any time. The brakes are applied when it is travelling at 20 ms^{-1}.

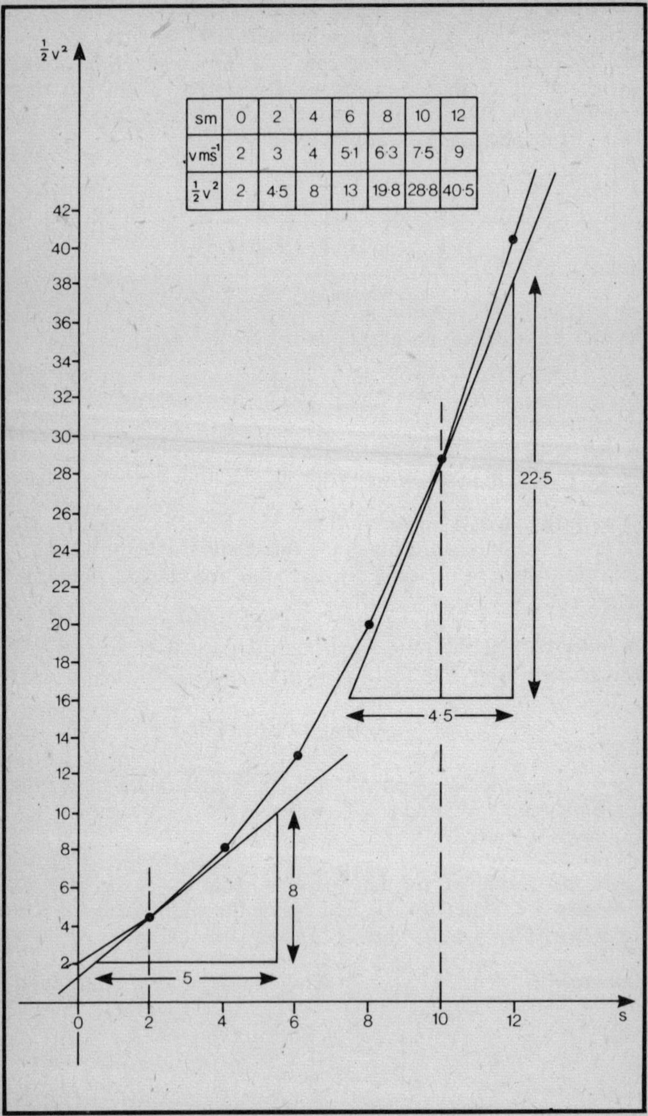

sm	0	2	4	6	8	10	12
vms^{-1}	2	3	4	5·1	6·3	7·5	9
$\frac{1}{2}v^2$	2	4·5	8	13	19·8	28·8	40·5

Figure 35.

sm	0	2	4	6	8	10	12
$\frac{1}{v}$	0·5	0·33	0·25	0·2	0·16	0·13	0·11

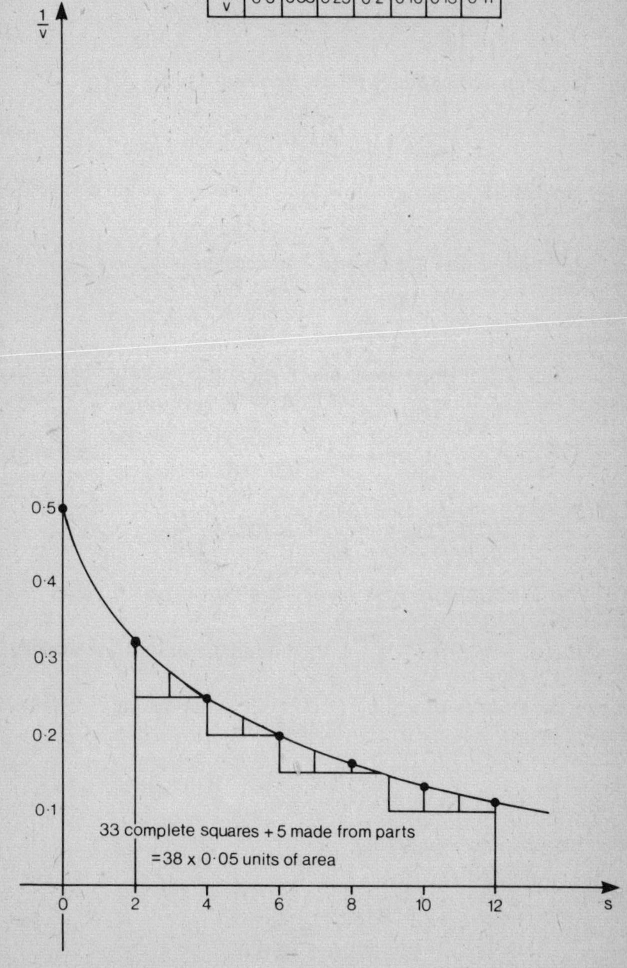

Figure 36.

(i) Show that it takes just over 18 s to come to rest.

(ii) Show that it will travel approximately 177 m after the brakes are applied. (It may prove useful to know that

$$\frac{v}{v+100} \text{ can be written as } 1 - \frac{100}{v+100}.)$$

(iii) If it then accelerates with acceleration

$$\left(1 + \frac{v^2}{100}\right) \text{ms}^{-2},$$

find, correct to 3 sig. figs., how long it takes to attain a speed of 20 ms^{-1}. (S.U.J.B.)

(i) When the brakes are applied, the acceleration,

$$\frac{dv}{dt}, \text{ is } -\left(1 + \frac{v}{100}\right).$$

If the train is travelling at 20 ms^{-1} when the braking begins, the time it takes, t, to come to rest is given by

$$\int_0^t dt = -\int_{20}^0 \frac{100}{100+v}\,dv. \qquad \text{Integrating,}$$

$$\mathbf{t} = -100\Big[\ln(100+v)\Big]_{20}^{0} = -100\ln\frac{100}{120} = \mathbf{18{\cdot}23\ s}$$

i.e. the train takes just over 18 s to come to rest.

(ii) Since $v\dfrac{dv}{dx} = -\left(1 + \dfrac{v}{100}\right)$, then after the brakes are applied

(when the train is travelling at 20 ms^{-1}), the distance, x, that it travels before coming to rest is given by

$$-\int_{20}^0 \frac{100v}{100+v}\,dv = \int_0^x dx$$

Rearranging and integrating,

$$x = -100\int_{20}^0 1 - \frac{100}{v+100}\,dv = -100\Big[v - 100\ln(v+100)\Big]_{20}^{0}$$

$$= -100(100\ln 1{\cdot}2 - 20) = \mathbf{176{\cdot}8\ m}$$

Hence the train travels approximately 177 m before coming to rest.

(iii) If the train accelerates from rest to 20 ms^{-1} with acceleration

$$\left(1+\frac{v^2}{100}\right),$$

then the time, t, taken to achieve this is given by

$$\frac{dv}{dt}=\left(1+\frac{v^2}{100}\right), \quad \text{so that} \quad \int_0^{20}\frac{100}{100+v^2}\,dv=\int_0^t dt.$$

$$\mathbf{t}=\left[\arctan\frac{v}{10}\right]_0^{20}=\arctan 2=\mathbf{1{\cdot}11\ s} \text{ correct to 3 sig. figs.}$$

Example 6 Two particles A and B of mass m_1 and m_2 respectively, where $m_1 > m_2$, are connected by a light inextensible string passing over a smooth fixed pulley. They are released from rest with the hanging parts of the string vertical. Find the force on the pulley.

After time, t_1, the string breaks just as the particles are passing each other. The particle A strikes the ground after a further time t_2. Find the height of particle B at this time.

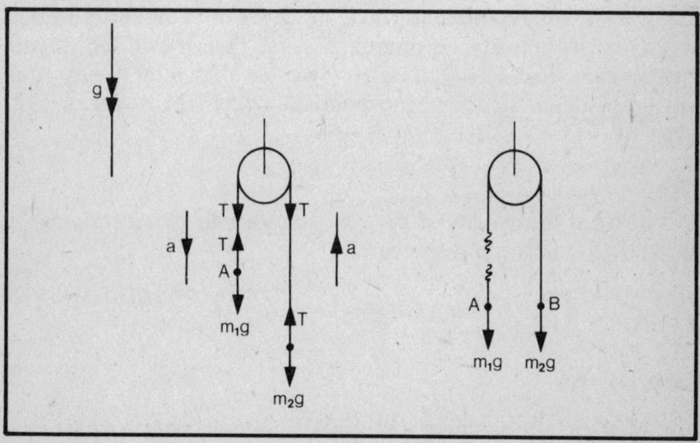

Figure 37.

Since the pulley is smooth and the string light, the tension, T, will be transmitted throughout the string. Applying Newton's second law, '$F = ma$', to the system we have for A, $m_1 g - T = m_1 a$ and for B, $T - m_2 g = m_2 a$ since the acceleration, a, will have the same magnitude for both particles.

Eliminating T from these equations,

$$a = \frac{(m_1 - m_2)g}{(m_1 + m_2)}$$

Thus $\mathbf{T} = m_1 g - m_1 \dfrac{(m_1 - m_2)g}{(m_1 + m_2)} = \dfrac{2m_1 m_2 g}{(m_1 + m_2)}$

The **force on the pulley** is $2T$ as shown in figure 37, which is

$$\frac{4m_1 m_2 g}{(m_1 + m_2)}$$

If the particles are travelling for time t_1 before the string breaks, since they have **constant** acceleration, a, then we can find the speed, v, at which they are travelling from '$v = u + at$'. (They must both have the same speed, v, but in opposite directions, since the length of the string is constant.) Hence $v = at_1$.

In time t_2, A will fall a distance, s_1, given by '$s = ut + \frac{1}{2}at^2$', and remembering that the acceleration is now only g, we see that

$$s_1 = at_1 t_2 + \tfrac{1}{2}gt_2^2.$$

B, however, will continue to travel upwards until, under the force of gravity, its velocity becomes zero, and then it will fall downwards again. From '$s = ut + \frac{1}{2}at^2$' we see that after time t_2, B will be distance s_2 above the point at which the string broke, given by

$$s_2 = at_1 t_2 - \tfrac{1}{2}gt_2^2.$$

B will be a distance of $s_1 + s_2$ above the ground when A hits the ground, i.e. a distance of

$$at_1 t_2 + \tfrac{1}{2}gt_2^2 + at_1 t_2 - \tfrac{1}{2}gt_2^2 = \frac{2(m_1 - m_2)gt_1 t_2}{(m_1 + m_2)}$$

Key terms

For a particle moving in a straight line with **acceleration** a, **velocity** v, and **displacement** s, after time t, the following relationships hold:

$$a = \frac{dv}{dt}\left(\frac{ds}{dt}\frac{dv}{ds}\right) = \frac{d}{ds}\left(\frac{1}{2}v^2\right);$$

$$v = \int a\, dt; \quad v = \frac{ds}{dt}; \quad s = \int v\, dt; \quad t = \int \frac{1}{v} ds.$$

Average velocity = $\dfrac{\text{increase in displacement}}{\text{time taken}}$

Average speed = $\dfrac{\text{Total distance travelled}}{\text{time taken}}$

We have derived certain equations of motion for **constant** acceleration: $v^2 = u^2 + 2as$; $s = ut + \frac{1}{2}at^2$; $s = vt - \frac{1}{2}at^2$; $v = u + at$; $s = \frac{1}{2}(u + v)t$.

These may be quoted when solving problems involving constant accelerations.

When particles are connected to both ends of a **light inextensible string** passing over a **smooth** fixed pulley, the tension is transmitted throughout the string, and both particles move with velocities and accelerations of the same magnitudes (but of opposite directions) since the string is of constant length. They move under the constant forces of **tension** and **gravity**.

Chapter 4
Friction

Friction is the name given to that force which comes into play between two rough surfaces in contact, in order to resist motion between them. The following **experimental laws** about friction have been formed:

i) Friction opposes motion.

ii) Up to a **limiting value**, the magnitude of the frictional force is just sufficient to prevent motion.

iii) The **limiting value** of the frictional force is directly proportional to the normal reaction between the surfaces, the constant of proportionality being the **coefficient of friction**, μ. The **coefficient of friction** is determined by the nature of the surfaces in contact.

iv) After motion has started the frictional force may be considered to be equal to the limiting value.

Consider a body of mass *m*, resting on a horizontal plane and being subjected to a gradually increasing force, *P*.

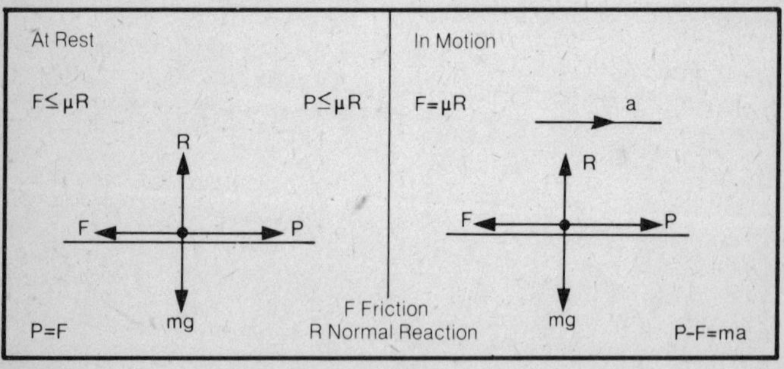

Figure 38.

As *P* is increased in magnitude, *F* will increase accordingly in order to prevent motion. However there comes a point at which motion begins: this is when *P* just exceeds μR so that *F* has reached its limiting value and can increase no more to prevent the motion. F will now remain at μR no matter how much P is increased.

When friction is **limiting**, the **angle** between the **resultant reaction** and the **normal reaction** of one surface on the other is called the **angle of friction**, usually denoted by λ.

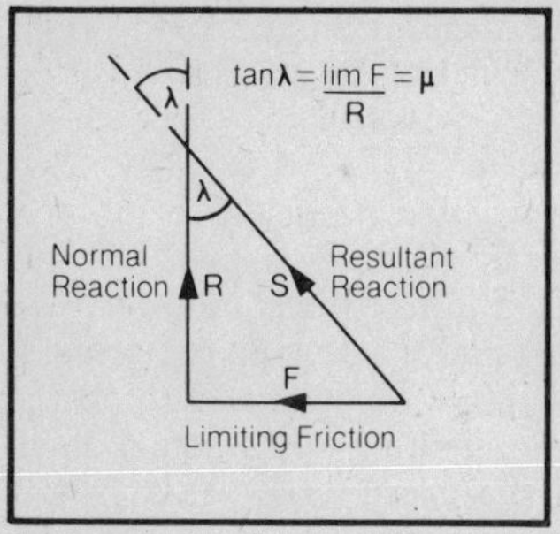

Figure 39.

We see from the triangle of vectors that:

$$\tan \lambda = \frac{\text{limiting friction}}{\text{normal reaction}}$$

$$= \frac{\mu R}{R} = \mu,$$

i.e., **the tangent of the angle of friction is the coefficient of friction.**

Consider now a particle in **rough contact** with an inclined plane as shown in figure 40. If it is not moving, then clearly **the resultant force acting on it must be zero.**

Let us then consider the forces acting on the particle. We have its weight, mg, acting vertically downwards; the normal reaction, R, acting perpendicular to the plane; and since, if the contact were smooth, the particle would slide down the plane, we must have friction, F, acting up the plane to prevent motion.

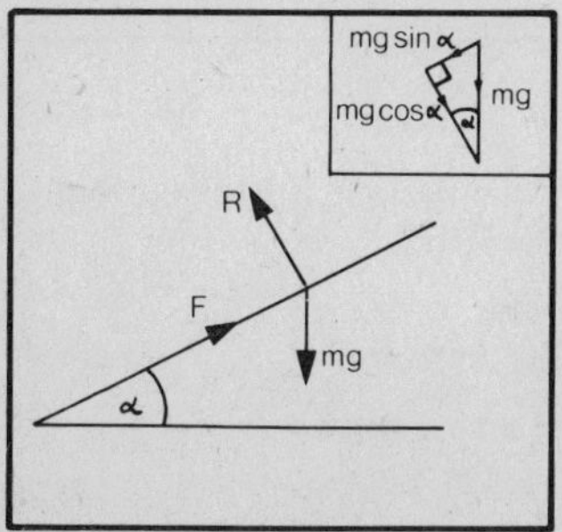

Figure 40.

It is convenient to **resolve the forces into their components parallel and perpendicular to the plane** and equate them to zero.

Parallel to the plane,

$$F - mg \sin \alpha = 0$$

and, perpendicular to the plane,

$$R - mg \cos \alpha = 0.$$

If the particle were on the **point of slipping down the plane** we would be in the situation of **limiting friction**, $\boldsymbol{F = \mu R}$. Since $R = mg \cos \alpha$ and $F = mg \sin \alpha$, then in **limiting friction**, $\boldsymbol{\mu = \tan \alpha}$.

This means that if a particle were placed on an inclined plane and the angle of inclination to the horizontal were increased until the particle just began to slide, then the tangent of the angle of inclination would give the coefficient of friction for that contact.

Let us now look at the situation where a particle is being prevented from sliding down the plane by a force P inclined at angle θ to the plane. P may be such that it just prevents motion down the plane so that friction is limiting and is acting up the plane. P may be such that the particle is on the point of sliding up the plane, in which case friction is again limiting but is acting down the plane this time. Any value of P between those values means that the particle is at rest and friction is not limiting.

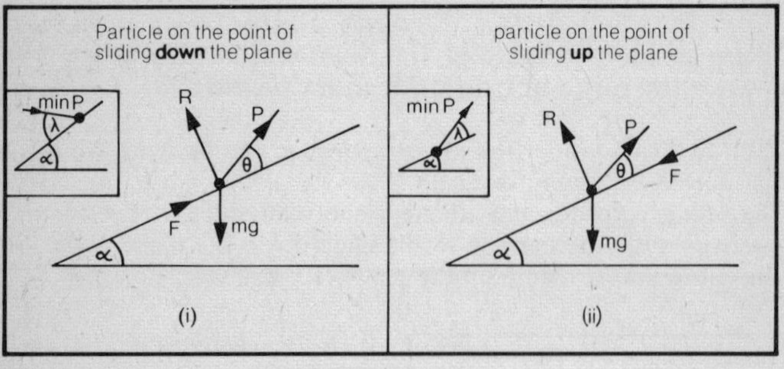

Figure 41.

Let us first consider the case when the particle is **just** prevented from sliding **down** the plane (figure 41(i)).

Resolving parallel to the plane,

$$P \cos \theta + F - mg \sin \alpha = 0$$

and **resolving perpendicular to the plane,**

$$R + P \sin \theta - mg \cos \alpha = 0$$

Since we are in **limiting friction**, then

$$F = \mu R = \mu(mg \cos \alpha - P \sin \theta)$$

Thus $$P = \frac{mg\,(\sin\alpha - \mu\cos\alpha)}{(\cos\theta - \mu\sin\theta)}$$

Since $\mu = \tan\lambda$, where λ is the angle of friction, then

$$P = \frac{mg\sin(\alpha - \lambda)}{\cos(\theta + \lambda)}.$$

If we were to vary θ in order to give the minimum force needed to prevent the particle from sliding down the plane, then we can see that P will be minimum when $\cos(\theta + \lambda) = 1$, i.e., when $\theta = -\lambda$. We can say that the **minimum force** needed to prevent the particle from sliding **down** the plane is $mg\sin(\alpha - \lambda)$ and occurs when the line of action of P makes a clockwise angle of λ with the upward direction of the slope.

If we now consider the case when the particle is on the point of sliding up the plane (figure 41(ii)) we can see that,

resolving parallel to the plane,

$$P\cos\theta - F - mg\sin\alpha = 0$$

and resolving perpendicular to the plane,

$$R + P\sin\theta - mg\cos\alpha = 0.$$

Again, we are in **limiting friction**, so that $F = \mu R$. From these three equations we see that

$$P = \frac{mg(\sin\alpha + \mu\cos\alpha)}{(\cos\theta + \mu\sin\theta)}.$$

Substituting $\mu = \tan\lambda$, $$P = \frac{mg\sin(\alpha + \lambda)}{\cos(\theta - \lambda)}.$$

This time, minimising P, to give the **least force** necessary to cause sliding **up** the plane, we see that this value occurs when $\cos(\theta - \lambda) = 1$ and $\theta = \lambda$; i.e., the **least force** necessary to cause sliding **up** the plane is $mg\sin(\alpha + \lambda)$ at an angle of λ with the upward direction of the slope.

If we are told that a particle is in **limiting equilibrium**, then it is at rest but on the point of motion and **friction** is **limiting**.

Worked examples

Example 1 Three particles A, B, C are of masses 4, 4, 2 kg respectively. They lie at rest on a horizontal table in a straight line, with particle B attached to the midpoint of a light inextensible string, and A and B attached to both ends. The string is taut. A force of 60 N is applied to A in the direction of CA produced, and a force of 15 N is applied to C in the opposite direction. Find the acceleration of the particles and the tension in each part of the string if the coefficient of friction is $\frac{1}{4}$. (Take g to be 10 ms^{-2}).

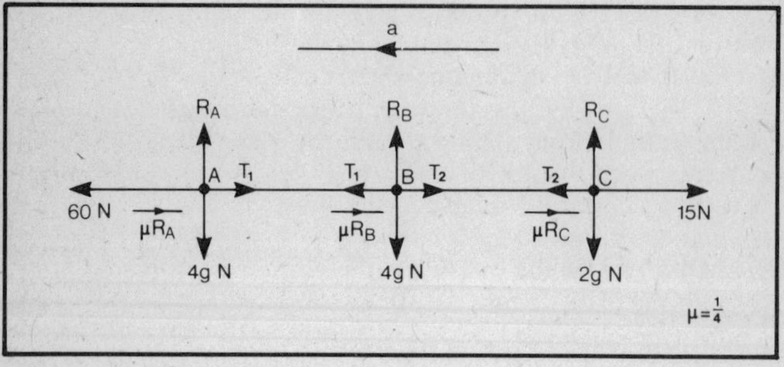

Figure 42.

Tension cannot be transmitted across a break in the string, so that tension, T_1, will be transmitted through part AB of the string and T_2 will be transmitted through part BC. Each part of the system will have the same acceleration, a, since the string is a fixed length, and we assume that it does not break. We shall apply Newton's second law to each particle in turn, and use the fact that, since the system is in motion, friction will have reached its limiting value μR. Thus for A, $60 - T_1 - \mu R_A = 4a$; for B, $T_1 - \mu R_B - T_2 = 4a$; for C, $T_2 - \mu R_C - 15 = 2a$.

Since there is no vertical motion, there must be no resultant vertical component of force, i.e., $R_A = 4g$, $R_B = 4g$ and $R_C = 2g$. Using these values for the normal reactions, $g = 10$, and $\mu = \frac{1}{4}$, gives $50 - T_1 = 4a$, $T_1 - T_2 = 4a + 10$, and $T_2 - 20 = 2a$.

Eliminating T_1 from the first two equations, $T_2 = 40 - 8a$, and substituting this into the third equation, $a = 2$ ms^{-2}.

Hence, $\boldsymbol{T_1 = 42}$ **N**, $\boldsymbol{T_2 = 24}$ **N**.

Example 2 A particle of weight W rests on a rough horizontal plane, the coefficient of friction being $\frac{12}{5}$. The particle is acted on by a force W inclined at an angle of arctan $(\frac{4}{3})$ to the vertical and by a variable force P inclined to the vertical at an angle of arctan ($\frac{12}{5}$). These forces act in the same vertical plane but on different sides of the vertical through the particle. Determine whether, as the magnitude of P increases from zero, the particle will slip before it is lifted from the plane.

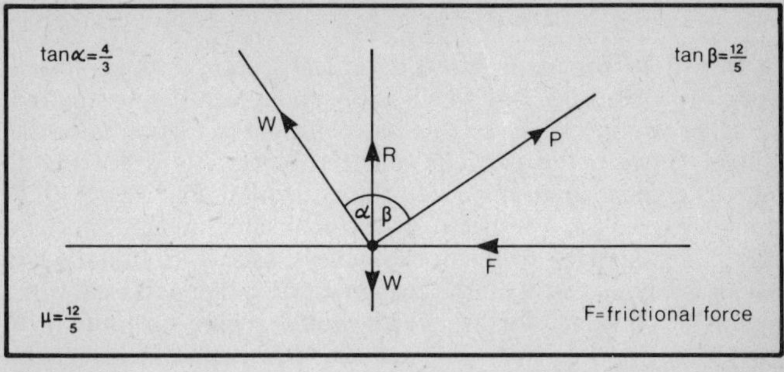

Figure 43.

Taking the forces as shown in figure 43 and resolving parallel to the plane, and perpendicular to the plane, for equilibrium:

$$P \sin \beta - F - W \sin \alpha = 0,$$

and $$R + P \cos \beta + W \cos \alpha - W = 0.$$

The maximum value of F is

$$\mu R = \frac{12}{5}(W - W \cos \alpha - P \cos \beta)$$

$$= \frac{12}{5}\left(W - \frac{3}{5}W - \frac{5}{13}P\right) = \frac{24}{25}W - \frac{12}{13}P.$$

For sliding to take place

$$P \sin \beta - W \sin \alpha > \mu R, \quad \text{i.e.,} \quad \frac{12}{13}P - \frac{4}{5}W > \frac{24}{25}W - \frac{12}{13}P$$

Therefore, $\boldsymbol{P > \frac{143}{150} W}$ **for sliding.**

For lifting to occur, contact will be lost between the particle and the plane so at the instant of lifting, $R = 0$, and

$$P \cos \beta + W \cos \alpha > W,$$

$$\text{i.e.,} \quad \frac{5}{13}P + \frac{3}{5}W > W, \quad \text{so that } P > \frac{26}{25}W.$$

Comparing the conditions for lifting and sliding tells us that **sliding will take place first when**

$$P = \frac{143}{150} W.$$

Example 3 Figure 44 shows a vertical section through a fixed block of wood *ABCD* in which the vertical face is smooth, and the other two faces are rough with coefficient of friction μ for the contact between the particles and the block. *BC* is horizontal and *CD* is inclined at angle α to the horizontal. Particles *P*, *Q*, *R* of mass m_1, m_2, m_3, respectively are connected by light inextensible strings passing over smooth pulleys, as shown. Initially the system is held at rest with the string taut. It is then released. Find the range of values for m_1 which would ensure no motion. If $\alpha = \arcsin \frac{3}{5}$, $\mu = \frac{1}{2}$, $m_1 = 3m_2$ and $m_3 = 2m_2$, find the acceleration of the system and the tensions in the strings.

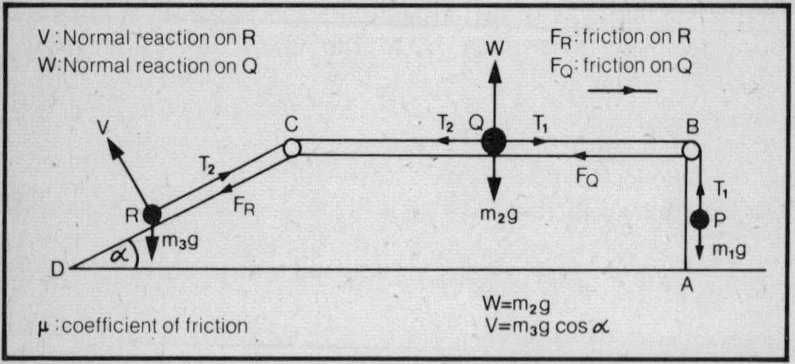

Figure 44.

Tension will be transmitted across **smooth** pulleys but not across the point of attachment of *Q*. Hence the tensions are as shown in figure 44. When the system is released from rest, there are three possibilities for it: it may stay at rest; it may begin to move in the direction *RQP* so that friction will be limiting for *R* and *Q*, acting in the direction which opposes motion; it may begin to move in the direction *PQR* so that friction will again be limiting but in the opposite sense to the previous case.

If no motion takes place on release of the system there must be no resultant force acting on the system. The extreme values for m_1 will be those for the system to be on the point of motion in either of the two directions mentioned. We shall obtain the equations for equilibrium in both cases when there is a tendency to move in the direction RQP and when there is a tendency to move in the direction PQR. The normal reactions are as shown in figure 44.

Consider the equilibrium conditions for the system, parallel to the surface, when it has a tendency to move in the direction RQP: $m_1 g = T_1$ for P; $T_1 = T_2 + F_Q$ for Q; and $T_2 = m_3 g \sin\alpha + F_R$ for R.

Eliminating T_1 and T_2, $F_R + F_Q = m_1 g - m_3 g \sin\alpha$.

Since $F_R \leq \mu m_3 g \cos\alpha$ and $F_Q \leq \mu m_2 g$, then for equilibrium,

$$m_1 g - m_3 g \sin\alpha \leq \mu m_3 g \cos\alpha + \mu m_2 g$$

$$\boldsymbol{m_1 \leq \mu m_3 \cos\alpha + \mu m_2 + m_3 \sin\alpha.}$$

Let us now consider the equilibrium conditions for the system parallel to the surface, when there is a tendency for motion in the direction PQR: $T_1 = m_1 g$ for P; $T_2 = T_1 + F_Q$ for Q; $m_3 g \sin\alpha = T_2 + F_R$ for R (friction is now acting in the opposite sense).

Eliminating T_1 and T_2, $F_R + F_Q = -m_1 g + m_3 g \sin\alpha$.

Again, $F_R \leq \mu m_3 g \cos\alpha$ and $F_Q \leq \mu m_2 g$, so that

$$-m_1 g + m_3 g \sin\alpha < \mu m_3 g \cos\alpha + \mu m_2 g$$

$$\text{and } \boldsymbol{m_1 \geq m_3 \sin\alpha - \mu m_3 \cos\alpha - \mu m_2.}$$

Hence, the range of values for m_1 such that no motion will take place is:

$$\boldsymbol{m_3 \sin\alpha - \mu m_3 \cos\alpha - \mu m_2 \leq m_1 \leq m_3 \sin\alpha + \mu m_3 \cos\alpha + \mu m_2.}$$

If $\alpha = \arcsin\frac{3}{5}$, $\mu = \frac{1}{2}$ and $m_1 = 3m_2$, $m_3 = 2m_2$, by inspecting the above inequalities, we see that

$$m_1 > m_3 \sin\alpha + \mu m_3 \cos\alpha + \mu m_2$$

thus motion will take place in the direction RQP.

Let a be the acceleration of the system. The equations of motion are: $m_1 g - T_1 = m_1 a$ for P; $T_1 - T_2 - \mu m_2 g = m_2 a$ for Q; $T_2 - m_3 g \sin\alpha - \mu m_3 g \cos\alpha = m_3 a$ for R (clearly, friction is at its limiting value).

Substituting the given values into these equations:

$3m_2 g - T_1 = 3m_2 a$; $\quad T_1 - T_2 - \frac{1}{2}m_2 g = m_2 a$;

$T_2 - 2m_2 g\frac{3}{5} - m_2 g\frac{4}{5} = 2m_2 a.$

Eliminating T_1 and T_2, $\boldsymbol{a = \frac{g}{12}}$.

Thus $\boldsymbol{T_1 = \frac{11}{4} m_2 g}$ and $\boldsymbol{T_2 = \frac{13}{6} m_2 g}$.

Key terms

When two surfaces in **rough contact** tend to move relative to each other, frictional forces come into play to oppose the potential motion.

Friction can increase in magnitude to be just sufficient to prevent motion up to a **limiting value** of $F = \mu R$, where R is the normal reaction and μ is the coefficient of friction for that contact. Hence at all times, $\boldsymbol{F \leq \mu R}$.

When the system is on the **point** of motion, $\boldsymbol{F = \mu R}$.

The **resultant reaction** of one surface on another is made up of two components: the frictional force parallel to the surface of contact and the normal reaction perpendicular to the surface of contact. The angle the resultant reaction makes with the normal reaction when friction is limiting is called the **angle of friction**, λ, and $\boldsymbol{\tan \lambda = \mu}$.

Chapter 5
Relative Motion

The position of a particle is defined by its displacement from a point which is regarded as fixed. The **frame of reference** of this point is an essential feature of the definition. The point which we regard as stationary in one frame of reference may well not be stationary in another frame of reference. A person sitting in a train moving with velocity $\mathbf{v}$ is stationary relative to the train, but relative to an observer on the earth's surface, the passenger will be moving with velocity $\mathbf{v}$.

Hence, any velocity which we observe, is a velocity relative to some frame of reference. In the absence of a stated frame, we assume the earth's surface.

Resultant velocity

Consider a person moving with velocity $\boldsymbol{u}$, relative to a train moving with velocity $\boldsymbol{v}$, relative to the earth's surface. To an observer on the ground, the person will have a **resultant velocity** which is the **vector addition of the two velocities**.

We are often asked to deal with the situations where we are told that a boat can travel at a certain speed in **still water** and asked to find its actual i.e., resultant velocity in the presence of a **current**. Similarly we may be told the speed of a plane in **still air** and asked to find the resultant velocity in the presence of a **wind**.

Relative velocity

If a moving object A is viewed from another moving object B, the motion of A as viewed from B will be the vector difference of the velocity of A and the velocity of B, and is called the **velocity of *A* relative to *B***.

Let $\mathbf{V}_A$ be the velocity of A and $\mathbf{V}_B$ be the velocity of B, then the velocity of A relative to B is $\mathbf{V}_A - \mathbf{V}_B$, often written as ${}_A\mathbf{V}_B$, and the velocity of B relative to A is $\mathbf{V}_B - \mathbf{V}_A$ also written as ${}_B\mathbf{V}_A$; see figure 45.

The velocity of A relative to B is the velocity which A appears to have when viewed from B. This may be considered to be the resultant velocity of A when a velocity of $-V_B$ is imposed on the whole system, making B 'stationary'; see figure 45(i).

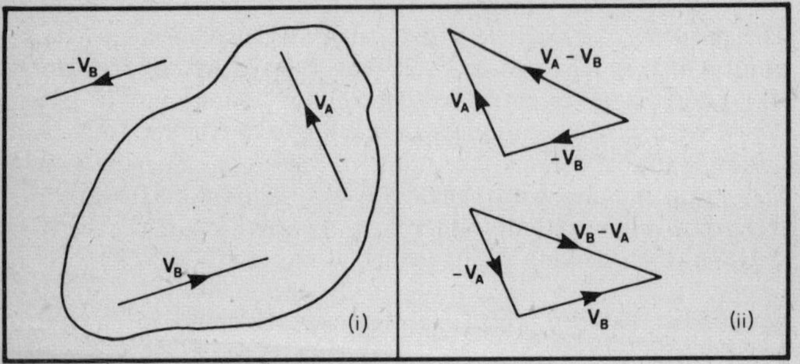

Figure 45.

Relative displacement

When one moving object A is viewed from another moving object B, the **relative displacements** of A from B depend on their **initial displacement** and on their **velocities**. Consider the vector polygon below:

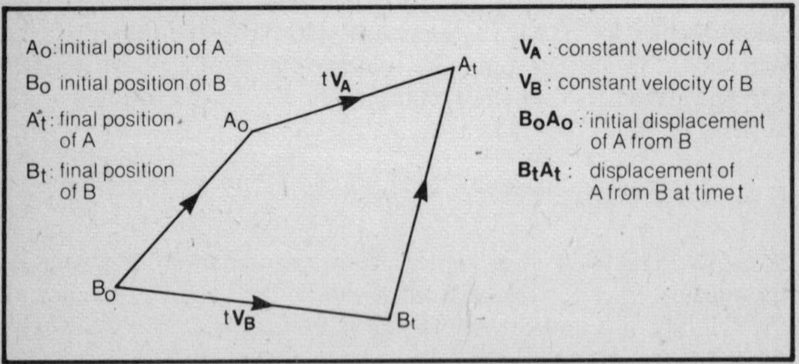

Figure 46.

The velocities are **constant** so that after time t, A will have travelled distance $V_A t$ in the direction of its velocity, and B will have travelled distance $V_B t$ in the direction of its velocity.

Hence, from the vector polygon we see that:

$$\mathbf{B_tA_t} = \mathbf{B_tB_o} + \mathbf{B_oA_o} + \mathbf{A_oA_t}$$

i.e., $\mathbf{B_tA_t} = -\mathbf{V_B}t + \mathbf{B_oA_o} + \mathbf{V_A}t = \mathbf{B_oA_o} + (\mathbf{V_A} - \mathbf{V_B})t$

Thus, **the displacement vector of *A* relative to *B* after time *t*, is the sum of the initial relative displacement vector, and the product of the time travelled with the velocity vector of *A* relative to *B*.**

The **shortest distance between *A* and *B*** may be found from the above relationship, that $\mathbf{B_tA_t} = \mathbf{B_oA_o} + (\mathbf{V_A} - \mathbf{V_B})t$, by considering the vector triangle; see figure 47(i).

Clearly the shortest distance will occur when $\mathbf{B_tA_t}$ is at right angles to $(\mathbf{V_A} - \mathbf{V_B})t$ and is represented by the length of this perpendicular.

We can also find the shortest distance if we make B 'stationary' by saying that the path of A viewed from B is in the direction of the relative velocity. If we know the positions A_o and B_o of A and B at a particular instant, then by drawing the relative path of A to B through A_o, the shortest distance apart will be given by the length represented by the perpendicular from B_o to the relative path; see figure 47(ii).

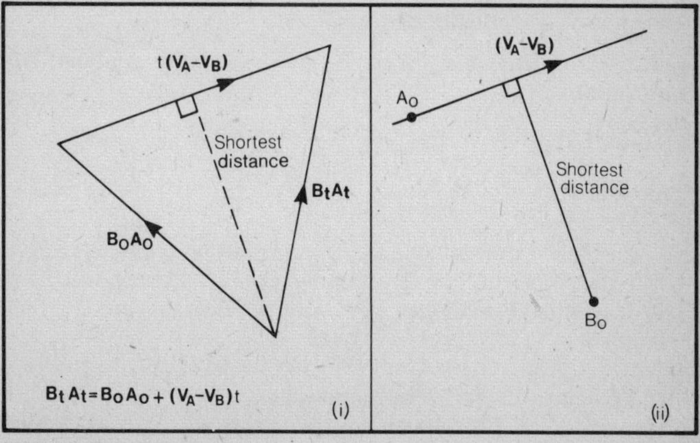

Figure 47.

The **line of closest approach** is the direction which a body of **fixed speed** but with the ability to **choose course**, would choose in order to approach as closely as possible to another body travelling with **fixed velocity**.

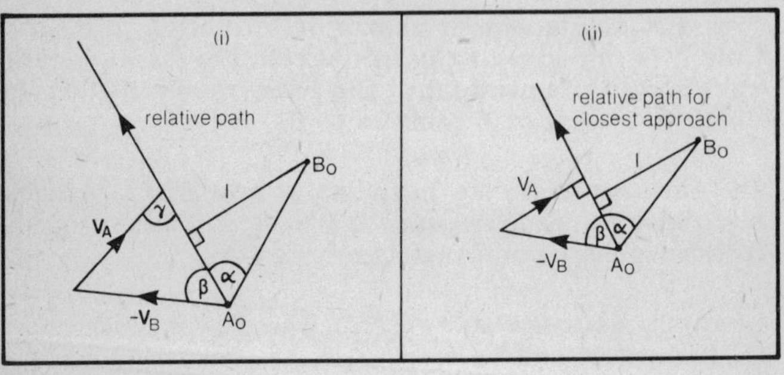

Figure 48.

Consider particle B moving with fixed velocity $\mathbf{V_B}$ and particle A having fixed speed $|\mathbf{V_A}|$ and choosing direction to approach B as closely as possible. Once the relative path is known, then as we have seen earlier, the shortest distance between A and B is $|\mathbf{A_o B_o}| \sin \alpha = l$.

In figure 48(i), the fixed quantities are:

$$(\alpha + \beta), \quad \mathbf{V_B}, \quad |\mathbf{V_A}|, \quad \mathbf{A_o B_o},$$

and the variable quantities are α, l, γ.

Since $l = |\mathbf{A_o B_o}| \sin \alpha$ and $\mathbf{A_o B_o}$ is fixed, then to minimise l we must minimise $\sin \alpha$.

$(\alpha + \beta)$ is constant, so we need to maximise β.

We see that $\dfrac{|\mathbf{V_B}|}{\sin \beta} = \dfrac{|\mathbf{V_A}|}{\sin \gamma}$.

In order to maximise β, we must maximise $\sin \gamma$, which is clearly when γ is 90°. Hence the direction which A should choose is such that it will make an angle of 90° with the relative path.

Often, problems on resultant velocity, relative velocity and relative path may be solved using graphical methods. The alternatives are to use trigonometry on the diagrams derived to solve the problems, or to represent the velocities and displacements in vector notation (usually cartesian coordinates are the most convenient) and then use vector algebra on them. We shall illustrate all of these methods in the following examples.

Worked examples

Example 1 A river flows at 5 ms^{-1} from west to east between parallel banks which are at a distance of 300 metres apart. A man rows a boat at a speed of 3 ms^{-1} in still water.

(i) State the direction in which the boat must be steered in order to cross the river from the southern bank to the northern bank in the shortest possible time. Find the time taken and the actual distance covered for this crossing.

(ii) Find the direction in which the boat must be steered in order to cross the river from the southern bank to the northern bank by the shortest possible route. Find the time taken and the actual distance covered by the boat for this crossing.

(A.E.B.)

(i) The resultant velocity of the boat will have components of $5 - 3\cos\alpha$ in the easterly direction, and $3\sin\alpha$ in the northerly direction. The boat must travel 300 m in the northerly direction in order to cross the river (see figure 49).

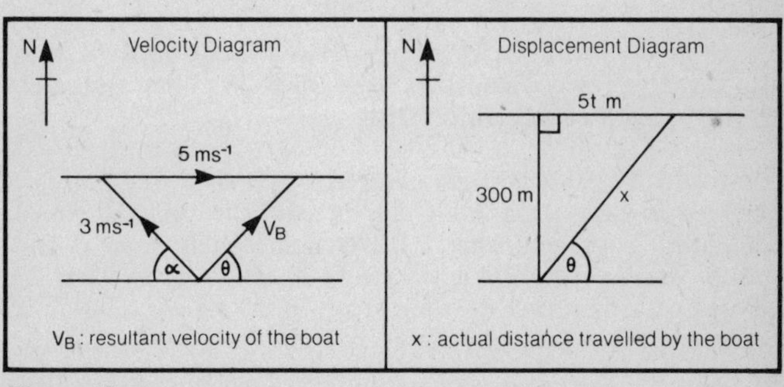

Figure 49.

If the boat takes time t to cross the river:

$$3\sin\alpha\, t = 300, \qquad t = \frac{100}{\sin\alpha}.$$

In order to minimise t, we must maximise $\sin\alpha$, i.e., $\alpha = 90°$.

Hence the boat must be steered in the northerly direction to cross the river in the shortest time possible of 100 s.

The **actual distance covered** is the resultant of a displacement of 300 m due north, and 500 m east i.e., $\sqrt{(500)^2 + (300)^2} =$ **583·1 m.**

(ii) The time taken to cross the river is $\dfrac{100}{\sin \alpha}$.

To minimise the actual distance travelled, we must minimise the easterly component of the journey, since the northerly component is fixed at 300 m.

The easterly component is $(5 - 3 \cos \alpha)t$,

$$\text{i.e. } (5 - 3 \cos \alpha)\frac{100}{\sin \alpha}$$

We must minimise this, subject to the constraint that $\sin \alpha \neq 0$ (otherwise the time would be infinite).

$$\frac{d}{d\alpha}\left(\frac{5}{\sin \alpha} - 3 \cot \alpha\right) = -5 \operatorname{cosec} \alpha \cot \alpha + 3 \operatorname{cosec}^2 \alpha.$$

Stationary values occur when $-\operatorname{cosec} \alpha\,(5 \cot \alpha - 3 \operatorname{cosec} \alpha) = 0$. Therefore, either $\operatorname{cosec}^2 \alpha = 0$ which is impossible, or $\cos \alpha = \frac{3}{5}$. Clearly the latter will give the minimum value.

Hence **the minimum distance travelled occurs when the boat takes a direction of N 36·9° W. This distance is 400 m and takes time 125 s.**

Example 2 At 12.00 h an aeroplane *A* is sighted 20 km due north of an airfield, travelling in a direction due east with speed 300 km h^{-1}. An aeroplane *B* starts immediately from *O* and travels with speed 400 km h^{-1} in a direction due north. By calculation, or by scale drawing in which the scales used should be stated:

(i) find the magnitude and direction of *B*'s velocity relative to *A*, the closest distance apart and the bearing of *B* from *A* when they are closest;

(ii) determine what course (at speed 400 km h^{-1}) *B* ought to have taken in order to intercept *A* and the time at which they would then have met.

(S.U.J.B.)

(i)

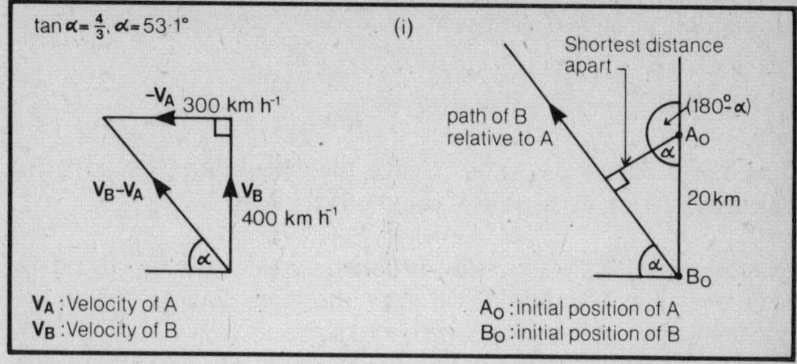

Figure 50.

We see from figure 50, that the velocity of B relative to A, i.e., $\mathbf{V_B - V_A}$, is **500 km h^{-1} in a direction of N 36·9° W.**

($\tan \alpha = \frac{4}{3}, \alpha = 53{\cdot}1°$).

The closest distance apart is found by 'reducing A to rest' when A and B are in their initial positions with A 20 km due north of B, and then finding the shortest distance from A_o to the relative path of B with respect to A. Hence **the closest distance between the planes is 20 cos α km, i.e., 12 km.**

The **bearing of *B* from *A* when they are closest** is N (180° − α) W, i.e., **N 126·9° W.**

(ii)

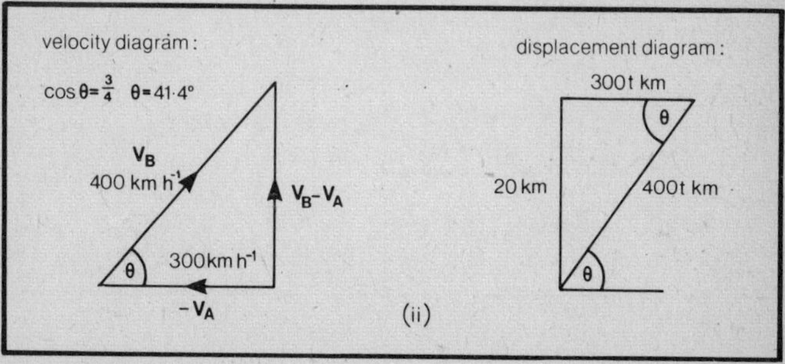

Figure 51.

If plane B travelling at 400 km h^{-1} is to **intercept** A, then the shortest distance apart must be zero. Hence the relative path of

B to A must be in the direction of the original displacement of A from B, i.e., due north. We see from figure 51(ii) that B must take a direction of N 48·6° E. From the displacement diagram, we see that

$$\tan 41{\cdot}4° = \frac{20}{300t},$$

thus $t = 4$ **mins 34 secs.** Hence the **interception should take place at 4 mins 34 secs past 12.00 h.**

(As the question states, this problem could also be solved by scale drawing. We should still draw the same diagrams as above, but instead of using trigonometry to solve them, we should draw them out to scale, using as large a scale as possible and clearly stating the scales used.)

Example 3 A cyclist observes that when his velocity is $u\mathbf{i}$ the wind appears to come from the direction $\mathbf{i} + 2\mathbf{j}$, but when his velocity is $u\mathbf{j}$ the wind appears to come from the direction $-\mathbf{i} + 2\mathbf{j}$. Prove that the true velocity of the wind is $(\frac{3}{4}\mathbf{i} - \frac{1}{2}\mathbf{j})u$. Find the speed and the direction of motion of the cyclist when the wind velocity appears to be $u\mathbf{i}$.

(A.E.B.)

To solve the first part of this question we must find a way of putting both sets of the given information together. The true velocity of the wind, $\mathbf{V}_w$, is constant and so we can draw the two velocity triangles using the same $\mathbf{V}_w$, as shown in figure 52.

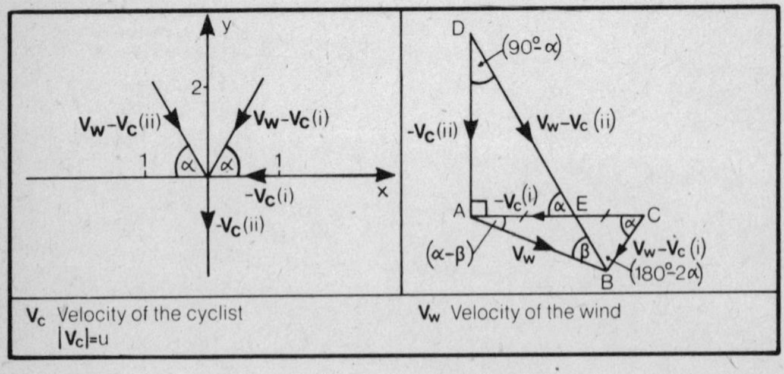

Figure 52.

We apply trigonometry to these triangles. The directions of $\mathbf{V}_w - \mathbf{V}_c$(i) and $\mathbf{V}_w - \mathbf{V}_c$(ii) are arctan 2 and arctan(−2).

Taking $A\hat{B}E = \beta$, then the other angles as marked follow.

Applying the sine rule to triangle ABC:

$$\frac{|\mathbf{V}_w|}{\sin\alpha} = \frac{|\mathbf{V}_c(\mathrm{i})|}{\sin(180 - (2\alpha - \beta))}, \quad \text{i.e., } |\mathbf{V}_w| = \frac{u \sin\alpha}{\sin(2\alpha - \beta)}$$

and to triangle ABD:

$$\frac{|\mathbf{V}_w|}{\sin(90 - \alpha)} = \frac{|\mathbf{V}_c(\mathrm{ii})|}{\sin\beta}, \quad \text{i.e., } |\mathbf{V}_w| = \frac{u\cos\alpha}{\sin\beta}$$

Equating these two values for $|\mathbf{V}_w|$ and rearranging:

$$\tan\beta = \frac{\sin 2\alpha}{\tan\alpha + \cos 2\alpha}$$

Since $\tan\alpha = 2$, $\sin 2\alpha = \frac{4}{5}$ and $\cos 2\alpha = -\frac{3}{5}$, thus $\tan\beta = \frac{4}{7}$.

Hence $|\mathbf{V}_w| = \dfrac{\sqrt{13}}{4}u$.

Since $\tan\beta = \frac{4}{7}$ and $\tan\alpha = 2$, then $\tan(\alpha - \beta) = \frac{2}{3}$ and the unit direction vector for $\mathbf{V}_w$ is

$$\frac{1}{\sqrt{13}}(3\mathbf{i} - 2\mathbf{j}). \quad \text{Thus} \quad \mathbf{V}_w = \left(\frac{3}{4}\mathbf{i} - \frac{1}{2}\mathbf{j}\right)u$$

If the wind velocity appears to be $u\mathbf{i}$, the velocity triangle is as shown in figure 53.

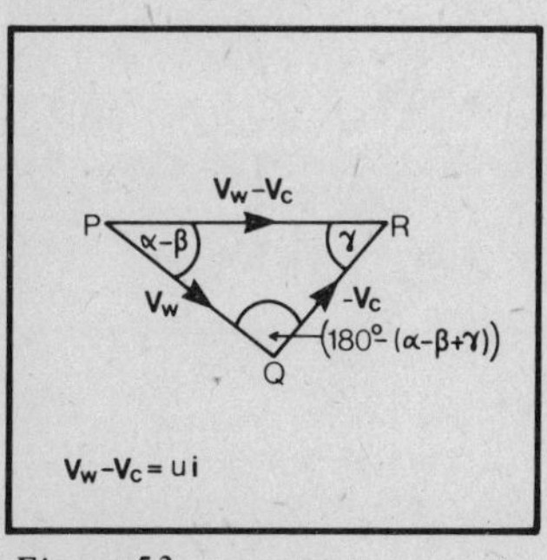

Figure 53.

Clearly, triangle PQR is congruent to triangle ABC in figure 52. Therefore $\gamma = \alpha$ and the unit direction vector for the velocity of the cyclist is

$$-\frac{1}{\sqrt{5}}(\mathbf{i} + 2\mathbf{j}).$$

Applying the sine rule to triangle PQR:

$$\frac{\sqrt{13}\,u}{4\sin\alpha} = \frac{|\mathbf{V}_c|}{\sin(\alpha - \beta)}.$$

Thus $|\mathbf{V}_c| = \dfrac{\sqrt{5}\,u}{4}$, and $\mathbf{V}_c = \left(\dfrac{\mathbf{i}}{4} + \dfrac{\mathbf{j}}{2}\right)u.$

Example 4 A particle A starts from a point P whose position vector is $(r\mathbf{i} + 3\mathbf{j} + 2\mathbf{k})$ with constant velocity vector $(\mathbf{i} - \mathbf{j} + \mathbf{k})$. At the same time another particle, B, starts from a point, Q, with position vector $(7\mathbf{i} - 2\mathbf{j} + 4\mathbf{k})$. Q travels with constant velocity vector $(-2\mathbf{i} + \mathbf{j} + 4\mathbf{k})$. Find the velocity of B relative to A and the position of B relative to A after time t. Find also the distance between A and B at this time. If the distance AB is least when $t = 2$ find r.

Let $\mathbf{V}_A$ be the velocity of A and $\mathbf{V}_B$ be the velocity of B. **The velocity of B relative to A is**

$$\mathbf{V}_B - \mathbf{V}_A = (-2\mathbf{i} + \mathbf{j} + 4\mathbf{k}) - (\mathbf{i} - \mathbf{j} + \mathbf{k}) = -3\mathbf{i} + 2\mathbf{j} + 3\mathbf{k}.$$

Let A_t be the position of A at time t and B_t be the position of B at time t.

The position of B relative to A at time t is

$\mathbf{A}_t\mathbf{B}_t = \mathbf{PQ} + (\mathbf{V}_B - \mathbf{V}_A)t$ (see figure 54(i)). Therefore

$$\mathbf{A}_t\mathbf{B}_t = (7\mathbf{i} - 2\mathbf{j} + 4\mathbf{k} - (r\mathbf{i} + 3\mathbf{j} + 2\mathbf{k})) + (-3\mathbf{i} + 2\mathbf{j} + 3\mathbf{k})t$$
$$= (7 - r - 3t)\mathbf{i} + (-5 + 2t)\mathbf{j} + (2 + 3t)\mathbf{k}$$

The **least distance** $|\mathbf{A}_t\mathbf{B}_t| = AB$ occurs when $\mathbf{A}_t\mathbf{B}_t$ is perpendicular to the relative path $(\mathbf{V}_B - \mathbf{V}_A)t$ as shown in figure 54(ii).

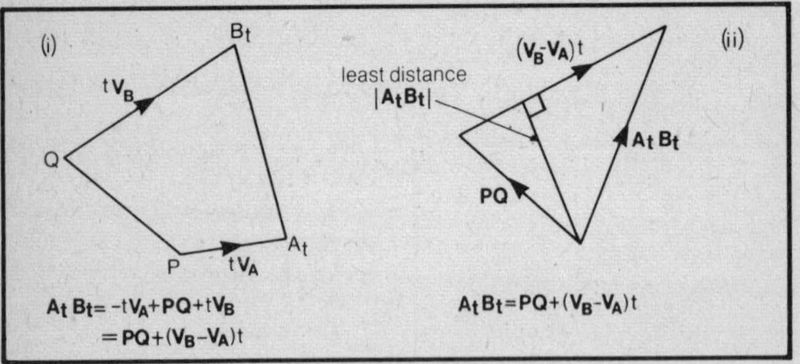

Figure 54.

We are told that the least distance occurs at time $t = 2$. At this time $\mathbf{A}_t\mathbf{B}_t = (1 - r)\mathbf{i} - \mathbf{j} + 8\mathbf{k}$ which must be perpendicular to $-3\mathbf{i} + 2\mathbf{j} + 3\mathbf{k}$, ($(\mathbf{V}_B - \mathbf{V}_A)$ has the same direction as $(\mathbf{V}_B - \mathbf{V}_A)t$).

Hence $((1 - r)\mathbf{i} - \mathbf{j} + 8\mathbf{k}) \cdot (-3\mathbf{i} + 2\mathbf{j} + 3\mathbf{k}) = 0$

$$\text{i.e., } -3 + 3r - 2 + 24 = 0,$$

$$r = -\frac{19}{3}.$$

Key terms

The **velocity of *A* relative to *B*** is the vector difference of their respective velocities, i.e. $\mathbf{V}_A - \mathbf{V}_B$, and is the velocity with which *A* appears to move when viewed from *B*.

The **displacement of *A* relative to *B*** is the sum of the initial relative displacement vector and the product of the velocity of *A* relative to *B* with the time travelled, i.e.,

$$\mathbf{B}_t\mathbf{A}_t = \mathbf{B}_o\mathbf{A}_o + (\mathbf{V}_A - \mathbf{V}_B)t.$$

The **shortest distance between them,** $|\mathbf{B}_t\mathbf{A}_t|$ occurs when $\mathbf{B}_t\mathbf{A}_t$ is perpendicular to the relative path.

Interception or collision occurs when the shortest distance between *A* and *B* is zero.

The line of closest approach for *A* to *B* where *A* has fixed speed, but may vary direction, and *B* has fixed velocity, occurs when *A* chooses the direction to be perpendicular to the relative path,

$$\text{i.e., } \mathbf{V}_A \cdot (\mathbf{V}_A - \mathbf{V}_B) = 0.$$

Chapter 6
General Motion of Particles in Planes

We have considered the motion in a straight line of a particle, we shall now look at the motion of a particle in a plane. Before we can do this we need to look at the quantities **angular velocity** and **angular acceleration**.

Angular velocity

Consider a particle moving on the circumference of a circle, centre O and radius r. It can only move in two directions: clockwise and anticlockwise. By convention anticlockwise is taken as the positive direction, and clockwise as the negative direction. If initially the particle is at rest at A, and θ is the angle the radius to the particle makes with OA, then the angular velocity is defined as the rate of change of θ, i.e., $d\theta/dt$, usually denoted by ω. θ is measured in radians and the unit of angular velocity is 1 radian per second (rs^{-1}).

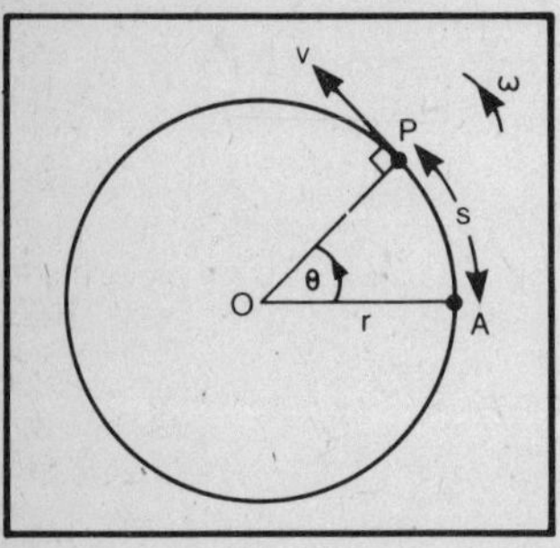

Figure 55.

We know that the arc length, s, may be written as $r\theta$, so that

$$\frac{ds}{dt} = r\frac{d\theta}{dt}$$

i.e., $v = \omega r$ where v is the tangential linear velocity at an instant.

If ω is constant then $\omega = \dfrac{\theta}{t}$

and $v = \dfrac{r\theta}{t}$ is also constant.

Angular acceleration

The angular acceleration, α, of a body is defined to be the rate of change of its angular velocity, i.e. $\alpha = d\omega/dt$, and is measured in radians per second per second. By considering the graph of angular velocity against time for **constant angular acceleration**, α, we can make comparisons with the corresponding graph for **constant linear acceleration**. We see that by substituting initial and final **angular** velocities, for initial and final **linear** velocities, **angular** displacement for **linear** displacement, and constant **angular** acceleration for constant **linear** displacement, the same basic equations of motion hold.

Hence, '$v^2 = u^2 + 2as$' becomes '$\omega_2{}^2 = \omega_1{}^2 + 2\alpha\theta$';

'$v = u + at$' becomes '$\omega_2 = \omega_1 + \alpha t$';

'$s = ut + \frac{1}{2}t^2$' becomes '$\theta = \omega_1 t + \frac{1}{2}\alpha t^2$';

'$s = vt - \frac{1}{2}at^2$' becomes '$\theta = \omega_2 t - \frac{1}{2}\alpha t^2$';

'$s = \frac{1}{2}(u + v)t$' becomes '$\theta = \frac{1}{2}(\omega_1 + \omega_2)t$'.

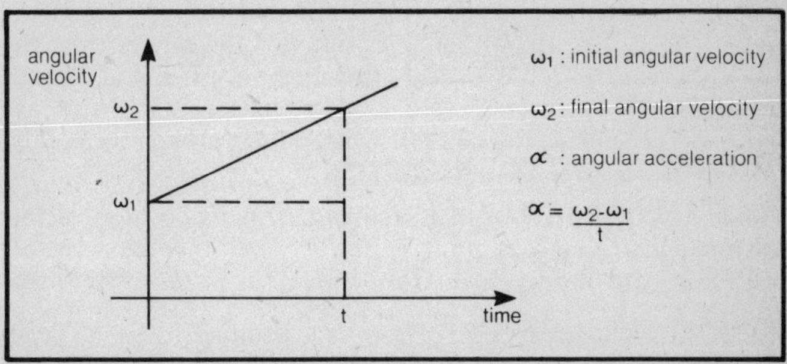

Figure 56.

We shall consider circular motion in detail in Chapter 10. Let us now consider the position, velocity and acceleration vectors of particles moving in planes. We shall use two systems of reference: Cartesian coordinates and polar coordinates.

Motion of particles in planes using Cartesian coordinates

We have seen that in general, if vector **a** is a function of a scalar where $\mathbf{a}(\lambda) = f(\lambda)\mathbf{i} + g(\lambda)\mathbf{j} + h(\lambda)\mathbf{k}$

then $$\frac{d\mathbf{a}}{d\lambda} = \frac{d}{d\lambda}(f(\lambda))\mathbf{i} + \frac{d}{d\lambda}(g(\lambda))\mathbf{j} + \frac{d}{d\lambda}(h(\lambda))\mathbf{k}$$

Let us now consider the position vector **r** of a particle P, moving on a curve where **r** is a function of the scalar variable time, t, such that $\mathbf{r}(t) = f(t)\mathbf{i} + g(t)\mathbf{j} + h(t)\mathbf{k}$. We write $\mathbf{r}(t + \delta t)$ as

$\mathbf{r} + \delta\mathbf{r}$, and we can see from the diagram below that $\delta\mathbf{r}$ [the small change in $\mathbf{r}$ over time δt, i.e., $(\mathbf{r} + \delta\mathbf{r}) - \mathbf{r}$] is represented by PP'. As $\delta t \to 0$, PP' will tend towards the tangent to the curve at P.

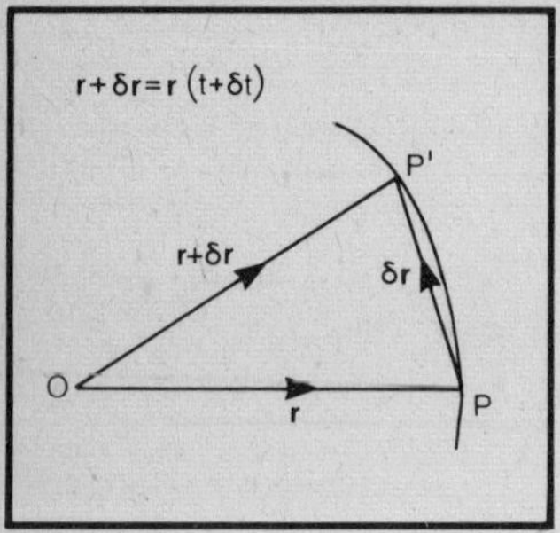

Figure 57.

Hence, $\lim\limits_{\delta t \to 0} \dfrac{\delta\mathbf{r}}{\delta t} = \dfrac{d\mathbf{r}}{dt}$

(which is, of course, the velocity vector) will be in the direction of the tangent to the curve at P.

If a point is moving on a **circle**, and the position vector is with reference to the centre of the circle, then the position vector, $\mathbf{r}$, and the velocity vector, $\mathbf{v} = d\mathbf{r}/dt$ are **perpendicular** to each other.

When a point is moving in a **straight line**, its position vector, with reference to a fixed point on the line, and its velocity vector, will be in **the same direction**, since $\delta\mathbf{r}$ is in the same direction as $\mathbf{r}$.

The acceleration vector,

$$\mathbf{a} = \frac{d\mathbf{v}}{dt} = \frac{d^2\mathbf{r}}{dt^2}$$

$$= \frac{d^2(f(t))}{dt^2}\mathbf{i} + \frac{d^2(g(t))}{dt^2}\mathbf{j} + \frac{d^2(h(t))}{dt^2}\mathbf{k}.$$

Before we consider these vectors in polar coordinates, we shall derive a useful result concerning unit vectors: that **a unit vector and its derivative with respect to a scalar variable are perpendicular**.

Let the unit vector $\hat{\mathbf{r}}$ be a function of the scalar variable t, so that $\hat{\mathbf{r}}(t) = l(t)\mathbf{i} + m(t)\mathbf{j} + n(t)\mathbf{k}$ where l, m, n, are the **variable** direction cosines.

$$\frac{d\hat{\mathbf{r}}}{dt} = \frac{dl}{dt}\mathbf{i} + \frac{dm}{dt}\mathbf{j} + \frac{dn}{dt}\mathbf{k}.$$

Thus $\hat{\mathbf{r}} \cdot \dfrac{d\hat{\mathbf{r}}}{dt} = l\dfrac{dl}{dt} + m\dfrac{dm}{dt} + n\dfrac{dn}{dt} = \dfrac{1}{2}\dfrac{d}{dt}(l^2 + m^2 + n^2) = 0$

since $l^2 + m^2 + n^2 = 1$.

Hence $\hat{\mathbf{r}}$ and $\dfrac{d\hat{\mathbf{r}}}{dt}$ must be perpendicular.

Motion of a particle in a plane using polar coordinates

Let us consider the motion of a particle P moving in the xy plane such that at time t it has polar coordinates (r, θ). The position vector $\mathbf{r}$ of P at time t is given by $\mathbf{r} = r(\cos\theta\,\mathbf{i} + \sin\theta\,\mathbf{j})$.

Let $\hat{\mathbf{r}}$ be the unit vector for $\mathbf{r}$ at time t and $\hat{\mathbf{s}}$ be the unit vector in the direction given by rotating $\hat{\mathbf{r}}$ through a positive right angle.

We can see that $\hat{\mathbf{r}} = \cos\theta\,\mathbf{i} + \sin\theta\,\mathbf{j}$

and $\hat{\mathbf{s}} = -\sin\theta\,\mathbf{i} + \cos\theta\,\mathbf{j}$.

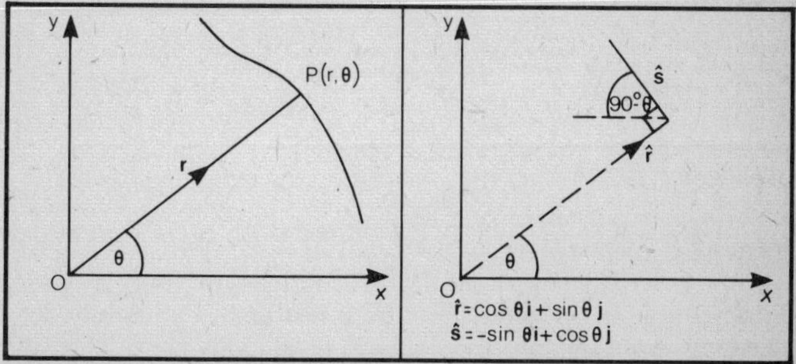

Figure 58.

The velocity vector $\mathbf{v}$ is given by:

$$\mathbf{v} = \frac{d\mathbf{r}}{dt} = \dot{r}(\cos\theta\,\mathbf{i} + \sin\theta\,\mathbf{j}) + r\dot{\theta}(-\sin\theta\,\mathbf{i} + \cos\theta\,\mathbf{j}) = \dot{r}\hat{\mathbf{r}} + r\dot{\theta}\hat{\mathbf{s}}.$$

(The 'dot' notation indicates differentiation with respect to time

i.e., $$\dot{r} = \frac{dr}{dt}, \quad \ddot{\theta} = \frac{d^2\theta}{dt^2}, \text{ etc.)}$$

The acceleration vector $\mathbf{a}$ is given by

$$\mathbf{a} = \frac{d\mathbf{v}}{dt} = \ddot{r}\hat{\mathbf{r}} + \dot{r}\dot{\hat{\mathbf{r}}} + (\dot{r}\dot{\theta} + r\ddot{\theta})\hat{\mathbf{s}} + r\dot{\theta}\dot{\hat{\mathbf{s}}}$$

Since $\dot{\hat{\mathbf{r}}} = \dot{\theta}(-\sin\theta\,\mathbf{i} + \cos\theta\,\mathbf{j}) = \dot{\theta}\hat{\mathbf{s}}$

and $\dot{\hat{\mathbf{s}}} = \dot{\theta}(-\cos\theta\,\mathbf{i} - \sin\theta\,\mathbf{j}) = -\dot{\theta}\hat{\mathbf{r}}$ then,

$$\mathbf{a} = \ddot{r}\hat{\mathbf{r}} + \dot{r}\dot{\theta}\hat{\mathbf{s}} + \dot{r}\dot{\theta}\hat{\mathbf{s}} + r\ddot{\theta}\hat{\mathbf{s}} - r\dot{\theta}^2\hat{\mathbf{r}} = (\ddot{r} - r\dot{\theta}^2)\hat{\mathbf{r}} + (r\ddot{\theta} + 2\dot{r}\dot{\theta})\hat{\mathbf{s}}$$

Hence both the velocity and acceleration vectors may be considered as the **sum of two component vectors**, one in the direction of OP called the **radial component** and one in the direction given by rotating OP through a positive right angle, called the **transverse component**. $\dot{\theta}$ is the angular velocity of the particle, and $\ddot{\theta}$ is the angular acceleration.

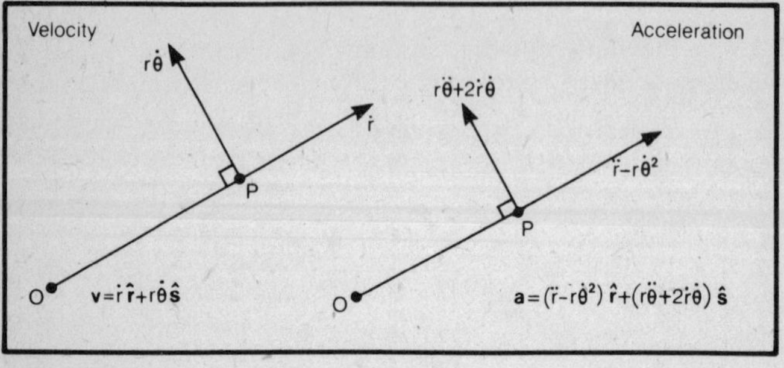

Figure 59.

Worked examples

Example 1 A flywheel starts from rest and is uniformly accelerated to an angular speed of 120 revolutions per minute. It maintains this speed until it is uniformly retarded to rest again. The magnitude of the retardation is three times the value of the starting acceleration. Between starting and coming to rest again, the flywheel completes N revolutions in five minutes. Sketch the angular speed-time graph and hence find, in terms of N, the time for which the flywheel is travelling at the maximum speed. Show that $300 < N < 600$.

If $N = 480$, find the starting acceleration and the number of revolutions completed in the first two minutes.

(A.E.B.)

The angular speed-time graph is as shown in figure 60. x minutes is the time taken to come to rest from the maximum speed. Hence the time taken to accelerate from rest to this speed will be $3x$ minutes, since the retardation is three times the acceleration.

We know that the total time taken is 5 minutes, so that the time travelling at the constant speed of 120 rev/min is $(5 - 4x)$ mins.

Since $\int \omega \, dt = \theta$, where ω is the angular velocity and θ the angular displacement, we can see that the area under the angular speed-time graph gives the angle turned through, which is given as N revolutions.

Hence, $\frac{1}{2} \times 3x \times 120 + (5 - 4x)120 + \frac{1}{2} \times x \times 120 = N$

i.e., $$\boldsymbol{x = \frac{600 - N}{240}}.$$

The time travelled at 120 rev/min is $(5 - 4x) = \boldsymbol{\frac{N - 300}{60}}$ **mins**

We know that this time must be greater than 0 mins and less than 5 mins.

Therefore, $0 < \frac{N - 300}{60} < 5$ and $\boldsymbol{300 < N < 600}$.

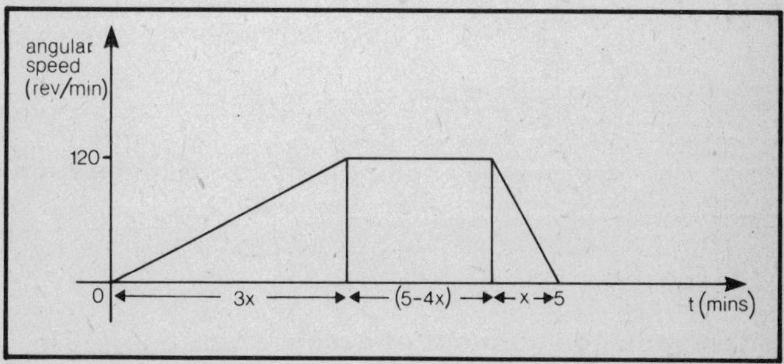

Figure 60.

The starting acceleration is

$$\frac{120}{3x} \text{ so that when } N = 480, \; x = \frac{1}{2}$$

and the starting acceleration is **80 rev/min²**. To find the number of revolutions completed in the first two minutes, we inspect the graph and see that for $1\frac{1}{2}$ mins the fly wheel is accelerating to 120 revs/min and turns through $\frac{1}{2} \times 3 \times \frac{1}{2} \times 120 = 90$ revolutions.

It then travels at 120 revs/min for a further $\frac{1}{2}$ min covering 60 revolutions. Hence the **total number of revolutions turned through in the first 2 minutes is 150**.

(If we had not been asked to solve this problem by drawing the angular speed-time graph, we could have applied the standard equations for constant angular acceleration to the problem.)

Example 2 A particle describes the curve $r = 3e^{\theta}$ so that the radial velocity of the particle, when it is at a distance r from the pole, is $2/r$. Show that the acceleration of the particle is $8/r^3$ directed towards O.

We know that the radial component of acceleration of a particle is $(\ddot{r} - r\dot{\theta}^2)$ and the transverse component is $(r\ddot{\theta} + 2\dot{r}\dot{\theta})$. (In order to answer this question, we must show that the transverse component of acceleration is zero and the radial component is $-8/r^3$). We are told that $r = 3e^{\theta}$ and that the radial velocity of the particle, $\dot{r}$, is $2/r$.

Since $r = 3e^{\theta}$ then $\dot{r} = 3e^{\theta}\dot{\theta} = \dfrac{2}{r}$ and $\dot{\theta} = \dfrac{2}{r^2}$.

Thus $\ddot{\theta} = -\dfrac{4}{r^3}\dot{r} = -\dfrac{8}{r^4}$. Also $\ddot{r} = -\dfrac{2}{r^2}\dot{r} = -\dfrac{4}{r^3}$.

Hence the **transverse component of acceleration**, $(r\ddot{\theta} + 2\dot{r}\dot{\theta})$, is

$$-r\,.\,\frac{8}{r^4} + 2\,.\,\frac{2}{r}\,.\,\frac{2}{r^2} = 0$$

and the **radial component of acceleration**, $(\ddot{r} - r\dot{\theta}^2)$, is

$$-\frac{4}{r^3} - r\frac{4}{r^4} = -\frac{8}{r^3},$$

which is the required result.

Example 3 The position vector of a particle P is given by $\mathbf{r} = (a_1 t^2 + a_2 t + 1)\mathbf{i} + (b_1 t^2 + b_2 t - 2)\mathbf{j}$. At time $t = 0$, the particle has velocity vector $3\mathbf{i} + \mathbf{j}$, and at time $t = 2$, the velocity vector is $11\mathbf{i} - 7\mathbf{j}$. Calculate the values of a_1, a_2, b_1, b_2. Show that the particle is moving under the action of a constant force and find its magnitude, given that the mass of the particle is 4 kg, (distances are measured in metres, and time in seconds).

Since $\mathbf{r} = (a_1 t^2 + a_2 t + 1)\mathbf{i} + (b_1 t^2 + b_2 t - 2)\mathbf{j}$, the velocity vector

$$\mathbf{v} = \frac{d\mathbf{r}}{dt} = (2a_1 t + a_2)\mathbf{i} + (2b_1 t + b_2)\mathbf{j}.$$

When $t = 0$, we are told that $\mathbf{v} = 3\mathbf{i} + \mathbf{j}$.

Comparing coefficients of vectors **i** and **j**, $\boldsymbol{a_2} = \mathbf{3}$ and $\boldsymbol{b_2} = \mathbf{1}$.

When $t = 2$, $\mathbf{v} = 11\mathbf{i} - 7\mathbf{j} = (4a_1 + 3)\mathbf{i} = (4b + 1)\mathbf{j}$,

therefore $\boldsymbol{a_1} = \mathbf{2}$ and $\boldsymbol{b_2} = \mathbf{-2}$

The acceleration vector,

$$\mathbf{a} = \frac{d\mathbf{v}}{dt} = 2a_1\,\mathbf{i} + 2b_1\,\mathbf{j} = 4\mathbf{i} - 4\mathbf{j},$$

which is clearly a constant vector. From Newton's second law, '$f = ma$'; **the particle is moving under a constant force of magnitude** $4|4\mathbf{i} - 4\mathbf{j}| = \mathbf{16\sqrt{2}\ N}$.

Example 4 A disc has its centre at the origin O of fixed rectangular axes Ox, Oy, and rotates about O in the xy plane. Fixed points A and B on the disc, initially have coordinates $(a, 0)$, $(a, 8a)$ respectively. The disc rotates in the clockwise sense with constant angular speed ω. A particle P starts from A and moves towards B at the same time as the disc begins to rotate. The particle moves with constant speed $2a\omega$ relative to the disc. Find the coordinates of P after it has been in motion for time t, and find its actual speed. Find also the magnitude of the particle's acceleration when it passes through the midpoint of AB and also at time $t = \pi/\omega$.

After time t, the disc will have rotated through a clockwise angle of ωt, and the particle will have moved a distance of $2a\omega t$ relative to the disc in the direction $A_t B_t$.

From figure 61, we can see that the **x coordinate of P_t** is $a \cos \omega t + 2a\omega t \sin \omega t$, and the **$y$ coordinate** is $-a \sin \omega t + 2a\omega t \cos \omega t$.

Hence the position vector, $\mathbf{r}$, of P after time t is: $\mathbf{r} = a(\cos \omega t + 2\omega t \sin \omega t)\mathbf{i} + a(-\sin \omega t + 2\omega t \cos \omega t)\mathbf{j}$. Differentiating with respect to t,

$$\mathbf{v} = \frac{d\mathbf{r}}{dt} = a\omega(\sin \omega t + 2\omega t \cos \omega t)\mathbf{i} + a\omega(\cos \omega t - 2\omega t \sin \omega t)\mathbf{j}.$$

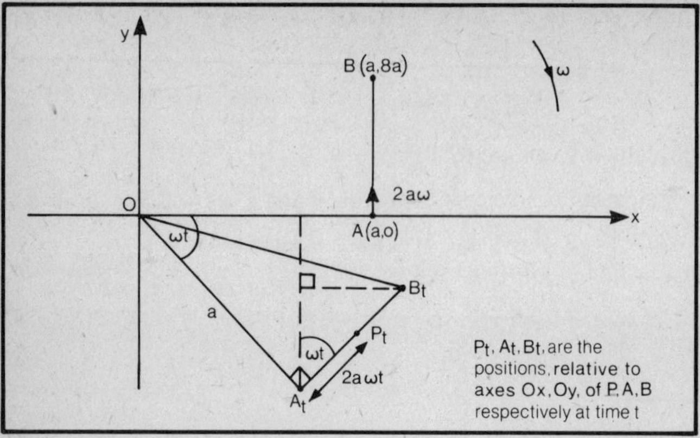

Figure 61.

Hence the **actual speed** of the particle is:

$$|\mathbf{v}| = a\omega(\sin^2 \omega t + 4\omega t \sin \omega t \cos \omega t + 4\omega^2 t^2 \cos^2 \omega t + \cos^2 \omega t - 4\omega t \cos \omega t \sin \omega t + 4\omega^2 t^2 \sin^2 \omega t)^{1/2}$$
$$= \boldsymbol{a\omega(1 + 4\omega^2 t^2)^{1/2}}.$$

The acceleration, **a**, of the particle is $\frac{d\mathbf{v}}{dt}$. Hence

$$\mathbf{a} = a\omega^3(3 \cos \omega t - 2\omega t \sin \omega t)\mathbf{i} - a\omega^2(3 \sin \omega t + 2\omega t \cos \omega t)\mathbf{j},$$
$$|\mathbf{a}| = a\omega^2(9 + 4\omega^2 t^2)^{1/2}.$$

When the particle is midway between A and B it has travelled $4a$ relative to the disc so that

$$2a\omega t = 4a, \ \boldsymbol{t} = \frac{\mathbf{2}}{\boldsymbol{\omega}}.$$

Hence the magnitude of the particle's **acceleration at the midpoint of AB is $5a\omega^2$**. When $t = \pi/\omega$ the magnitude of the acceleration is $\boldsymbol{a\omega^2(9 + 4\pi^2)^{1/2}}$.

Key terms

Angular velocity, ω, is the rate of change of angular displacement

i.e.,
$$\omega = \frac{d\theta}{dt} = \dot{\theta}$$

Angular acceleration, α, is the rate of change of angular velocity,

i.e., $$\alpha = \frac{d\omega}{dt} = \ddot{\theta} = \dot{\omega}$$

When a particle is rotating with **constant angular acceleration**, we have **quotable basic equations of motion** corresponding to those for linear motion. With the usual notation:

$$\omega_2 = \omega_1 + \alpha t, \qquad \theta = \omega_1 t + \tfrac{1}{2}\alpha t^2,$$

$$\theta = \omega_2 t - \tfrac{1}{2}\alpha t^2, \qquad \omega_2{}^2 = \omega_1{}^2 + 2\alpha\theta$$

$$\theta = \tfrac{1}{2}(\omega_1 + \omega_2)t.$$

The derivative of a unit vector with respect to a scalar variable is perpendicular to the unit vector. In particular, $\dot{\hat{\mathbf{r}}} = \dot{\theta}\hat{\mathbf{s}}$ where $\hat{\mathbf{s}}$ is the unit vector in the direction given by rotating $\hat{\mathbf{r}}$ through 90° and $\dot{\theta}$ is the angular velocity.

In the polar coordinate system the coordinates of a point are (r, θ), and the velocity and acceleration vectors may be given as the sum of two component vectors, one in the direction of **r**, called the **radial component**, and one in the direction given by rotating **r** through a **positive** angle of 90°, called the **transverse component**, i.e., $\mathbf{v} = \dot{r}\hat{\mathbf{r}} + r\dot{\theta}\hat{\mathbf{s}}$

and $\mathbf{a} = (\ddot{r} - r\dot{\theta}^2)\hat{\mathbf{r}} + (r\ddot{\theta} + 2\dot{r}\dot{\theta})\hat{\mathbf{s}}$.

where $\hat{\mathbf{r}}$ is the **unit radial vector** and $\hat{\mathbf{s}}$ is the **unit transverse vector**.

Chapter 7
Projectiles

A **projectile** is a particle which is given an initial velocity and then moves entirely under the action of its own weight. (Air resistance is taken to be negligible, unless otherwise stated).

Projectiles on a horizontal plane

Consider a particle of mass m projected from a horizontal plane with initial velocity $\boldsymbol{u}$ at an angle θ to the horizontal.

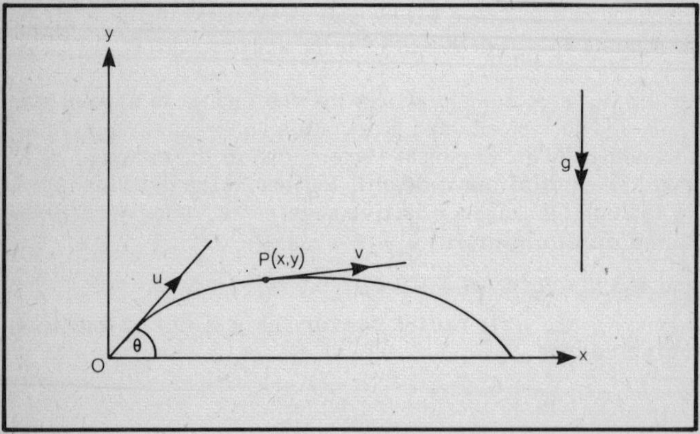

Figure 62.

Since the only force acting on the particle is its own weight, it will have an acceleration of $\boldsymbol{g}$ vertically downwards. Hence, if we consider the vertical and horizontal components of motion we can apply the equations of motion for **constant acceleration vertically** and for **constant velocity horizontally**.

Taking the axes as shown in figure 62:

$\ddot{y} = -g$ (i)

$\dot{y} = u \sin\theta - gt$ (ii) (From $v = u + at$)

$y = u \sin \theta \, t - \frac{1}{2}gt^2$ (iii) (From $s = ut + \frac{1}{2}at^2$)

$\ddot{x} = 0$ (iv)

$\dot{x} = u \cos \theta$ (v) (i.e., constant velocity)

$x = u \cos \theta \, t$ (vi) (From $s = ut + \frac{1}{2}at^2$, where $a = 0$)

Equations (iii) and (vi) are parametric equations of the path of the projectiles which combine to give:

$$y = x \tan \theta - \frac{gx^2}{2u^2} \sec^2 \theta$$

which is the equation of the path of the projectile. This is a parabola with its axis vertically downwards.

At any instant t and velocity v:

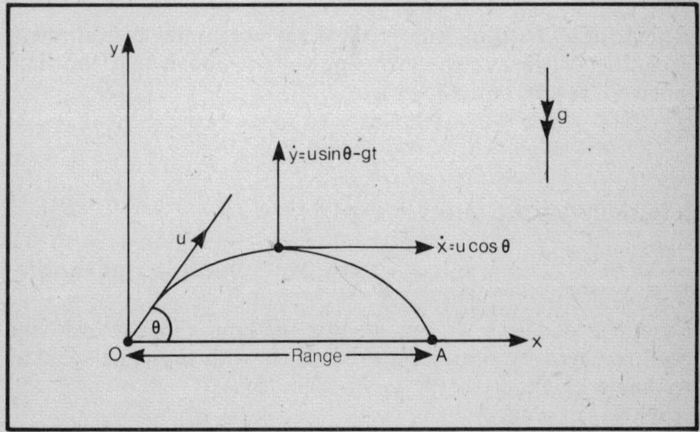

Figure 63.

We can see that the particle's velocity v is given by:

$v = \sqrt{\dot{x}^2 + \dot{y}^2}$ at a direction of $\arctan \frac{\dot{y}}{\dot{x}}$ to the horizontal.

Since $\frac{\dot{y}}{\dot{x}} = \frac{dy/dt}{dx/dt} = \frac{dy}{dx}$

then we can see that the direction of velocity is along the tangent, which is what we would expect.

We now have all the general information about the motion of a particle which we need in order to look at the following particular properties of its flight.

Time of flight: this is the time for which the particle is in the air, i.e., the time taken for it to travel along its parabolic path from O to A.

We know $y = u \sin\theta\, t - \frac{g}{2} t^2$, but at A, $y = 0$

so $$0 = u \sin\theta\, t - \frac{g}{2} t^2$$

Either $t = 0$ (this is the initial position),

or $\boldsymbol{t = \frac{2u}{g} \sin\theta}$ which is the **time of flight**.

Horizontal range: this is the final horizontal displacement from the origin i.e., OA. We know from above that the time taken to travel from O to A is

$$\frac{2u}{g} \sin\theta,$$

so substituting this into $x = u \cos\theta\, t$, we get

$$\boldsymbol{OA} = \frac{2u^2}{g} \cos\theta \sin\theta = \boldsymbol{\frac{u^2}{g} \sin 2\theta}, \quad \textbf{which is the range.}$$

From this equation we can see that the same range would have been achieved by projecting the particle with the same speed at an angle of elevation $(90° - \theta)$, since $\sin(90° - \theta) = \cos\theta$ and $\cos(90° - \theta) = \sin\theta$.

Substituting these values in the expression for the range would have given the same result;

$$\text{i.e., Range} = \frac{2u^2}{g} \sin\theta \cos\theta$$

$$= \frac{2u^2}{g} \cos(90° - \theta) \sin(90° - \theta) = \frac{u^2}{g} \sin 2\theta$$

However, the different trajectories would have produced different times of flight.

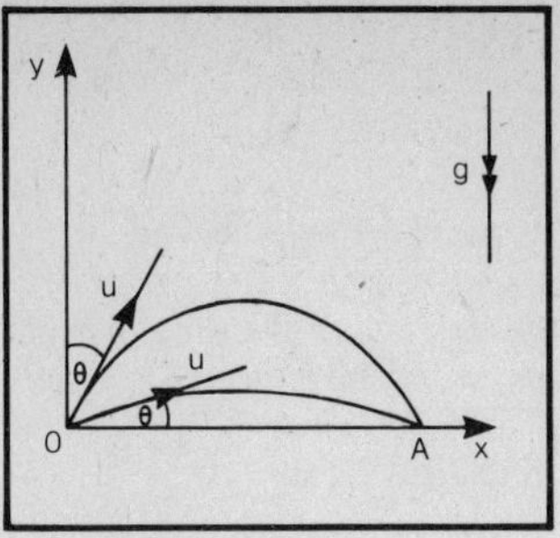

Figure 64.

For angle of projection, θ,

$$t = \frac{2u}{g}\sin\theta,$$

and for angle of projection $(90° - \theta)$,

$$t = \frac{2u}{g}\sin(90° - \theta)$$

$$= \frac{2u}{g}\cos\theta.$$

The smaller angle will give the faster time of flight since it will give the larger horizontal component of velocity. (The smaller the angle, the larger the cosine of it.)

Maximum horizontal range: this is the maximum horizontal range which can be achieved for a fixed speed u by varying the angle of projection. We know the range for θ is given by:

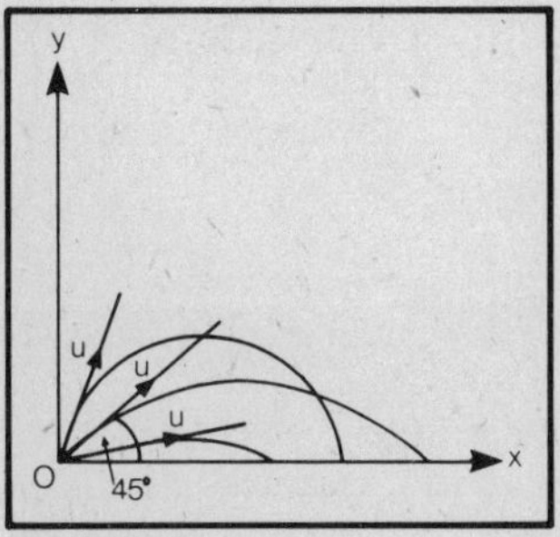

Figure 65.

$$R = \frac{u^2}{g}\sin 2\theta$$

where u^2 and g are constant.

So to maximise R, we need to maximise $\sin 2\theta$.

i.e., $\sin 2\theta = 1$

so $2\theta = 90°$

and $\theta = 45°$

Therefore the **maximum horizontal range is u^2/g** and occurs when the angle of projection is 45°.

Greatest height: this is the maximum vertical displacement above the horizontal, i.e., h.

When the particle reaches its maximum height, its vertical component of velocity will be zero (it will have stopped travelling upwards and be on the point of travelling downwards).

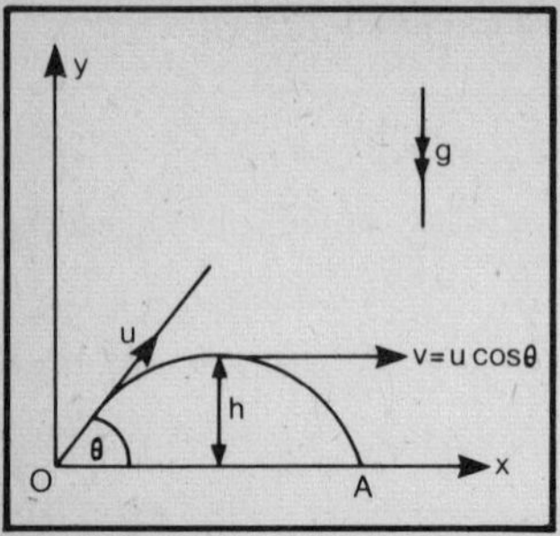

Figure 66.

$\dot{y} = u \sin\theta - gt$ and at the greatest height, $\dot{y} = 0$, so $0 = u \sin\theta - gt$,

$$t = \frac{u}{g} \sin\theta$$

(which is half the time of flight, as would be expected).

Substituting in

$$y = u \sin\theta\, t - \frac{g}{2} t^2,$$

we see that $h = u \sin\theta \frac{u}{g} \sin\theta - \frac{gu^2 \sin^2\theta}{2g^2}$,

i.e., the **maximum height is** $\boldsymbol{\dfrac{u^2 \sin^2\theta}{2g}}$.

Angle of projection to pass through a particular point: this is the angle of projection for the particle to pass through a given point $P(a, b)$ for a given initial speed, u. The equation of the parabolic path for the particle is:

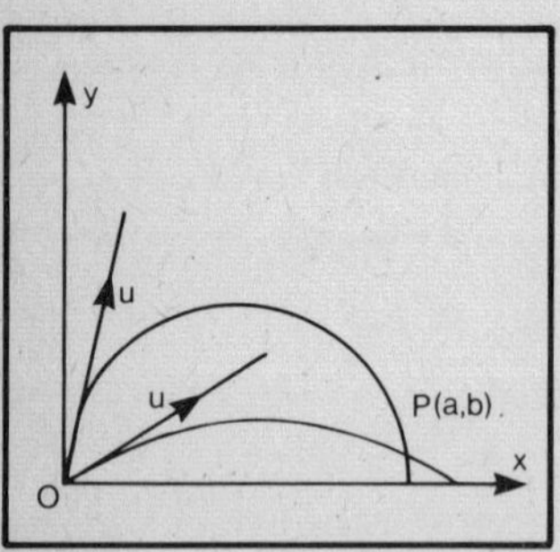

Figure 67.

$$y = x \tan\theta - \frac{gx^2}{2u^2} \sec^2\theta.$$

Substituting $y = b$, and $x = a$ into this equation,

$$b = a \tan\theta - \frac{ga^2}{2u^2} \sec^2\theta.$$

Using $\tan^2\theta + 1 = \sec^2\theta$ and rearranging gives:

$$\frac{a^2 g}{2u^2} \tan^2\theta - a \tan\theta + \left(b + \frac{a^2 g}{2u^2}\right) = 0,$$

which is a quadratic equation in $\tan\theta$ giving two solutions. Providing that these are both real and positive there will be two possible angles of projection for the particle to pass through $P(a, b)$ with given initial speed u.

Projectiles on an inclined plane

We can also consider the motion of a projectile on an **inclined plane** using the techniques outlined so far, but in most cases it is better to analyse the motion in its **components, parallel and perpendicular to the plane**.

Consider a particle projected up a plane inclined at an angle α to the horizontal with initial speed u at an angle of elevation θ from the plane.

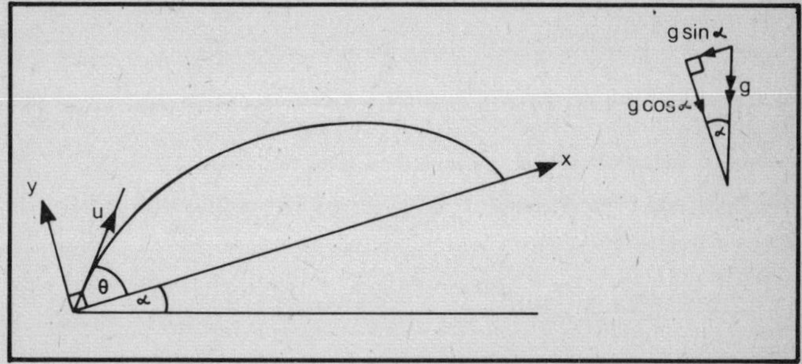

Figure 68.

Taking the y-axis perpendicular to the plane and the x-axis parallel to the plane we have:

$$\ddot{y} = -g \cos \alpha$$

(which is a constant acceleration, so we apply the standard equations of motion for constant acceleration.)

$$\dot{y} = u \sin \theta - g \cos \alpha \, t$$

$$y = u \sin \theta \, t - \tfrac{1}{2} g \cos \alpha \, t^2$$

$$\ddot{x} = -g \sin \alpha \text{ (constant acceleration)}$$

$$\dot{x} = u \cos \theta - g \sin \alpha \, t$$

$$x = u \cos \theta \, t - \tfrac{1}{2} g \sin \alpha \, t^2$$

As in the previous section, these general equations of motion enable us to look at particular properties of the flight of the projectile.

Time of flight. The projectile hits the plane again when $y = 0$, i.e., when $0 = u \sin\theta\, t - \frac{1}{2} g \cos\alpha\, t^2$, so neglecting $t = 0$,

the **time of flight** $= \dfrac{\mathbf{2u \sin \theta}}{\mathbf{g \cos \alpha}}$.

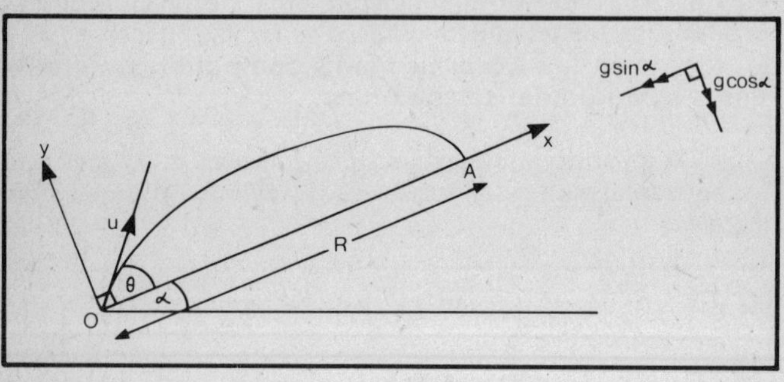

Figure 69.

Range up the plane, R, is given by the value of x when t is the time of flight, i.e.,

$$\boldsymbol{R} = u \cos\theta \frac{2u \sin\theta}{g \cos\alpha} - \frac{g}{2} \sin\alpha \left(\frac{2u \sin\theta}{g\cos\alpha}\right)^2$$

$$= \frac{2u^2 \sin\theta}{g \cos^2\alpha} (\cos\theta \cos\alpha - \sin\theta \sin\alpha)$$

$$= \frac{\mathbf{2u^2 \sin\theta \cos(\theta + \alpha)}}{\mathbf{g \cos^2 \alpha}}.$$

Maximum range up the plane for fixed angle of inclination of the plane, α, and for fixed speed of projection, u, (i.e., only θ, the angle of projection of the particle, can vary) is given by maximising

$$\frac{2u^2 \sin\theta \cos(\theta + \alpha)}{g \cos^2\alpha}.$$

In this expression

$$\left(\frac{\mathbf{2u^2}}{\mathbf{g \cos^2 \alpha}}\right) \textbf{ is constant}$$

and only **$\sin\theta \cos(\theta + \alpha)$ can vary**. Hence, to **maximise R** we need to **maximise $\sin\theta \cos(\theta + \alpha)$**.

We can do this by finding when

$$\frac{d}{d\theta}(\sin\theta\cos(\theta+\alpha)) = 0$$

i.e., when $\cos\theta\cos(\theta+\alpha) - \sin\theta\sin(\theta+\alpha) = 0$.

Thus for maximum R, $\cos(2\theta+\alpha) = 0$, $(2\theta+\alpha) = 90°$, and

$$\boldsymbol{\theta = 45° - \frac{\alpha}{2}}.$$

From physical considerations this is clearly a maximum and not a minimum or point of inflection. We see from figure 70(i) that to achieve the **maximum range** then the **angle of projection of the particle must bisect the angle between the upward slope of the plane and the vertical.** The **maximum range** is

$$\frac{2u^2}{g\cos^2\alpha}\left(\sin\left(45° - \frac{\alpha}{2}\right)\cos\left(45° + \frac{\alpha}{2}\right)\right)$$

$$= \frac{2u^2}{g\cos^2\alpha}\left(\tfrac{1}{2}(\sin 90° - \sin\alpha)\right)$$

$$= \boldsymbol{\frac{u^2}{g\cos^2\alpha}(1 - \sin\alpha)}.$$

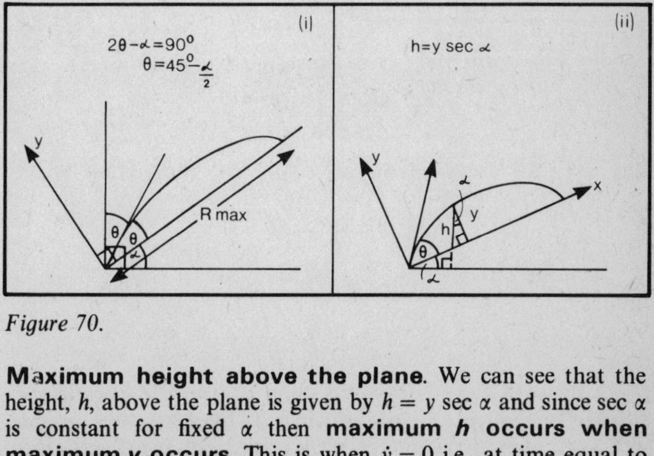

Figure 70.

Maximum height above the plane. We can see that the height, h, above the plane is given by $h = y\sec\alpha$ and since $\sec\alpha$ is constant for fixed α then **maximum h occurs when maximum y occurs.** This is when $\dot{y} = 0$ i.e., at time equal to **half the time of flight**.

We can also consider the flight of a particle projected **down** an inclined plane. Taking the axes as shown in figure 71, we can see that the only difference from the analysis of the motion for projection **up** the plane, is that the **component of acceleration in the direction of increasing x is positive**. As before:

$\ddot{y} = -g \cos \alpha,$

$\dot{y} = u \sin \theta - gt \cos \alpha,$

$y = u \sin \theta \, t - \frac{1}{2}gt^2 \cos \alpha,$ and so the **time of flight** is still given by

$$\boldsymbol{t = \frac{2u \sin \theta}{g \cos \alpha}}, \quad \text{but now:}$$

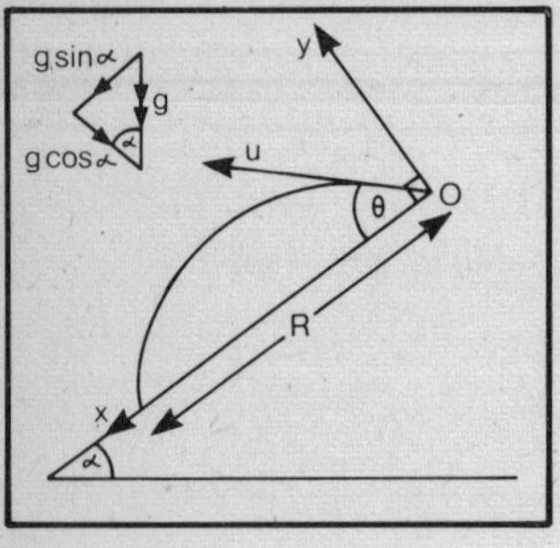

Figure 71.

$\ddot{x} = +g \sin \alpha,$

$\dot{x} = u \cos \theta + gt \sin \alpha,$

$x = u \cos \theta \, t + \frac{1}{2}gt^2 \sin \alpha,$

and the **range down the plane** is given by

$$\boldsymbol{R} = u \cos \alpha \frac{2u \sin \theta}{g \cos \alpha} + \frac{g \sin \alpha}{2}\left(\frac{2u \sin \theta}{g \cos \alpha}\right)^2$$

$$= \boldsymbol{\frac{2u^2 \sin \theta \cos(\theta - \alpha)}{g \cos^2 \alpha}}.$$

This may be **maximised** as before by considering when

$$\frac{d}{d\theta}(\sin \theta \cos(\theta - \alpha)) = 0,$$

i.e., when $\cos \theta \cos(\theta - \alpha) - \sin \theta \sin(\theta - \alpha) = 0$.

Hence, $\cos(2\theta - \alpha) = 0$, $2\theta - \alpha = 90°$ and $\boldsymbol{\theta = 45° + \frac{\alpha}{2}}$.

This tells us that the **range down the plane is maximum when the angle of projection θ bisects the angle between the downward slope and the vertical**, and that the **maximum range down the plane is:**

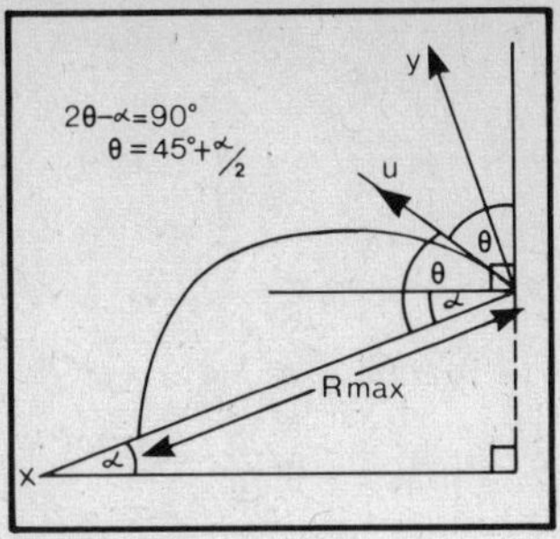

Figure 72.

$$\frac{2u^2 \sin\left(45^\circ + \frac{\alpha}{2}\right)\cos\left(45^\circ - \frac{\alpha}{2}\right)}{g\cos^2\alpha}$$

$$= \frac{2u^2}{g\cos^2\alpha}\left(\tfrac{1}{2}(\sin 90^\circ + \sin\alpha)\right)$$

$$= \frac{u^2}{g\cos^2\alpha}(1 + \sin\alpha).$$

We could also analyse the motion in terms of its **horizontal** and **vertical components** as shown in figure 73, but unless a question is specifically on horizontal and vertical components, then it is very rarely simpler than the above methods.

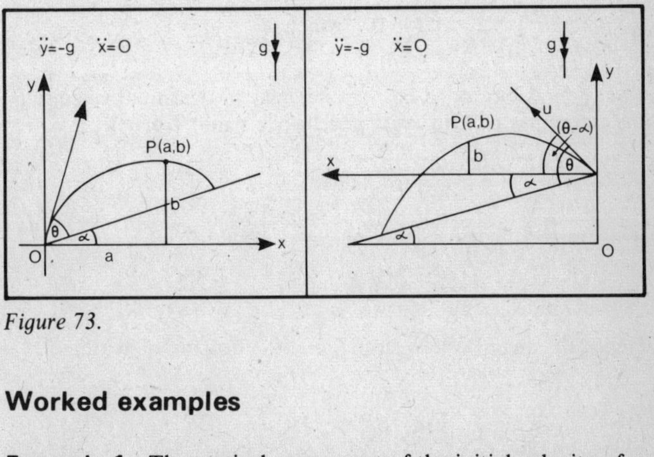

Figure 73.

Worked examples

Example 1 The vertical component of the initial velocity of a projectile, P, is twice the horizontal component. Find the range of the projectile on the horizontal plane through the point of projection O. Find also the angle between the horizontal and OP when the velocity is at right angles to its initial velocity.

If we let the horizontal component of velocity be u, then the vertical component will be $2u$.

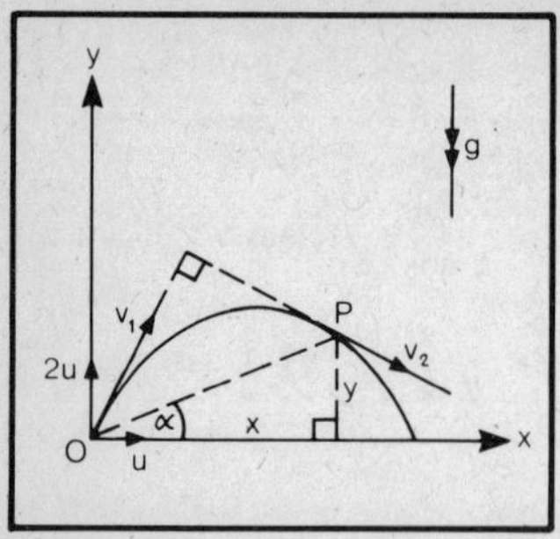

Figure 74.

Taking the axes as shown in the diagram:

$\ddot{y} = -g \qquad \ddot{x} = 0$

$\dot{y} = 2u - gt \qquad \dot{x} = u$

$y = 2ut - \frac{1}{2}gt^2 \qquad x = ut$

$y = 0$, when $0 = 2ut - \frac{1}{2}gt^2$,

$$\text{thus } t = 0 \text{ or } t = \frac{4u}{g}.$$

The value of x at this time is $\mathbf{\frac{4u^2}{g}}$ which is the **range**.

To find the **direction** of v_1, we need

$$\frac{\dot{y}}{\dot{x}} \text{ at } t = 0 \quad \text{which is} \quad \frac{2u}{u} = 2.$$

For the velocity v_2 to be perpendicular to the initial velocity v_1, the **product of the two gradients must be −1**.

At time t, $\frac{\dot{y}}{\dot{x}} = \frac{2u - gt}{u}$.

Thus v_2 will be achieved after time t given by

$$2\left(\frac{2u - gt}{u}\right) = -1 \quad \text{i.e., } t = \frac{5u}{2g}.$$

Hence the angle α, between OP and the horizontal at time

$$t = \frac{5u}{2g} \text{ is given by } \tan\alpha = \frac{y}{x} = \frac{2u \cdot \frac{5u}{2g} - \frac{g}{2} \cdot \frac{25u^2}{4g^2}}{\frac{5u^2}{2g}}$$

i.e., **angle α = arctan $\frac{3}{4}$.**

Example 2 A particle is projected with speed v ms^{-1} from a point A at an angle of elevation α and moves freely under gravity. The highest point in the path of the particle is B and AB is inclined at an acute angle θ to the horizontal. Show that $\tan\alpha = 2\tan\theta$.

If $\alpha = 45°$ and the vertical height of B above A is 50 m, find the value of v. Find also the time that elapses from the instant of projection until the instant when the velocity is parallel to AB. (Take the acceleration due to gravity to be 10 ms^{-2}.)

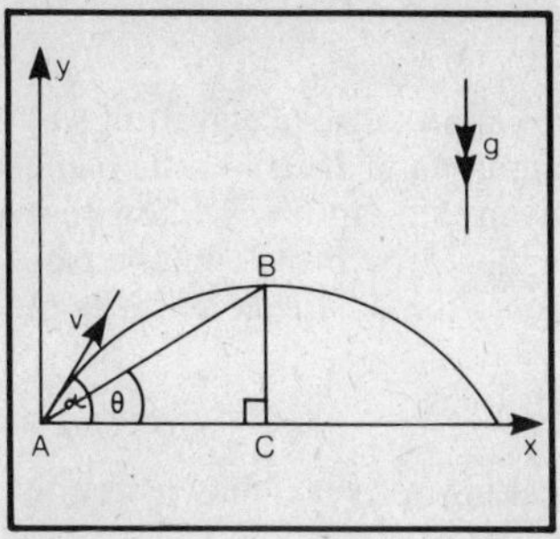

Figure 75.

Taking axes as shown:

$\ddot{y} = -g,$

$\dot{y} = v \sin \alpha - gt,$

$y = v \sin \alpha\, t - \frac{1}{2}gt^2,$

$\ddot{x} = 0,$

$\dot{x} = v \cos \alpha,$

$x = v \cos \alpha\, t.$

BC is the maximum height and is achieved when $\dot{y} = 0$, i.e., when

$$t = \frac{v \sin \alpha}{g}.$$

Substituting this value of t into the expressions for y and x,

$$BC = v \sin \alpha \,.\, \frac{v \sin \alpha}{g} - \frac{g}{2}\left(\frac{v \sin \alpha}{g}\right)^2, \quad \text{i.e., } BC = \frac{v^2 \sin^2 \alpha}{2g} \quad \text{(i)}$$

$$\text{and } AC = \frac{v \cos \alpha\, v \sin \alpha}{g} = \frac{v^2 \cos \alpha \sin \alpha}{g} \quad \text{(ii)}.$$

$$\text{From (i) and (ii), } \tan \theta = \frac{BC}{AC} = \frac{(v^2 \sin^2 \alpha)/2g}{(v^2 \cos \alpha \sin \alpha)/g}$$

i.e., $\tan \alpha = 2 \tan \theta$.

Substituting $BC = 50$ m, $g = 10\ \text{ms}^{-2}$ and $\alpha = 45°$ into (i),

$50 = \dfrac{v^2}{2 \times 2 \times 10}$. Therefore, $v^2 = 2000$ and $\mathbf{v = 20\sqrt{5}\ ms^{-1}}$.

When the **velocity is parallel to *AB***, $\dot{y}/\dot{x}$ will have the same value as $\tan \theta$,

i.e., $\dfrac{\dot{y}}{\dot{x}} = \tan \theta = \dfrac{\tan \alpha}{2} = \dfrac{1}{2}.$

However $\frac{\dot{y}}{\dot{x}} = \frac{v \sin \alpha - gt}{v \cos \alpha} = \frac{20\sqrt{5}/\sqrt{2} - 10t}{20\sqrt{5}/\sqrt{2}}$.

Thus $\frac{1}{2} = \frac{20\sqrt{5}/\sqrt{2} - 10t}{20\sqrt{5}/\sqrt{2}}$ and $\boldsymbol{t = \frac{\sqrt{5}}{\sqrt{2}}}$ **secs.**

Example 3 A particle is projected at an angle of elevation θ at speed v from a point O on a horizontal plane. It strikes the plane at point A. Find the distance OA given that $\tan \theta = \frac{2}{3}$. Show that the same range would have been achieved by projecting the particle at an angle of arctan $(\frac{3}{2})$. Find also the difference in times of flight for the two projectiles.

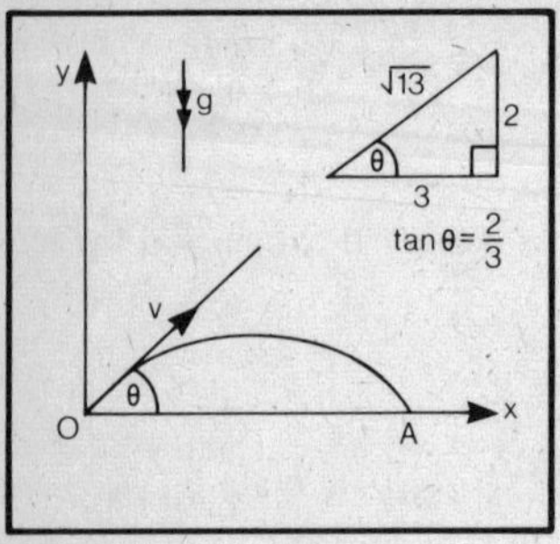

Figure 76.

Taking axes as shown in the diagram:

$\ddot{y} = -g,$

$\dot{y} = v \sin \theta - gt,$

$y = v \sin \theta \, t - \frac{1}{2}gt^2,$

$\ddot{x} = 0,$

$\dot{x} = v \cos \theta,$

$x = v \cos \theta \, t,$

When $y = 0$, $v \sin \theta \, t - \frac{1}{2}gt^2 = 0$,

i.e., $t = 0$, which is obviously when the particle is at O,

or, $\boldsymbol{t = \frac{2u}{g} \sin \theta}$.

Substituting this value of t into the expression for x will give us the **range**,

i.e., $OA = v \cos \theta \frac{2u}{g} \sin \theta.$

Since $\tan \theta = \frac{2}{3}$, then $\sin \theta = \frac{2}{\sqrt{13}}$ and $\cos \theta = \frac{3}{\sqrt{13}}$.

Therefore $\boldsymbol{OA = \frac{12v^2}{13g}}$.

If the angle of projection were arctan $(\frac{3}{2})$ then

$$\sin\theta = \frac{3}{\sqrt{13}} \quad \text{and} \quad \cos\theta = \frac{2}{\sqrt{13}},$$

which on substitution into the expression for OA will give the same value of

$$\frac{12v^2}{13g}.$$

The times of flight for the two angles of projection will be different since

$$t = \frac{2v}{g}\sin\theta$$

and the **difference is** $\frac{2v}{g}\left(\frac{3}{\sqrt{13}} - \frac{2}{\sqrt{13}}\right) = \frac{\mathbf{2v}}{\sqrt{\mathbf{13g}}}$ **secs.**

Example 4 A ball is hit with initial speed 25 ms^{-1} at an elevation of 50°. Find its greatest height, range on level ground and the time of flight.

Show that it will clear a wall 10 m high which is 50 m from where the ball was hit. (Take g to be 10 ms^{-2}).

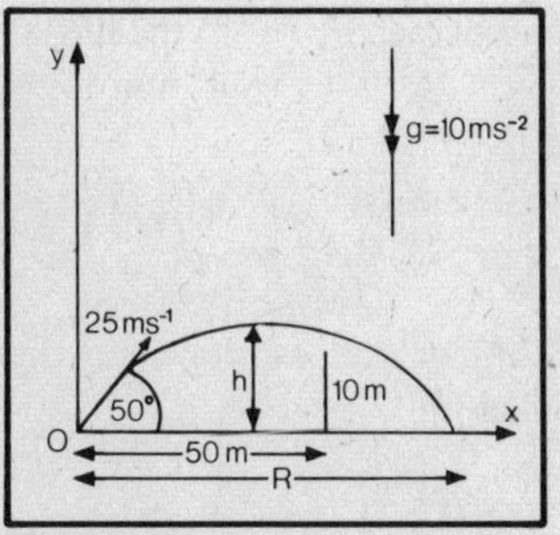

Figure 77.

Taking axes as shown:

$\ddot{y} = -10,$

$\dot{y} = 25 \sin 50° - 10t,$

$y = 25t \sin 50° - 5t^2,$

$\ddot{x} = 0,$

$\dot{x} = 25 \cos 50°,$

$x = 25t \cos 50°.$

At the greatest height, h, $\dot{y} = 0$.

$$\text{Thus } 0 = 25 \sin 50° - 10t \text{ and } t = \frac{25 \sin 50°}{10}.$$

Substituting this value of t into the expression for y,

$$h = 25 \sin 50° \frac{25}{10} \sin 50° - 5\left(\frac{25 \sin 50°}{10}\right)^2 = \mathbf{18{\cdot}3\ m}.$$

The **time of flight** is twice that taken to reach the greatest height,

i.e., $$2\left(\frac{25 \sin 50°}{10}\right) = \mathbf{3{\cdot}8\ s}.$$

The **range** is the value of x when $\boldsymbol{t} = \mathbf{3{\cdot}8\ s}$.

i.e., $\boldsymbol{R} = 25 \cos 50° \times 3{\cdot}8 = \mathbf{61\ m}$.

To find if the ball will clear the wall we must find the value of y when $x = 50$ m,

i.e., when $t = \dfrac{50}{25 \cos 50°}$. At this time,

$$\boldsymbol{y} = 25 \sin 50° \left(\frac{50}{25 \cos 50°}\right) - 5\left(\frac{50}{25 \cos 50°}\right)^2 = \mathbf{11{\cdot}18\ m}$$

and **the ball will clear the wall**.

Example 5 A shell is fired with speed v at an elevation 2α to the horizontal from the foot of a plane of inclination α to the horizontal, the direction of projection being in the vertical plane, through the line of greatest slope of the inclined plane. Find the components of its velocity along and perpendicular to the line of greatest slope after time T. Find its range on the plane and show that it strikes the plane at an angle

$$\tan^{-1}\left\{\frac{\tan \alpha}{1 - 2 \tan^2 \alpha}\right\}. \qquad \text{(W.J.E.C.)}$$

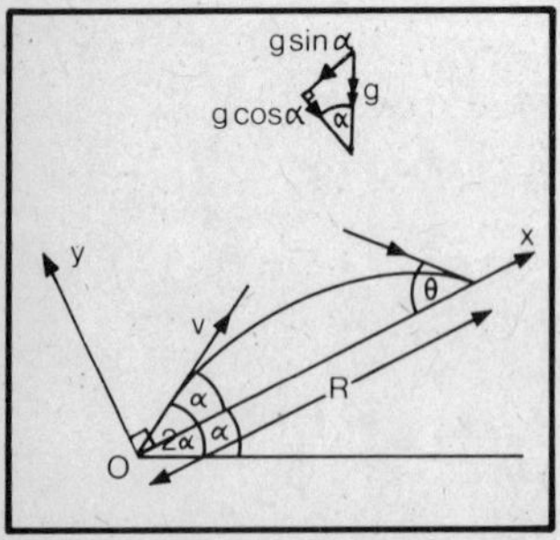

Figure 78.

Taking axes as shown in the diagram:

$\ddot{y} = -g \cos \alpha$,

$\dot{y} = v \sin \alpha - g \cos \alpha\, T$,

$y = v \sin \alpha\, T - \frac{1}{2} g \cos \alpha\, T^2$,

and

$\ddot{x} = -g \sin \alpha$,

$\dot{x} = v \cos \alpha - g \sin \alpha\, T$,

$x = v \cos \alpha\, T - \frac{1}{2} g \sin \alpha\, T$.

The component of velocity along the line of greatest slope is $x = v\cos\alpha - g\sin\alpha\, T$, and perpendicular to the line of greatest slope is $y = v\sin\alpha - g\cos\alpha\, T$. The particle hits the plane when $y = 0$,

i.e., when $0 = v\sin\alpha\, T - \frac{1}{2}g\cos\alpha\, T^2$.

Thus $T = 0$, which is the starting point, or $T = \dfrac{2v\sin\alpha}{g\cos\alpha}$,

and the range, R, is given by the value of x at this time. Hence,

$$\begin{aligned} \boldsymbol{R} &= v\cos\alpha\left(\frac{2v\sin\alpha}{g\cos\alpha}\right) - \frac{g}{2}\sin\alpha\left(\frac{2v\sin\alpha}{g\cos\alpha}\right)^2 \\ &= \boldsymbol{\frac{2v^2}{g}\sin\alpha\,(1 - \tan^2\alpha)}. \end{aligned}$$

The angle with which the particle strikes the plane, θ, is given by

$$\tan(180° - \theta) = \frac{\dot{y}}{\dot{x}} \text{ at time } T = \frac{2v\sin\alpha}{g\cos\alpha},$$

i.e., $$-\tan\theta = \frac{v\sin\alpha - g\cos\alpha\left(\dfrac{2v\sin\alpha}{g\cos\alpha}\right)}{v\cos\alpha - g\sin\alpha\left(\dfrac{2v\sin\alpha}{g\cos\alpha}\right)} = \frac{-\tan\alpha}{1 - 2\tan^2\alpha}.$$

Therefore $$\boldsymbol{\theta = \tan^{-1}\left\{\frac{\tan\alpha}{1 - 2\tan^2\alpha}\right\}}.$$

Example 6 A particle is projected from a point on a plane which is inclined at an angle α to the horizontal. Its initial speed is u at an angle θ to the horizontal in a vertical plane through a line of greatest slope of the plane. Show that the range up the plane is

$$\frac{2u^2\sin(\theta - \alpha)\cos\theta}{g\cos^2\alpha}.$$

Deduce that the range up the plane is maximum when

$$\theta = \frac{\pi}{4} + \frac{\alpha}{2}.$$

If the maximum range up the plane is half the maximum range down the plane find α.

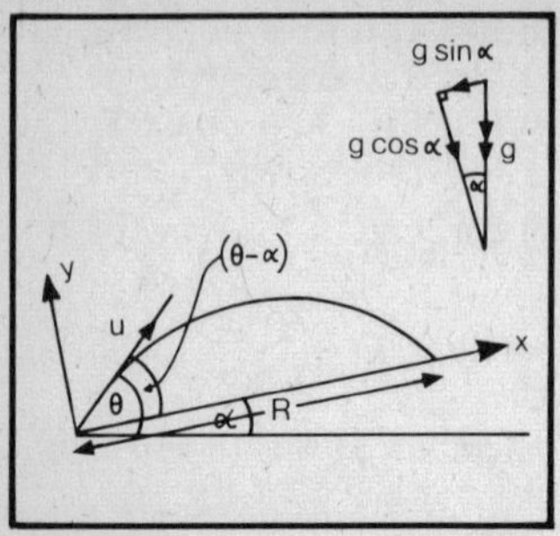

Figure 79.

Considering projection **up** the plane first, and taking axes as shown in the diagram:

$\ddot{y} = -g \cos \alpha,$

$\dot{y} = u \sin(\theta - \alpha) - g \cos \alpha \, t,$

$y = u \sin(\theta - \alpha)\, t - \tfrac{1}{2} g \cos \alpha \, t^2,$

$\ddot{x} = -g \sin \alpha,$

$\dot{x} = u \cos(\theta - \alpha) - g \sin \alpha \, t,$

$x = u \cos(\theta - \alpha)\, t - \tfrac{1}{2} g \sin \alpha \, t^2,$

The particle hits the plane when $y = 0$,

i.e., $0 = u \sin(\theta - \alpha)\, t - \tfrac{1}{2} g \cos \alpha \, t^2.$

Therefore $t = 0$ or $t = \dfrac{2u \sin(\theta - \alpha)}{g \cos \alpha}$.

The latter value gives the time of flight.

R is given by the value of x at this time, i.e.,

$$\boldsymbol{R = u \cos(\theta - \alpha)\left(\frac{2u \sin(\theta - \alpha)}{g \cos \alpha}\right) - \frac{g}{2} \sin \alpha \left(\frac{2u \sin(\theta - \alpha)}{g \cos \alpha}\right)^2}$$

$$= \frac{2u^2 \sin(\theta - \alpha)}{g \cos^2 \alpha}(\cos(\theta - \alpha)\cos \alpha - \sin(\theta - \alpha)\sin \alpha)$$

$$\boldsymbol{= \frac{2u^2}{g \cos^2 \alpha} \sin(\theta - \alpha)\cos \theta.}$$

In this expression $\dfrac{2u^2}{g \cos^2 \alpha}$

is constant and only $\sin(\theta - \alpha)\cos \theta$ may vary. To maximise R we need to find when

$$\frac{d}{d\theta}(\sin(\theta - \alpha)\cos \theta) = 0 \quad \text{i.e., when}$$

$$\cos(\theta - \alpha)\cos \theta - \sin(\theta - \alpha)\sin \theta = 0. \text{ So, } \cos(2\theta - \alpha) = 0$$

and $2\theta - \alpha = \frac{\pi}{2}$ giving $\boldsymbol{\theta = \frac{\pi}{4} + \frac{\alpha}{2}}$.

Substituting this value of θ into the expression for R, gives:

$$\begin{aligned}\text{maximum } R &= \frac{2u^2}{g\cos^2\alpha}\sin\left(\frac{\pi}{4} - \frac{\alpha}{2}\right)\cos\left(\frac{\pi}{4} + \frac{\alpha}{2}\right) \\ &= \frac{2u^2}{g\cos^2\alpha}\left(\frac{1}{2}\left(\sin\frac{\pi}{2} - \sin\alpha\right)\right) \\ &= \boldsymbol{\frac{u^2}{g\cos^2\alpha}(1 - \sin\alpha)}.\end{aligned}$$

We were asked to deduce that the maximum range occurs when $\theta = \pi/4 + \alpha/2$, which we have done, but we are not asked to do so for the maximum range down the plane so we may quote the result **that the maximum range down the plane occurs when the angle of projection with the downward slope bisects the angle between the downward slope and the upward vertical and is**

$$\boldsymbol{\frac{u^2}{g\cos^2\alpha}(1 + \sin\alpha)}.$$

We are given that the maximum range up the plane is half the maximum range down the plane and so

$$\frac{2u^2}{g\cos^2\alpha}(1 - \sin\alpha) = \frac{u^2}{g\cos^2\alpha}(1 + \sin\alpha),$$

giving $\boldsymbol{\sin\alpha = \frac{1}{3}}$ i.e., $\boldsymbol{\alpha = \arcsin(\frac{1}{3})}$.

Example 7 A particle is projected under gravity from a point A on an inclined plane which makes an angle α with the horizontal. The velocity of projection is V at an angle $\tan^{-1}(\frac{1}{2})$ to an upward line of greatest slope of the plane. The motion takes place in the vertical plane through a line of greatest slope and the particle hits the plane at the point B which is further up the plane than A. Find the time taken for the particle to travel between the points A and B in terms of V, g and α. Find also the possible values of $\tan\alpha$ if the particle is moving at an angle α to the horizontal when it strikes the plane at B.

(A.E.B.)

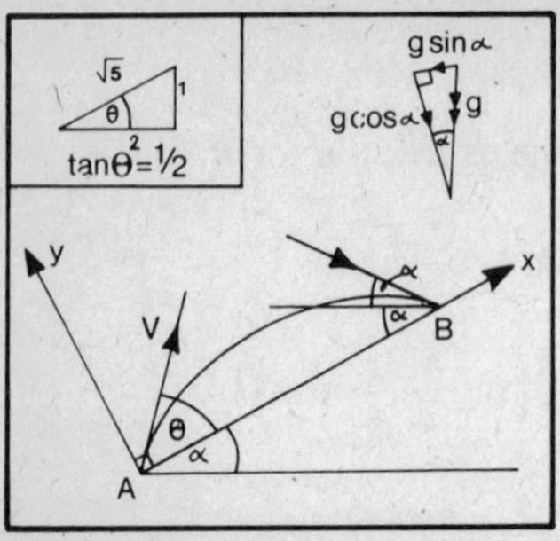

Figure 80.

Taking axes as shown in the diagram:

$\ddot{y} = -g \cos \alpha,$

$\dot{y} = V \sin \theta - g \cos \alpha\, t,$

$y = V \sin \theta\, t - \frac{1}{2} g \cos \alpha\, t^2,$

$\ddot{x} = -g \sin \alpha,$

$\dot{x} = V \cos \theta - g \sin \alpha\, t,$

$x = V \cos \theta\, t - \frac{1}{2} g \sin \alpha\, t^2.$

The particle strikes the plane at B when $y = 0$, i.e., when $0 = V \sin \theta\, t - \frac{1}{2} g \cos \alpha\, t^2$.

Therefore $t = 0$, or $t = \dfrac{2V \sin \theta}{g \cos \alpha}$. Hence the

time taken to travel between *A* and *B* is $\dfrac{2V}{\sqrt{5}\,g \cos \alpha}$

since $\sin \theta = 1/\sqrt{5}$.

The angle, 2α, at which the particle strikes the plane is given by

$$\tan(180° - 2\alpha) = \frac{\dot{y}}{\dot{x}} \text{ at time } t = \frac{2V \sin \theta}{g \cos \alpha},$$

$$\text{i.e.,}\quad \tan(180° - 2\alpha) = \frac{V \sin \theta - g \cos \alpha \left(\dfrac{2V \sin \theta}{g \cos \alpha}\right)}{V \cos \theta - g \sin \alpha \left(\dfrac{2V \sin \theta}{g \cos \alpha}\right)}$$

$$= \frac{-\sin \theta}{\cos \theta - 2 \tan \alpha \sin \theta} = \frac{-\tan \theta}{1 - 2 \tan \alpha \tan \theta}.$$

Since $\tan(180° - 2\alpha) = -\tan 2\alpha$ and $\tan \theta = \frac{1}{2}$, then

$$\tan 2\alpha = \frac{1}{2 - 2 \tan \alpha} \quad \text{and} \quad 2 \tan 2\alpha - 2 \tan 2\alpha \tan \alpha - 1 = 0.$$

Thus $$\frac{4 \tan \alpha}{1 - \tan^2 \alpha} - \frac{4 \tan^2 \alpha}{1 - \tan^2 \alpha} - 1 = 0,$$

$$3\tan^2\alpha - 4\tan\alpha + 1 = 0, \quad (3\tan\alpha - 1)(\tan\alpha - 1) = 0.$$

Therefore $\mathbf{\tan\alpha = 1}$ or $\mathbf{\tan\alpha = \frac{1}{3}}$.

Key terms

The only force acting on a projectile (unless otherwise stated) is its own weight. It therefore has a **constant acceleration of *g* vertically downwards, and no acceleration, hence constant velocity, in the horizontal direction.**

When solving a problem, show clearly on your diagram the axes which you are taking and first write down the equations of motion in both of these directions.

Choose your axes to suit the problem. It is nearly always more convenient to take axes parallel and perpendicular to the slope when considering projectiles on an **inclined** plane. In this case there will be components of constant acceleration in both of these directions.

Chapter 8
Work, Energy and Power

Work done by a constant force

When a body moves under the action of a **constant** force, **F**, through displacement, **r**, then the work done by **F** is defined to be the product of the magnitude of the component of force in the direction of motion with the distance moved by the point of application of the force, i.e., $F \cos \theta \, r = \mathbf{F} \cdot \mathbf{r}$, where θ is the angle between the line of action of the force and the displacement.

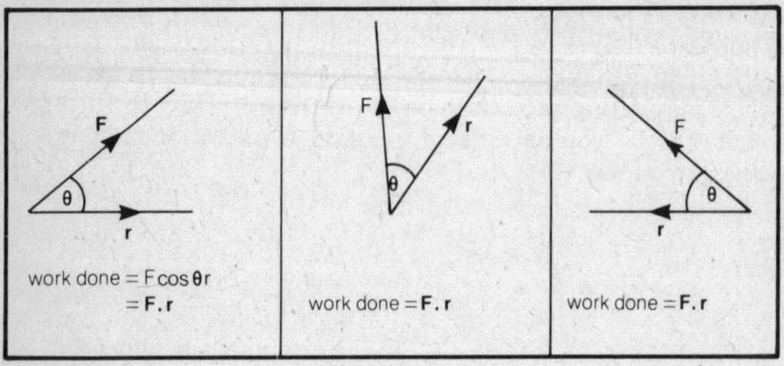

Figure 81.

Work is a scalar. The work done by **F** in moving its point of application through **r** is unaffected by the orientation of **F** and **r**, provided that their relative position remains the same, see figure 81.

When a number of forces $\mathbf{f}_1, \mathbf{f}_2, \ldots, \mathbf{f}_n$, with resultant **F**, act on a particle to give it a displacement **r**, then the work done is $\mathbf{f}_1 \cdot \mathbf{r} + \mathbf{f}_2 \cdot \mathbf{r} + \cdots \mathbf{f}_n \cdot \mathbf{r}$ which, by the distributive law for scalar products, is $(\mathbf{f}_1 + \mathbf{f}_2 + \cdots \mathbf{f}_n) \cdot \mathbf{r} = \mathbf{F} \cdot \mathbf{r}$. Hence the work done is the same as it would be if the system of forces were replaced by their resultant force.

When a **constant** force **F** acts on a particle which undergoes successive displacements $\mathbf{r}_1, \mathbf{r}_2, \ldots, \mathbf{r}_n$ the work done by **F** is

$\mathbf{F} . \mathbf{r}_1 + \mathbf{F} . \mathbf{r}_2 + \cdots \mathbf{F} . \mathbf{r}_n$, which, by the distributive law for scalar products is $\mathbf{F} . (\mathbf{r}_1 + \mathbf{r}_2 + \cdots \mathbf{r}_n)$. Hence the work done is equal to the scalar product of $\mathbf{F}$ and the vector sum of the displacements. If the point of application of the force $\mathbf{F}$ moves from a point with position vector $\mathbf{r}_1$ to a point with position vector $\mathbf{r}_2$, then the work done is $\mathbf{F} . (\mathbf{r}_2 - \mathbf{r}_1)$ and is **independent of the path taken.**

If the angle of inclination of the line of action of the force to the displacement is greater than 90°, then work is said to be being done **against** the force, e.g., when a body moves on a rough surface, work is said to be done **against friction**.

If the angle of inclination of the force to the displacement is 90°, then zero work is being done by that force, since $\mathbf{F} . \mathbf{r} = 0$.

Units of work

Since work is the product of a force and a distance, the unit of work is **1 Nm**, called a **joule** (J). 1 J is the amount of work done by a force of 1 N in moving its point of application a distance of 1 m in the direction of the force.

Work done by a variable force

The work done by a variable force, $\mathbf{F}$, may be found by assuming the force to be constant while moving the point of application a small distance δr in what can be assumed to be a straight line, and then summing over the total path.

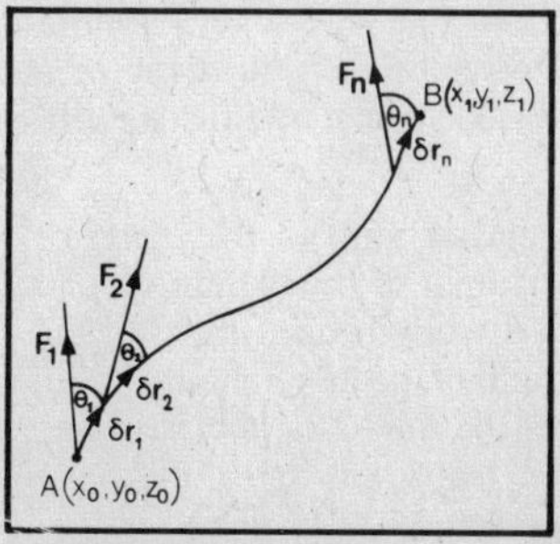

Figure 82.

Consider the force $\mathbf{F}$ to have the values $\mathbf{F}_1, \mathbf{F}_2, \ldots, \mathbf{F}_n$, while moving through displacements $\delta\mathbf{r}_1, \delta\mathbf{r}_2, \ldots, \delta\mathbf{r}_n$, along AB (see figure 82).

The work done is

$$\lim_{\delta r \to 0} \sum \mathbf{F} . \delta\mathbf{r}$$

$$= \lim_{\delta r \to 0} \sum F \cos\theta \, \delta r$$

$$= \int_{AB} F \cos\theta \, dr.$$

Since the work done by a force AB is equal to the work done by its components (letting X, Y and Z be the variable components of F) then

$$\int_{AB} F \cos\theta \, dr = \int_{x_0}^{x_1} X \, dx + \int_{y_0}^{y_1} Y \, dy + \int_{z_0}^{z_1} Z \, dz.$$

In particular, if the path is a **straight line**, say from x_0 to x_1, and the variable force is acting along that straight line, the work done is

$$\int_{x_0}^{x_1} F \, dx$$

We shall now look at the work done by the **variable** force of tension in a **stretched elastic string or spring**, and of **thrust in a compressed spring**. However, before we can do this we must examine some properties of elastic strings and of springs.

Springs and elastic strings: Hooke's Law

The **natural length** of a string or spring is its length in the absence of any forces being applied to it.

The **extension of a stretched string or spring**, or the **compression of a spring in thrust**, is the difference between its actual length and its natural length.

The **modulus of elasticity** is a measure of the '**stiffness**' of a spring and of the '**elasticity**' of a string, and is constant for that particular string or spring.

Hooke's Law is an experimental law and gives us a relationship between the **tension**, T, in a stretched elastic string or a spring, or the **thrust**, T, in a compressed spring, and the **extension or compression**, x.

It has been shown that, up to the elastic limit of the string or spring, the tension is directly proportional to the extension, and the thrust in a compressed spring is directly proportional to the compression, i.e., $\boldsymbol{T \propto x}$. The constant of proportionality is λ/l where l is the natural length and λ is the modulus of elasticity.

Hence, **Hooke's Law** may be expressed as $\boldsymbol{T = \dfrac{\lambda x}{l}}$.

We can see from this that since T has the dimensions of force, and x and l are distances, λ **must have the dimensions of force**.

The **elastic limit** of a string or spring, is the point after which it will not return to its natural length: it has become 'over-stretched' and no longer obeys Hooke's Law.

Work done in stretching an elastic string or spring (or in compressing a spring)

If the **natural length** of the string or spring is l, its modulus of elasticity, λ, its tension, T, and its extension, s, then the work done in stretching it a further infinitesimally small distance, δs,

$$\text{is } T\delta s = \frac{\lambda s}{l}\,\delta s,$$

assuming the tension to be constant over δs. **The total work done in stretching it from its natural length l to its final length, $(l+x)$ is therefore**

$$\int_0^x \frac{\lambda s}{l}\,\mathrm{d}s = \frac{\lambda x^2}{2l}.$$

This is also the work done in compressing a spring from its natural length, l, to $(l-x)$.

Energy

Energy is the capacity to do work. Work and energy have the same units, joules, and both are scalars.

We are concerned with **mechanical energy** in mechanics, as opposed to chemical, heat, electrical, etc., forms. **Mechanical energy** is of two types: that due to motion, called **kinetic energy**, and that due to position, called **potential energy**.

Kinetic energy is the capacity of a body to do work due to its motion. Consider a body of mass m moving with speed v. Work has been done by a force to get the body to this speed from rest, and work will have to be done on it to bring it to rest. Either of these quantities give us the kinetic energy of the body. We shall look at the work needed to accelerate the body to velocity v from rest.

Let the resultant force acting be a constant force of magnitude F, giving the body an acceleration of a, taking it from rest to velocity v over distance x. $F = ma$ (Newton's second law) and the work done by this force is $Fx = max$. From the equation of motion for constant acceleration '$v^2 = u^2 + 2as$', $v^2 = 2ax$. Hence the work done, and therefore the kinetic energy of a body of mass m travelling at v ms^{-1} is $\frac{1}{2}mv^2$.

If the body we have just considered was accelerated from u ms^{-1} to v ms^{-1} over distance x by this constant resultant force F, then the **work done** by it would be $Fx = max = \frac{1}{2}mv^2 - \frac{1}{2}mu^2$ i.e., **the change in kinetic energy**.

If the force had been **variable** but acting along the direction of motion, then the work done would be

$$\int_0^x F\,\mathrm{d}x = \int_0^x mv\frac{\mathrm{d}v}{\mathrm{d}x}\,\mathrm{d}x = \int_u^v mv\,\mathrm{d}v = \frac{1}{2}mv^2 - \frac{1}{2}mu^2,$$

i.e., **the change in K.E.**

This can also be shown to be the case for a **variable** force moving the particle along a path other than a straight line (using vector techniques more advanced than we have considered). Hence we arrive at the conclusion, called the **Principle of Work**, which states that **the work done by the resultant force in accelerating a body is equal to the change in kinetic energy it produces.**

Potential energy is the capacity of a body to do work by virtue of its position. We shall consider two types: **gravitational** and **elastic**.

Gravitational potential energy is the work that a body is capable of doing in moving under the action of its weight from its position to a standard level. Consider a particle of mass m at height h above a certain level. Its weight mg would do work of mgh in moving to the standard level. Hence we say that its **gravitational potential energy is *mgh*.** If it were at height h **below** the standard level, then work would have to be done against its weight to raise it to the standard level and we say that it has **negative potential energy**. In this case its gravitational potential energy is $-mgh$. We can say that the **work done against gravity** in lifting a body a height h is mgh, i.e., **its increase in potential energy**.

The reference level we choose for zero potential energy is arbitrary and so we must be careful to mark it clearly when using potential energy in the solution of a problem.

Elastic potential energy is the energy that a stretched string possesses by virtue of its extension. A stretched or compressed spring also possesses **e.p.e.** by virtue of its extension or compression. We have seen that the work done in stretching an elastic string, or in stretching or compressing a spring is $\lambda x^2/2l$ with the usual notation, and so the **elastic potential energy is also $\boldsymbol{\lambda x^2/2l}$**, since that is the capacity to do work. When at their natural lengths, springs and elastic strings possess no **e.p.e.**, since if released they would not gain kinetic energy, unlike stretched strings and springs and compressed springs which would immediately begin to move if released.

The **Principle of Conservation of Energy** states that energy can neither be created nor destroyed. This is a very important concept and its implications are enormous, but it is not very useful to us for solving problems. However, a modification of this, **the Principle of Conservation of Mechanical Energy**, is extremely useful. Before we state this we need to define forces called **conservative forces**.

A **conservative force** is such that the work it does in moving its point of application from one position to another is independent of the path it takes; if its point of application moves in a **closed path** then there is **no loss or gain of energy**. An example of a **non-conservative** force is **friction**. Examples of **conservative forces** are **gravity**, **tension in an elastic string**, **tension or thrust in a spring**, and any **constant** force.

We have seen that the work done by a **constant** force is $\mathbf{F} . (\sum \delta\mathbf{r})$ and in a **closed** path $\sum \delta\mathbf{r} = \mathbf{0}$.

The **Principle of Conservation of Mechanical Energy states that in a conservative system of forces the total mechanical energy remains constant**. It follows that if work is done by a **non-conservative force** then that work is equal to the **change in mechanical energy it produces**.

A typical conservative system of forces is one where the only external force causing work to be done is gravity and where no sudden change in motion takes place (otherwise mechanical energy may be converted into heat, light, sound, etc.).

Consider a mass m ascending a rough slope against constant frictional force F with initial velocity u ms^{-1}, reducing to v ms^{-1} after distance x m. From the **principle of work**, i.e.,

work done by the resultant force = change in K.E., we see that

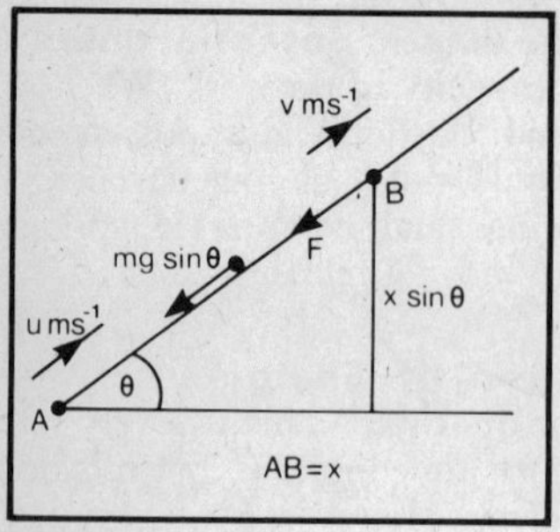

Figure 83.

$$(-F - mg\sin\theta)x = \tfrac{1}{2}mv^2 - \tfrac{1}{2}mu^2$$

giving

$$-Fx = (\tfrac{1}{2}mv^2 + mgx\sin\theta) - \tfrac{1}{2}mu^2,$$

= Final M.E. − Initial M.E.

confirming that the work done against friction is given by the change in mechanical energy it produces.

Power

Power is the rate at which a force does work. Its units are watts (W) which are joules per second. Power is a scalar.

Consider a body working with power P, exerting a force $\mathbf{F}$, causing a displacement $\mathbf{r}$, then **generally,**

$$P = \frac{\mathrm{d}}{\mathrm{d}t}(\mathbf{F}\,.\,\mathbf{r}).$$

If $\mathbf{F}$ is constant then $P = \mathbf{F}\,.\,\dfrac{\mathrm{d}\mathbf{r}}{\mathrm{d}t} = \mathbf{F}\,.\,\mathbf{v}$.

If, as is usually the case, **F is constant and in the same direction as the displacement**, and hence the velocity, then $\boldsymbol{P = Fv}$.

We often consider the power of an engine to pull a vehicle. The force that an engine exerts is called its **tractive force**, and will of course be P/v.

Efficiency

The efficiency of any mechanical system is the **ratio of the output of work to the input of work**, and so will always be less than 1.

Worked examples

Example 1 A lorry of mass 2000 kg is subject to a constant frictional resistance of 2400 N. Find in kW the power at which the engine is working when the lorry is travelling along a level road at a steady speed of 36 km h^{-1}.

If the engine continues to work at this rate, find the steady speed in km h^{-1} at which the lorry ascends a hill of inclination arcsin $(\frac{1}{50})$. (Take g to be 9·8 ms^{-2}.)

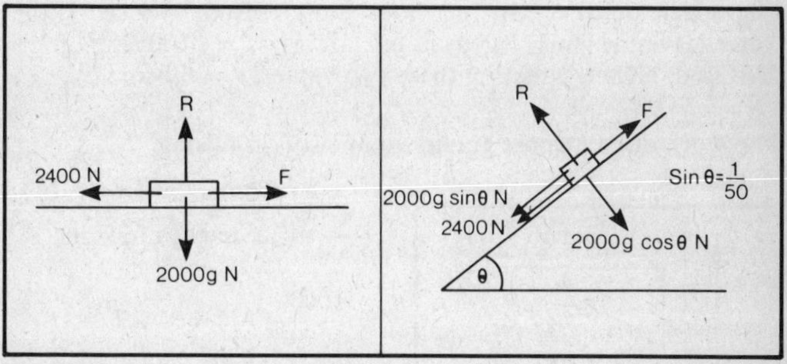

Figure 84.

The **equation of motion** for the lorry moving on the level road is $F - 2400 = 2000a$, where F is the **tractive force** of the engine and a is its **acceleration**. However, since it is travelling at a **steady speed** there will be **no acceleration** and the equation reduces to $\mathbf{F = 2400\ N}$.

The lorry is travelling at 36 km h^{-1} which must be converted to ms^{-1} before we can substitute it in $P = Fv$, so

$$\mathbf{P} = 2400\left(\frac{36 \times 1000}{60 \times 60}\right) = 24000 \text{ W} = \mathbf{24\ kW}.$$

The equation of motion for the lorry climbing the hill at a steady speed, and hence with no acceleration is $F - (2400 + 2000g \sin \theta) = 0$. Therefore

$$F = \left(2400 + \frac{2000 \times 9{\cdot}8}{50}\right) \text{N}.$$

The engine is working at the same rate as when it was on the level road, 24000 W. Since $P = Fv$, then

$$v = \frac{P}{F} = \frac{24000}{2792} = 8{\cdot}6\ \text{ms}^{-1} = \mathbf{31\ km\ h^{-1}}.$$

Example 2 A pump raises 3 m^3 of water per hour through a height of 3·7 m and delivers it in a horizontal jet through a circular nozzle of diamter 12·5 mm. Find the speed at which the water leaves the nozzle. If the pump works at 70% efficiency, calculate the power needed to drive it. (Take the density of water to be 1000 kg m^{-3} and $g = 9{\cdot}8$ ms^{-2}.)

The nozzle delivers 3 m^3 per hour, i.e., 1/1200 m^3 per sec. If the water leaves at v ms^{-1} then in one sec $v \times \pi \times (0{\cdot}00625)^2$ m^3 is delivered. Hence, equating these two values, $\mathbf{v = 6{\cdot}8\ ms^{-1}}$.

The work done against gravity in one sec, in raising

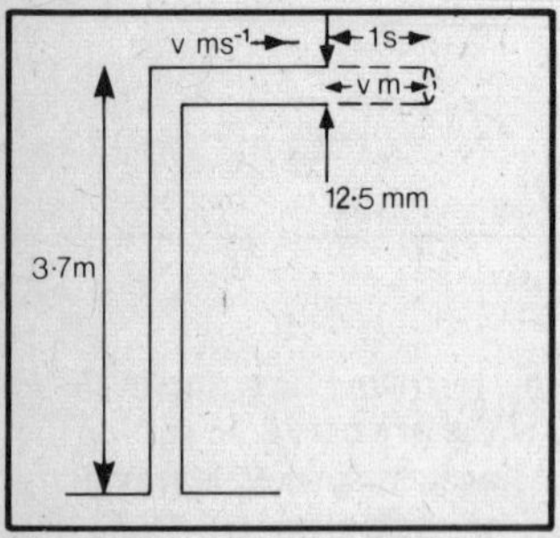

Figure 85.

$\frac{1}{1200}$ m^3 a height of 3·7 m

is $\frac{1000}{1200}g(3{\cdot}7) = 30{\cdot}2$ J,

and the work done in 1 s to give

$\frac{1}{1200}$ m^3 a velocity of 6·8 ms^{-1}

is $\frac{1}{2}\left(\frac{1000}{1200}\right)(6{\cdot}8)^2 = 19{\cdot}3$ J.

Hence the **total work done per second is 49·5 J** and 70% of the power of the pump must be **49·5 W**.

Thus the actual power of the pump is **70·7 W**.

Example 3 A particle of mass 1 kg is attached to one end of an elastic string of modulus $4g$ N and natural length 1 m. The other end of the string is fastened to a fixed point A. Find the point of equilibrium when hanging under gravity.

If the particle is released from rest at A, find the kinetic energy of the particle when it is at a depth of x m below its equilibrium position, provided the string does not exceed its elastic limit.

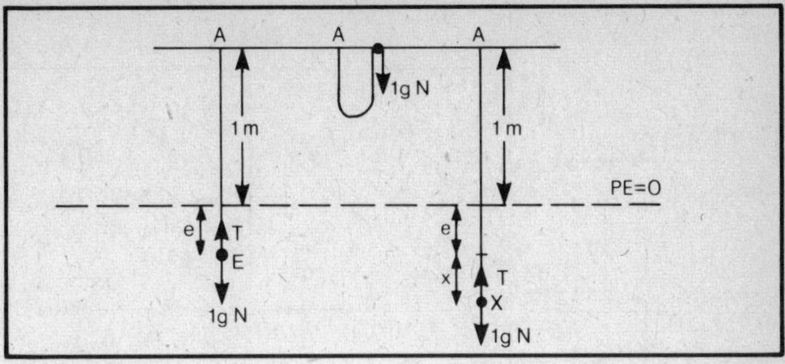

Figure 86.

Let the extension be e m, when the particle is at rest under gravity. For equilibrium, $T = 1g$.

$$\text{By Hooke's Law, } T = \frac{\lambda x}{1} = 4ge\text{; thus } \boldsymbol{e = \frac{1}{4}}.$$

Since the system of forces is **conservative**, we know that the total mechanical energy is constant at all times.

Taking the zero potential energy line as shown in figure 86, at A the only M.E. is the gravitational potential energy of $1g$ J and at X there is negative G.P.E. of $-(\frac{1}{4} + x)g$ J, elastic potential energy of $4g(\frac{1}{4} + x)^2/2$ J, and unknown kinetic energy, K.E.

Hence $g = -(\frac{1}{4} + x)g + 2g(\frac{1}{4} + x)^2 + \text{K.E.}$,

therefore

$$\text{K.E.} = g + \frac{g}{4} + xg - \frac{g}{8} - gx - 2gx^2 = \boldsymbol{g\left(\frac{9}{8} - 2x^2\right)} \textbf{ J}.$$

Example 4 The engine of a car of mass m kg develops constant power P watts when the car is in motion. The resistance to the motion of the car is constant. The maximum speed of the car on a level road is V ms^{-1}. Find the maximum speed of the car

(i) directly up a road inclined at angle arcsin $(1/n)$ to the horizontal;

(ii) directly down the same road.

Find the acceleration of the car on a level road when the speed is $\frac{1}{2}V$ ms^{-1}.

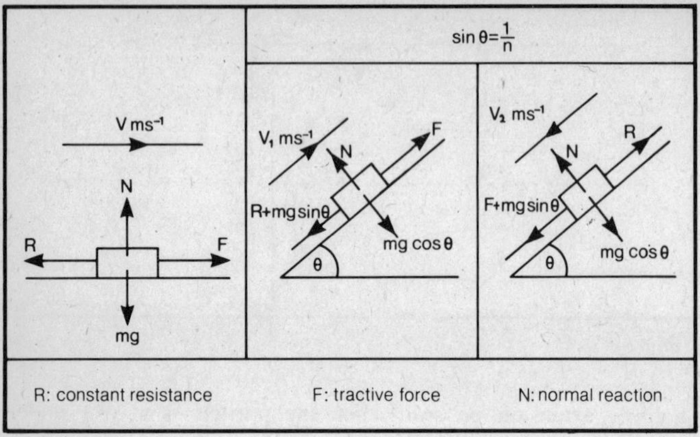

Figure 87.

Since $P = Fv$ then $F = P/v$, and the equation of motion for the car on level ground is (figure 87)

$$\frac{P}{v} - R = m\frac{dv}{dt}.$$

When the car has reached its maximum speed, V ms^{-1},

$$\text{then } \frac{dv}{dt} = 0 \quad \text{and} \quad R = \frac{P}{V}.$$

(i) The equation of motion for the car **ascending** the hill is

$$F - mg \sin\theta - R = m\frac{dv}{dt},$$

and when it has reached its **maximum speed,** $\boldsymbol{V_1}$ ms^{-1},

then $\frac{dv}{dt} = 0$, i.e., $\frac{P}{V_1} - \frac{mg}{n} - \frac{P}{V} = 0.$

Therefore, $\boldsymbol{V_1} = \dfrac{\boldsymbol{PVn}}{\boldsymbol{Pn + mgV}}.$

(ii) The equation of motion for the car **descending** the hill is

$$F + mg \sin\theta - R = m\frac{dv}{dt},$$

and when it has reached its **maximum speed,** $\boldsymbol{V_2}$ ms^{-1}, then $\frac{dv}{dt} = 0$, i.e. $\frac{P}{V_2} + \frac{mg}{n} - \frac{P}{V} = 0$.

Therefore, $\boldsymbol{V_2 = \frac{PVn}{Pn - mgV}}$.

To find the acceleration of the car when it is travelling at $\frac{1}{2}V$ ms^{-1} on level ground we use the equation of motion

$$\frac{P}{v} - \frac{P}{V} = m\frac{dv}{dt},$$

and substitute $\frac{1}{2}V$ ms^{-1} for the velocity v.

Thus $\frac{2P}{V} - \frac{P}{V} = m\frac{dv}{dt}$ and $\boldsymbol{\frac{dv}{dt} = \frac{P}{mV}}$.

Example 5 A particle is projected with kinetic energy E up a line of greatest slope of a rough plane inclined at angle α to the horizontal. If the constant coefficient of friction between the plane and the particle is μ, show that the work done against the frictional force before the particle comes to rest is

$$\frac{E\mu \cos \alpha}{\sin \alpha + \mu \cos \alpha}.$$

Show that if $\tan \alpha \geq \mu$ then the particle will return to the point of projection with kinetic energy

$$E\left(\frac{\tan \alpha - \mu}{\tan \alpha + \mu}\right).$$

(W.J.E.C.)

Since $R = mg \cos \alpha$ and $F = \mu R$, then $F = \mu mg \cos \alpha$. Let x be the distance the particle travels up the hill before coming to rest.

From the Principle of Work, which states that the work done by the resultant force = change in K.E.,

$-(mg \sin \alpha + F)x = -E$.

$$\text{Thus } x = \frac{E}{mg(\sin \alpha + \mu \cos \alpha)}.$$

Hence the **work done against the frictional force is**;

$$\boldsymbol{Fx = \frac{E\mu \cos \alpha}{\sin \alpha + \mu \cos \alpha}}.$$

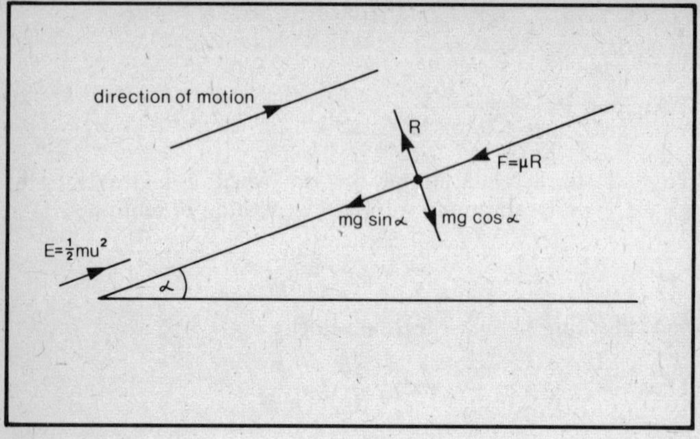

Figure 88.

The particle will then slip down the plane provided its component of weight down the plane, $mg \sin \alpha$, is sufficient to overcome the frictional resistance, i.e.,

$$mg \sin \alpha > \mu mg \cos \alpha, \quad \boldsymbol{\tan \alpha \geq \mu}.$$

If E_1 is the K.E. with which the particle will return to its point of projection, from the Principle of Work,

$$(mg \sin \alpha - \mu mg \cos \alpha)x = E_1.$$

$$\text{Thus, } E_1 = (mg \sin \alpha - \mu mg \cos \alpha)\frac{E}{mg(\sin \alpha + \mu \cos \alpha)}$$

$$= \boldsymbol{E\left(\frac{\tan \alpha - \mu}{\tan \alpha + \mu}\right)}.$$

Example 6 A pump raises water from an underground reservoir through a height of 10 m and delivers it at a speed of 9 ms^{-1} through a circular pipe of internal diameter 20 cm. Taking 1 litre of water to have a mass of 1 kg, and g to have the value 9·8 ms^{-2} find:

(i) the mass of water raised per second, correct to 3 sig. fig;
(ii) the kinetic energy imparted to the water each second, correct to 3 sig. fig;
(iii) the effective power of the pump correct to 2 sig. fig;
(iv) the actual power of the pump if it is 70% efficient.

(S.U.J.B.)

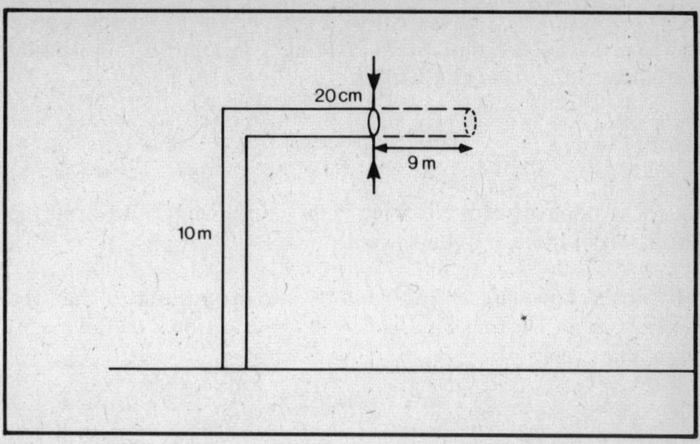

Figure 89.

(i) The volume of water raised per second is $9 \times \pi \times 0{\cdot}1^2$ m^3. Therefore the mass is $9 \times \pi \times 0{\cdot}1^2 \times 10^3$ kg = **283 kg correct to 3 significant figures**.

(ii) The kinetic energy imparted to the water per second is $\frac{1}{2} \times 283 \times 81 = 11\,462$ W = **11·5 kW correct to 3 significant figures**.

(iii) The work done against gravity per second in lifting 283 kg a height of 10 m is $283 \times 9{\cdot}8 \times 10$ W = 27 734 W = **27·7 kW**, correct to 3 sig. figs.

Therefore the effective power of the pump is **39 kW**, correct to 2 sig. figs.

(iv) The actual power of the pump, if it is 70% efficient, is

$$\frac{39}{70} \times 100 \text{ kW} = \mathbf{56\ kW}, \text{ correct to 2 sig. figs.}$$

Key terms

The **work** done by a **constant** force is the product of its component in the direction of motion and the distance moved by the point of application of the force.

Work is a scalar and its units are **joules** (J).

The work done by a **variable** force, **F**, acting in the **same direction** as the motion in moving its point of application distance r in a **straight line** is

$$\int_0^r F\,\mathrm{d}r.$$

Work is done **against** a force if the component of force in the direction of motion opposes motion.

Hooke's Law for springs and elastic strings states that the tension in an elastic string is directly proportional to the extension, i.e..

$$T \propto x \quad \text{and} \quad T = \frac{\lambda x}{l}$$

(similar results hold for springs).

The **work done** in giving an elastic string an extension of x is

$$\int_0^x T\,\mathrm{d}x = \frac{\boldsymbol{\lambda x^2}}{\mathbf{2l}}.$$

Energy is the capacity to do work, is measured in **joules**, and is a **scalar**. **Energy and work are mutually convertible**.

Mechanical energy is either **kinetic energy** (due to motion) or **potential energy** (due to position).

The **kinetic energy** of a body of mass m kg moving with velocity v ms^{-1} is $\frac{1}{2}\boldsymbol{mv^2}$ J.

The **gravitational potential energy** of a body of mass m kg at height h m **above** the level of zero potential energy is $\boldsymbol{mgh}$ J, and if at a depth of h m below the standard level then it has negative potential energy of $-mgh$ J.

Work of mgh J is said to be done **against** gravity in lifting a body of mass m kg a vertical height of h m.

The **elastic potential energy** of a spring or elastic string is the **work done** in stretching or compressing it x m from its natural length, l m, i.e.,

$$\frac{\lambda x^2}{2l}\ \mathrm{J}.$$

A **conservative force** is such that any **work done** by it in moving a body from one position to another is **independent of the path taken**, and such that in moving in a closed path there is no loss of energy. An example of a **non-conservative force is friction**.

The **principle of work** states that the **work done by the resultant force = change in kinetic energy** produced by that **resultant** force.

The principle of conservation of energy states that energy can neither be created nor destroyed.

The Principle of conservation of mechanical energy states that in a system of **conservative forces** the **total mechanical energy remains constant**, i.e., K.E. + P.E. = const. This modifies to give: 'the work that is done by a non-conservative force = change in mechanical energy it produces'.

Power is the **rate of doing work**, its **units** are **watts** and it is a **scalar**. Generally, $P = [d/dt(\mathbf{F} \cdot \mathbf{r})]$ and if $\mathbf{F}$ is constant $P = \mathbf{F} \cdot \mathbf{v}$. If F is constant and in the direction of motion then $\boldsymbol{P = Fv}$.

The **tractive force** of an engine is $F = \dfrac{P}{v}$.

The **efficiency** of any mechanical system is the **ratio of the output of work to the input of work** and so will be **less than unity**.

Chapter 9
Momentum, Impulse and Impact

Momentum

The momentum of a body is the product of its mass and velocity. It is therefore a vector quantity with the direction of its velocity. **Momentum** $= m\mathbf{v}$.

If a body of constant mass moves with constant velocity, it has constant momentum and an external force will have to act on it in order to change its momentum (Newton's first law).

Impulse

The **impulse** of a **constant** force, $\mathbf{F}$, is defined as the product of that force with the time for which it was acting. It is a vector quantity with direction of the velocity. The unit of impulse is 1 newton second (1 Ns). **Impulse** $= \mathbf{F}t$.

When a constant force $\mathbf{F}$ is acting on a particle of mass m, it produces a constant acceleration, $\mathbf{a}$. Let the initial velocity of the particle be $\mathbf{u}$, the final velocity be $\mathbf{v}$ and the time for which the force is acting be t. From '$\mathbf{F} = m\mathbf{a}$' (Newton's second law), and '$\mathbf{v} = \mathbf{u} + \mathbf{a}t$', $\mathbf{F}t = m\mathbf{v} - m\mathbf{u}$. Hence **the impulse of a constant force is equal to the change in momentum it produces**.

Momentum is measured in the same units as impulse, i.e., newton seconds.

The impulse of a **variable** force, $\mathbf{F}$, acting over an interval of time, t, is found by dividing the interval of time into a large number, n, of sub-intervals, $\delta t_1, \delta t_2, \ldots, \delta t_n$, and assuming $\mathbf{F}$ to be constant over these sub-intervals, taking the values $\mathbf{F}_1, \mathbf{F}_2, \ldots, \mathbf{F}_n$, respectively. The impulse $\mathbf{I}$ is approximately $\mathbf{F}_1\,\delta t_1 + \mathbf{F}_2\,\delta t_2 + \cdots \mathbf{F}_n\,\delta t_n$ and as $n \to \infty$ the approximation becomes more accurate. Hence

$$\mathbf{I} = \lim_{n\to\infty} \sum_{1}^{n} \mathbf{F}_r\,\delta t_r = \int_0^t \mathbf{F}\,\mathrm{d}t.$$

As with a constant force, the change in momentum produced by the action of a variable force gives the impulse:

$$\mathbf{I} = \int_0^t \mathbf{F}\,\mathrm{d}t = \int_0^t m\frac{\mathrm{d}\mathbf{u}}{\mathrm{d}t}\,\mathrm{d}t = m\int_{\mathbf{u}}^{\mathbf{v}} \mathrm{d}\mathbf{u} = m\mathbf{v} - m\mathbf{u}$$

When a force acts on a body, its momentum is changed in the direction of the force. Hence, **if in a certain direction no force affects the motion of the body, its momentum in that direction will remain constant**.

Impact

When two bodies collide, they exert equal and opposite forces of action and reaction on each other (Newton's third law). They are in contact with each other for exactly the same length of time. Hence two bodies in collision exert equal and opposite impulses on each other. Since impulse = change in momentum, we see that equal and opposite impulses will produce equal and opposite changes in momenta. Hence, the total momenta of the two objects immediately before and after the impact, will be equal. After the collision, any external forces acting on the system will affect the motion.

The Principle of Conservation of Linear Momentum states that if in a certain direction no external force affects the motion of the system, then the total linear momentum of the system remains constant in that direction.

Impulsive tension

When a string is jerked, equal and opposite tensions act instantaneously at each end of the string. They exert equal and opposite **impulses** called the **impulsive tensions**. If one end of the string is fixed, then the change in momentum of a particle attached to the free end, when the string is jerked, is equal to the impulsive tension, and is in the direction of that impulse. Perpendicular to the string, no impulse acts and so the momentum of the system in that direction remains unaltered by the jerk.

If there are particles attached to both ends of the string and they are both free to move, then the equal and opposite impulsive tensions, which occur when the string is jerked, produce equal and opposite changes in momenta in the direction of the string. Hence the total momentum of the system will remain unaltered by the jerk.

Elastic impact

When two objects **collide and bounce**, the impact is said to be **elastic**, whereas if they **collide and coalesce**, then the impact is said to be **inelastic**.

Newton's law of restitution

Experimental evidence suggests that when two bodies collide, their relative velocities along the line of impact before and after impact, are in a constant ratio for that particular pair of bodies.

Newton formulated his **law of restitution** which states that when two bodies collide:

$$\frac{\text{relative velocity along the line of impact after impact}}{\text{relative velocity along the line of impact before impact}} = -e,$$

where e is **the coefficient of restitution** for those two bodies.

If $e = 0$, then the collision is **inelastic** and the bodies must be moving as one after the collision. **If $e = 1$**, then there is no loss of kinetic energy due to impact, and we say that the impact is **perfectly elastic**: the bodies bounce back with the same speed as that with which they collided.

Hence the limits for e are 0 and 1: $\mathbf{0 \le e \le 1}$.

Direct and oblique impact

If the two bodies were moving **along the line of impact** just before impact, their impact is said to be **direct**.

If the two bodies were **not moving along the line of impact** just before impact, their impact is said to be **oblique**.

If the two bodies are smooth and are both free to move, then they will exert equal and opposite impulses along the line of impact which will result in equal and opposite changes in momenta. Hence the total momenta of the two bodies along the line of impact will remain unaltered by that impact.

Perpendicular to the line of impact, provided that the bodies make smooth contact, there will be no impulsive forces acting and their momenta in this direction will remain unchanged by the impact, i.e., the components of velocity perpendicular to the line of impact remain unaltered by the impact.

If impact occurs where one of the bodies is **fixed**, e.g., a ball striking a wall, then clearly there will be an external force along the line of impact: the reaction of the wall to the impact. Hence, momentum will not be conserved in this direction. However, Newton's law of restitution will still apply along the line of impact.

Worked examples

Example 1 An inelastic pile driver of mass 4000 kg falls freely from a height of 5 m on to a pile of mass 1000 kg driving the pile 20 cm into the ground. Find the speed with which the pile starts to move into the ground. Find the loss of kinetic energy due to the impact. Find the resistance to penetration of the ground, assuming it be constant. (Take $g = 10\ \text{ms}^{-2}$).

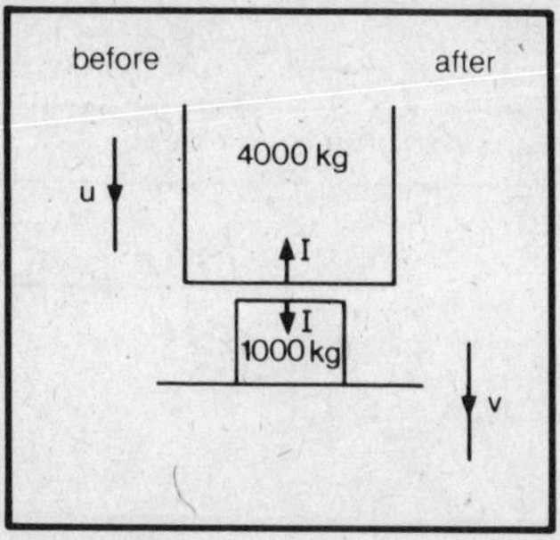

Figure 90.

Let us first find the velocity u with which the pile driver hits the pile. Applying '$v^2 = u^2 + 2as$', we see that $u = 10\ \text{ms}^{-1}$ since it is falling freely under gravity. At the moment of impact, equal and opposite impulses, I, will act and produce equal and opposite changes in momenta. Let v be the velocity of the pile and pile driver combined after the impact.

For the pile driver, $-I = 4000(v - 10)$, and for the pile $I = 1000(v - 0)$. **Therefore, $v = 8\ \text{ms}^{-1}$.**

The kinetic energy before impact was $\frac{1}{2} \cdot 4000 \cdot 10^2 = 200{,}000$ J and after impact was $\frac{1}{2}(4000 + 1000)8^2 = 160{,}000$ J.

Hence **the loss of kinetic energy due to the impact = 40,000 J**.

The equation of motion for the pile and pile driver as it penetrates the ground against a resistance of R N is

$$5000g - R = 5000a$$

where a is their acceleration, which will be constant. From '$v^2 = u^2 + 2as$', given that the penetration is 20 cm, **$a = -160\ \text{ms}^{-2}$**.

Substituting this value of a into the equation of motion above, $\boldsymbol{R = 850{,}000}$ **N**.

Example 2 A gun of mass M is free to move horizontally, but not vertically. The gun fires a bullet of mass m with velocity v.
(i) If the barrel is horizontal, find the velocity of recoil of the gun.
(ii) If the barrel is inclined at an angle θ to the horizontal, find the initial direction of motion of the bullet in terms of m, M, and θ.

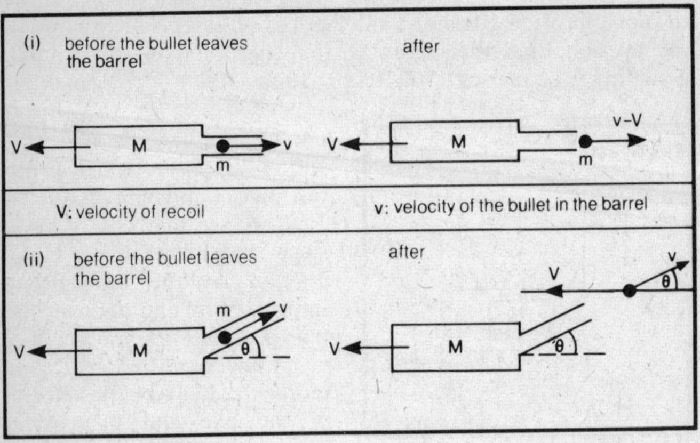

Figure 91.

(i) Before the gun is fired, both the gun and bullet are at rest, therefore the initial total momentum is zero. Taking the velocities as shown in figure 91, and applying the Principle of Conservation of Linear Momentum in the horizontal direction:

$$0 = m(v - V) - MV. \text{ Hence, } \boldsymbol{V = \frac{mv}{M + m}}.$$

(ii) Again, before the gun is fired, both the gun and bullet are at rest. Therefore the initial total momentum is zero. Taking the velocities as shown in the diagram, and applying Conservation of Momentum in the horizontal direction:

$$0 = m(v \cos \theta - V) - MV.$$

Hence, $V = \dfrac{mv \cos \theta}{m + M}$.

The bullet leaves the barrel with the velocity which is the resultant of the two inclined components, v and V, as shown in figure 91(ii). Hence it has component $v \cos \theta - V$ in the horizontal direction, and component $v \sin \theta$ in the vertical direction. Therefore, **the direction of motion of the bullet** is

$$\arctan\left(\frac{v \sin \theta}{v \cos \theta - V}\right) \text{ to the horizontal,}$$

i.e.
$$\boldsymbol{\arctan\left(\frac{(M + m)\tan \theta}{M}\right)}.$$

Example 3 A light inextensible string AB has a particle of mass $2m$ attached at A and a particle of mass m attached at B. The particles are placed on a smooth horizontal table with the string taut. A horizontal impulse is applied to the particle at A, of magnitude mu in the direction which makes an angle of 120° with AB.
(i) Show that B starts to move with speed $u/6$ along BA.
(ii) Find the component of the initial velocity of the particle at A along BA and perpendicular to BA and state the magnitude of the impulse of the tension in the string.

(A.E.B. part)

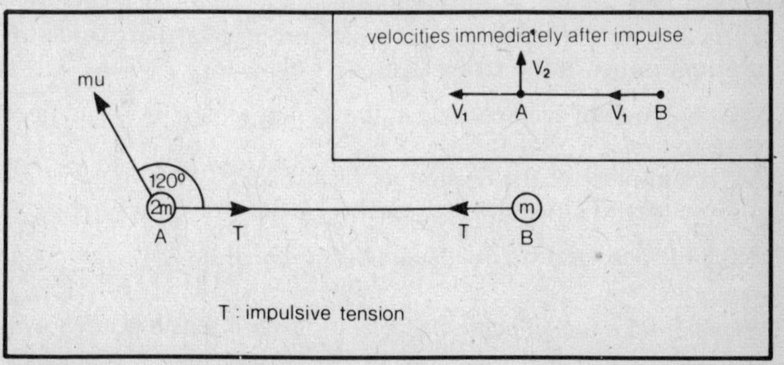

Figure 92.

(i) The external impulse mu causes the string to jerk, exerting internal impulsive tensions T, as shown.

Using 'impulse = change in momentum' in the direction parallel to BA for particle A, $mu \cos 60° - T = 2mv_1$, and for particle B, $T = mv_1$. Combining these equations: $v_1 = u/6$.

Particle B receives an impulse only in the direction of BA. **Hence it will begin to move with speed $u/6$ along BA.**

(ii) Using 'impulse = change in momentum' in the direction perpendicular to BA for A, $mu \sin 60° = 2mv_2$, therefore

$$v_2 = \frac{\sqrt{3}\,u}{4}.$$

Hence A begins to move with velocity

$$\frac{u}{6} \text{ along } BA \text{ and } \frac{\sqrt{3}\,u}{4} \text{ perpendicular to } BA.$$

The magnitude of the impulse of the tension is

$$\frac{mu}{6}.$$

Example 4 Particles A and B, masses 7m kg and 3m kg respectively are attached to the ends of a light inextensible string which passes over a smooth fixed pulley. If the system is released from rest, find the speed of the particles after 1 second of motion. At this instant the particle A hits an inelastic stop and is abruptly brought to rest. The particle B continues to move freely under gravity and comes to instantaneous rest before reaching the pulley. When the particle descends again it jerks the particle A into motion when the string tightens. Calculate:

(i) the speed of each particle at the instant after A is jerked into motion,
(ii) the impulse of the tension in the string.
(iii) the loss in kinetic energy caused by the jerk.

(Take the acceleration due to gravity to be 10 ms^{-2}).

We shall take the velocities, accelerations and tensions as shown in figure 93. To find the speed of the particle after 1 second of motion, let us look at the equations of motion for each particle.

Applying Newton's second law '$f = ma$': for A, $7mg - T_1 = 7ma$, and for B, $T_1 - 3mg = 3ma$. Eliminating T_1, $\boldsymbol{a} = \mathbf{4\ ms^{-2}}$. Hence applying '$v = u + at$' when $t = 1$ second, $\boldsymbol{u} = \mathbf{4\ ms^{-1}}$.

A hits an inelastic stop and is brought to rest from 4 ms^{-1}. B then moves upwards under gravity, comes to instantaneous rest, and then falls freely until the string again becomes taut. Its speed just before the string is jerked is clearly 4 ms^{-1}.

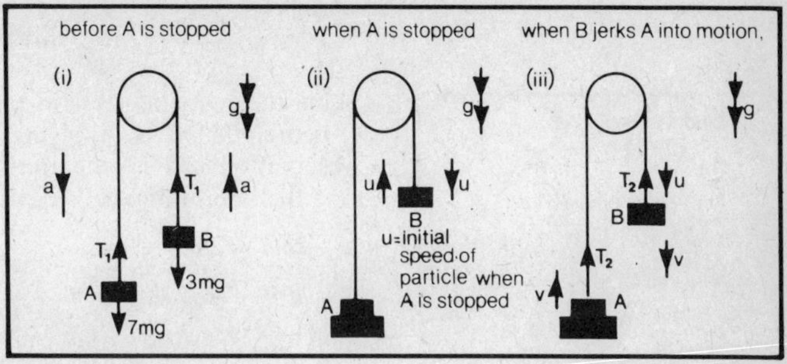

Figure 93.

(i) When A is jerked into motion, it is under the action of the impulsive tension T_2. Using 'impulse = change in momentum' for A, $T_2 = 7mv$ and for B, $-T_2 = 3mv - 3mu$; i.e., $-T_2 = 3mv - 12m$.

Combining these equations, $v = 1{\cdot}2$ ms^{-1}. Hence **the particles move with speed 1·2 ms^{-1} at the instant after A is jerked.**

(ii) The impulse, T_2, of the tension in the string is **8·4*m* Ns.**

(iii) Immediately before the jerk, A is at rest and B is moving with speed 4 ms^{-1}. Hence the kinetic energy before the jerk is $\frac{1}{2}(3m)16 = 24m$ J.

After the jerk A and B both move with speed 1·2 ms^{-1} so that the kinetic energy after the jerk is $\frac{1}{2}(10m)(1{\cdot}2)^2 = 7{\cdot}2m$ J and **the loss of kinetic energy is 16·8*m* J.**

Example 5 (In this question diagrams showing directions of the velocities 'before' and 'after' each impact must be drawn.)

Particles A and B of mass m and $2m$ respectively lie at rest on a smooth horizontal plane. They are projected towards each other both with speed u. Show that, if the coefficient of restitution between the particles is $\frac{3}{4}$, their directions of motion are reversed on impact and find their speeds after impact.

If A subsequently hits a smooth vertical plane normally, show that A and B will collide again, only if the coefficient of restitution between A and the vertical plane is greater than $\frac{1}{8}$.

(S.U.J.B.)

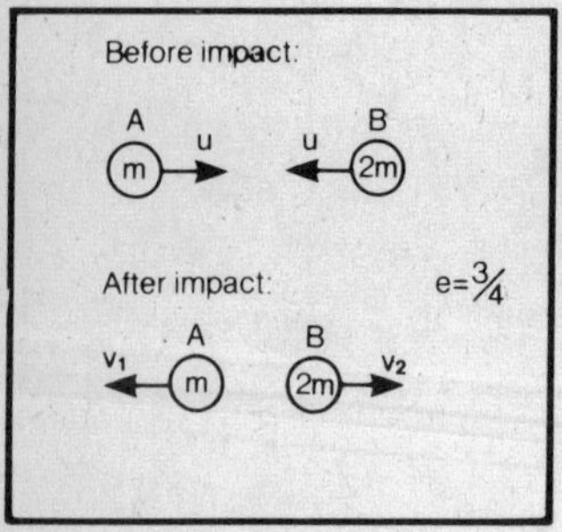

Figure 94.

Taking the velocities as shown in figure 94, and applying Conservation of Momentum along the line of impact,

$mu - 2mu = 2mv_2 - mv_1.$

Therefore $2v_2 - v_1 = -u.$

Applying Newton's law of restitution along the line of impact,

$$\frac{-v_1 - v_2}{2u} = -\frac{3}{4}.$$

Therefore, $v_1 + v_2 = \frac{3u}{2}$. Combining these equations:

$$\mathbf{v_2 = \frac{u}{6}} \quad \text{and} \quad \mathbf{v_1 = \frac{4u}{3}}.$$

Since these values are positive we must have the directions correct in the diagram. **Hence the directions of motion of A and B are reversed on impact. A has speed (4/3)u after impact and B has speed u/6 after impact.**

When the particle A hits the wall with speed $\frac{4}{3}u$, we cannot apply Conservation of Momentum to the situation since an external impulse in the form of the reaction of the wall will act.

However we can use Newton's law of restitution,

$$\frac{-v_3 - 0}{\frac{4}{3}u - 0} = -e.$$

Hence $\mathbf{v_3 = e\frac{4}{3}u}$.

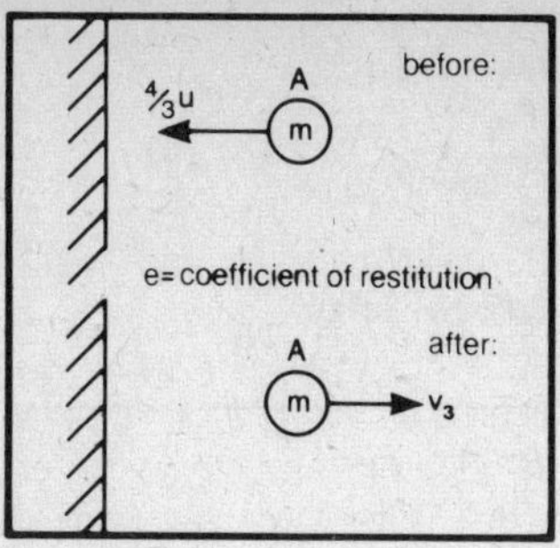

Figure 95.

If A and B are to collide again, A must bounce off the wall with a greater speed than B in order to catch it up,

i.e., $\frac{4ue}{3} > \frac{u}{6}$.

Hence $\boldsymbol{e} > \frac{1}{8}$.

Example 6 Two particles, A and B, of masses m and $3m$ respectively, lie at rest on a smooth horizontal plane. A is projected along the plane towards B with velocity u. Find the velocity of B after the impact if the coefficient of restitution between the particles is $\frac{1}{4}$.

Particle B goes on to strike a smooth vertical wall directly, and after rebounding from the wall, is brought to rest by a second impact with A. Find the coefficient of restitution between B and the wall.

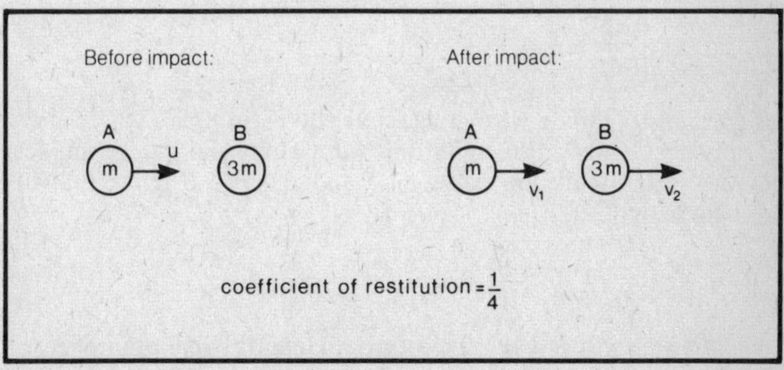

Figure 96.

Taking the velocities as shown in figure 96, and applying Conservation of Momentum along the line of impact, for the first impact of A and B, $mu = mv_1 + 3mv_2$. Applying the law of restitution along the line of impact,

$$\frac{v_1 - v_2}{u - 0} = -\frac{1}{4}, \text{ so that } v_2 - v_1 = \frac{u}{4}.$$

Combining these equations: $v_2 = \dfrac{5u}{16}$ and $v_1 = \dfrac{u}{16}$.

Hence **particle *B* moves off with velocity** $\boldsymbol{\dfrac{5u}{16}}$.

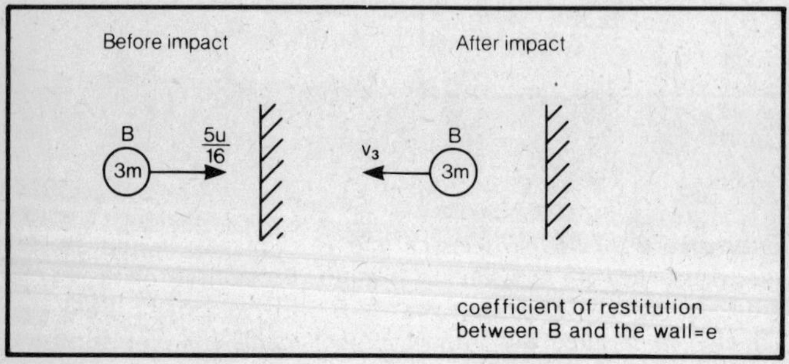

Figure 97.

Applying Newton's law of restitution along the line of impact of *B* and the wall (see figure 97),

$$\frac{-v_3 - 0}{\frac{5}{16}u - 0} = -e. \quad \text{Therefore, } \boldsymbol{v_3 = e\frac{5u}{16}}.$$

B goes on to collide with *A* again as shown in figure 98. Applying Conservation of Momentum along the line of impact of the second collision between *A* and *B*, since *B* is brought to rest by the collision,

$$\frac{mu}{16} - 3me\frac{5u}{16} = -mv_4.$$

Applying Newton's law of restitution along the line of impact;

$$-\frac{v_4 - 0}{\frac{u}{16} + e\frac{5u}{16}} = -\frac{1}{4} \quad \text{so that} \quad v_4 = \frac{u}{64}(1 + 5e).$$

Combining the two equations, $\boldsymbol{e = \dfrac{1}{11}}$.

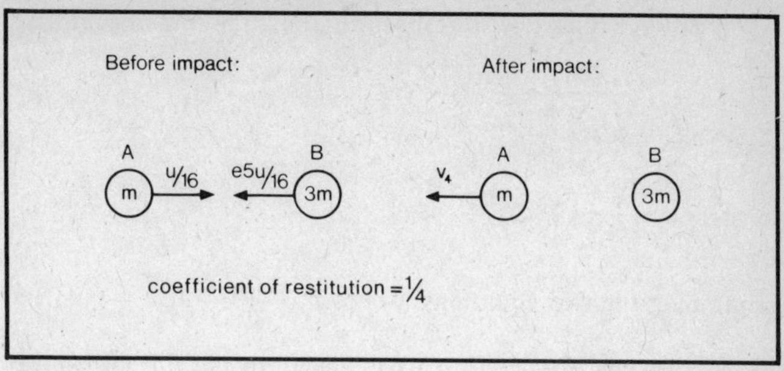

Figure 98.

Example 7 A smooth sphere A of mass m, moving with speed u collides with another stationary identical sphere B. The directions of motion of A before and after impact make angles of $\pi/3$ and arctan $10\sqrt{3}$ respectively with the line of impact as shown in figure 99. Find the coefficient of restitution between the spheres.

Find also the loss of kinetic energy due to the impact.

The line of impact is the line joining the centres of the spheres. Since the spheres are smooth there will be no impulse in the direction normal to the line joining the centres. Hence B will move off in the direction of the line joining the centres, as shown.

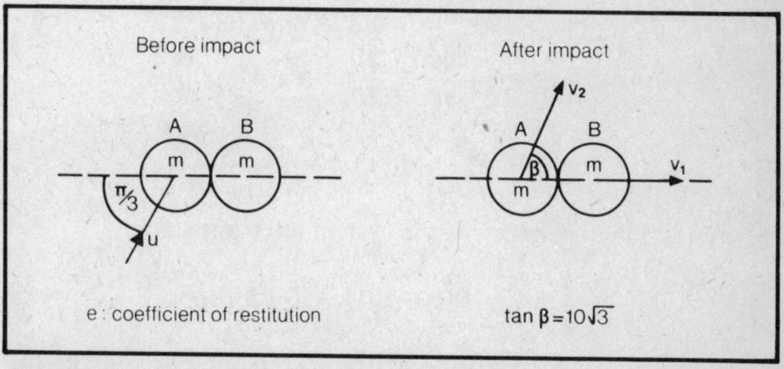

Figure 99.

Taking the velocities as shown in figure 99, and applying Conservation of Momentum along the line of centres,

$$mu \cos \frac{\pi}{3} = mv_2 \cos \beta + mv_1.$$

Applying Newton's law of restitution along the line of centres,

$$\frac{v_2 \cos \beta - v_1}{u \cos \frac{\pi}{3} - 0} = -e$$

so that $$v_2 \cos \beta - v_1 = -e\frac{u}{2}.$$

Combining the two equations: $v_2 \cos \beta = \frac{u}{4}(1 - e)$.

A's component of velocity perpendicular to the line of centres will be unchanged,

so that $$v_2 \sin \beta = u\frac{\sqrt{3}}{2}.$$

Dividing $v_2 \sin \beta$ by $v_2 \cos \beta$: $\tan \beta = \frac{2\sqrt{3}}{(1 - e)}$.

However, $\tan \beta = 10\sqrt{3}$, so that $e = \frac{4}{5}$. Hence the **coefficient of restitution between the spheres is**

$$\mathbf{\frac{4}{5}}.$$

Before impact, the kinetic energy of the system is $\frac{1}{2}\boldsymbol{mu^2}$. After impact the kinetic energy of the system is $\frac{1}{2}mv_2{}^2 + \frac{1}{2}mv_1{}^2$.

$$\mathbf{v}_1 = v_2 \cos \beta + \frac{4u}{10} = \frac{\mathbf{9u}}{\mathbf{20}}, \quad \mathbf{v}_2 = \frac{\boldsymbol{u}}{\mathbf{20 \cos \beta}}$$

$$\text{and } \cos \boldsymbol{\beta} = \frac{\mathbf{1}}{\sqrt{\mathbf{301}}} \text{ so that } \mathbf{v}_2 = \frac{\boldsymbol{u}\sqrt{\mathbf{301}}}{\mathbf{20}}.$$

Hence the kinetic energy of the system after impact is

$$\frac{1}{2}m\frac{u^2}{400}(81 + 301) = \mathbf{0{\cdot}48}\ \boldsymbol{mu^2}.$$

The loss of kinetic energy due to the impact is therefore 0·02 $\boldsymbol{mu^2}$ (correct to 2 dec. pl.).

Example 8 A smooth sphere A, of mass 1 kg, has velocity vector $5\mathbf{i} + 12\mathbf{j}$. A second sphere B, of mass 3 kg, has velocity $-3\mathbf{i} + 4\mathbf{j}$. Find the kinetic energy of each sphere.

The spheres collide and on impact their line of centres is in the direction **j**. If the sphere A moves in the direction $\mathbf{i}+\mathbf{j}$ immediately after impact, show that the coefficient of restitution is $\frac{1}{6}$. Find:

(i) the velocity vector of B after impact;

(ii) the kinetic energy lost by A due to the collision.

(A.E.B.)

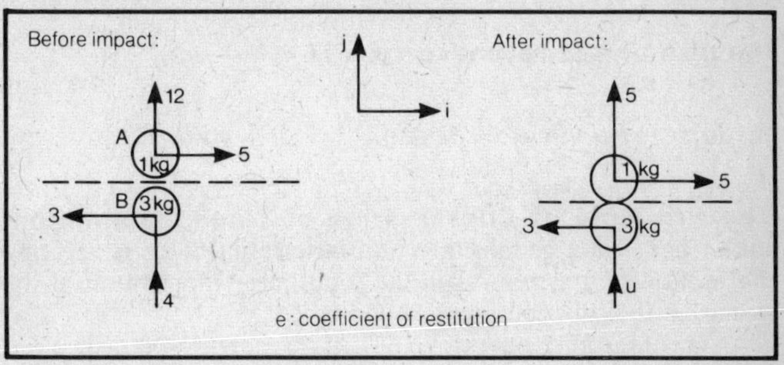

Figure 100.

Since the spheres are smooth there will be no impulse in the **i** direction. Hence the components of velocity in that direction will be unchanged. A moves off in the direction (**i** + **j**). Therefore, after impact, A will have components of velocity equal in magnitude in the **i** and **j** directions.

Hence ***A*'s velocity vector after impact** is 5**i** + 5**j**.

Applying Conservation of Momentum in the **j** direction, $12+12=3u+5$. Therefore,

$$u=\frac{19}{3}\ \mathrm{ms}^{-1}.$$

Applying Newton's law of restitution in this direction,

$$\frac{5-u}{12-4}=-e, \quad e=\frac{1}{6}.$$

(i) the velocity vector of B after the collision is

$$-3\mathbf{i}+\frac{19}{3}\mathbf{j}.$$

(ii) Before impact, A's kinetic energy is $\frac{1}{2}(12^2+5^2)=\frac{169}{2}$ J.

After impact, A's kinetic energy is $\frac{1}{2}(5^2 + 5^2) = 25$ J. **Hence *A*'s loss of kinetic energy due to the collision is 59·5 J.**

Key terms

Momentum = mass × velocity.

Impulse of a constant force = $\mathbf{F}t = m\mathbf{v} - m\mathbf{u}$.

Impulse of a variable force = $\int_0^t \mathbf{F}\,\mathrm{d}t = m\mathbf{v} - m\mathbf{u}$.

The Principle of Conservation of Linear Momentum states that if in a certain direction no external force is affecting the motion of a system, then the total linear momentum of the system in that direction will be conserved.

Two bodies in collision exert equal and opposite impulses on each other and **for the impact** are governed by 'impulse = change in momentum'. After the impact, the motion of the system is governed by the external forces acting on the system.

When a string is jerked, the **impulsive tension** will cause a change in momentum of particles attached to its end in the direction of the string. Perpendicular to the string no impulse will act and so the momentum will be conserved in that direction.

When two objects collide and bounce, the impact is said to be **elastic**. If they coalesce on collision, the impact is said to be **inelastic**.

Newton's (experimental) law of restitution states that the ratio of the relative velocity along the line of impact after impact, to the relative velocity along the line of impact before impact, is $-e$, where e is the coefficient of restitution for the two objects concerned.

If $e = 0$, the collision is **inelastic**. If $e = 1$, the collision is **perfectly elastic**. Generally, $0 \leq e \leq 1$.

Direct impact occurs when the bodies are travelling along the line of impact before the impact.

Indirect impact, or oblique impact occurs when the bodies are not travelling along the line of impact before impact.

When solving problems about collisions it is a good idea to draw separate diagrams for each collision both before and after the impact.

Chapter 10
Circular Motion of Particles

We shall now consider the accelerations required in order to move particles in circular paths. We shall first look at a particle moving in a circle with **constant** speed.

Circular motion at constant speed

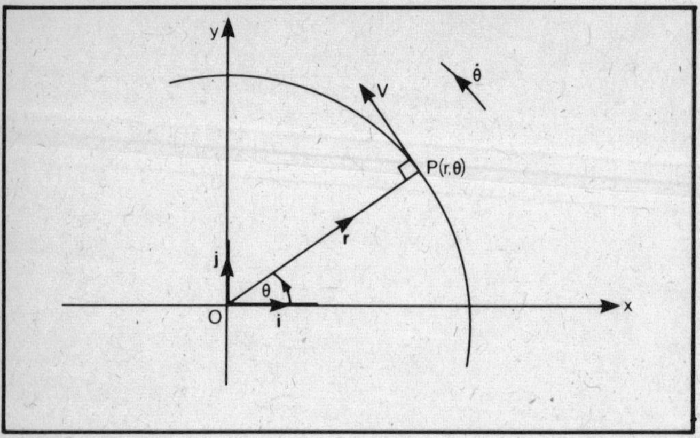

Figure 101.

Refer to figure 101.

Consider a particle P describing a circle centre O and radius r at constant speed v. Its position vector $\mathbf{r}$ is a function of the scalar variable, time, and may be written as:

$$\mathbf{r} = r(\cos\theta\,\mathbf{i} + \sin\theta\,\mathbf{j})$$

Since r is constant and θ is a function of time,

$$\mathbf{v} = \dot{\mathbf{r}} = r\dot{\theta}(-\sin\theta\,\mathbf{i} + \cos\theta\,\mathbf{j}).$$

Since $\dot{\theta}$ is constant ('$v = \dot{\theta}r$' and both v and r are constant)

then $$\mathbf{a} = \ddot{\mathbf{r}} = r\dot{\theta}^2(-\cos\theta\,\mathbf{i} - \sin\theta\,\mathbf{j}) = -\dot{\theta}^2\mathbf{r}.$$

Hence **the acceleration of *P* is directed towards *O* and has magnitude $r\dot{\theta}^2 = v^2/r$.**

Therefore, if a particle of mass m is moving in a circle of radius r with constant speed v, it must be doing so under the action of a resultant force, directed towards the centre of the circle, of magnitude $mr\omega^2 = mv^2/r$, where ω is its angular velocity. This force is not constant, although it is of constant magnitude, since its direction must vary.

Conical pendulum

Consider an inextensible string of length l fixed at one end A and with a particle, P, of mass m attached to the other end. P is rotating in a horizontal circle with constant angular velocity ω, and the string is at a constant angle θ to the vertical. This system is called a conical pendulum, for obvious reasons.

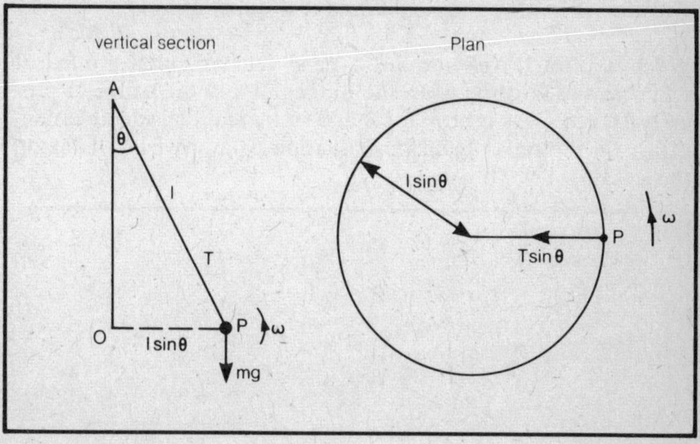

Figure 102.

Let us consider the forces acting on the particle in order that it should move in this way. Horizontally, we know that the resultant force on it must be directed towards the centre, O of the horizontal circle it describes, and be of magnitude $m\omega^2 l \sin\theta$.

The only horizontal force affecting the motion of the particle is $T\sin\theta$ directed towards the centre of the circle.

Hence, $T\sin\theta = m\omega^2 l\sin\theta$, i.e., $\boldsymbol{T = m\omega^2 l}$.

Vertically, since the particle is rotating in a horizontal circle, there is no motion. Thus, the resultant force in the vertical direction must be zero.

Hence, $T \cos \theta = mg$.

The **vertical depth** of P below A is

$$l \cos \theta = \frac{lmg}{T} = \frac{g}{\omega^2}.$$

Banked tracks

When a vehicle travels around a bend on horizontal ground, it relies entirely on the frictional force between it and the ground to provide the necessary central force. There is clearly a limit to that force (limiting friction) and hence a limit to the speed with which a vehicle may travel round the bend without slipping. The situation is improved by banking the ground.

When a train travels around a bend on horizontal ground, it relies on the force between the outer rail and the flange to provide the necessary central force and so there is considerable side-thrust on the rails. Again this situation is improved by banking the track.

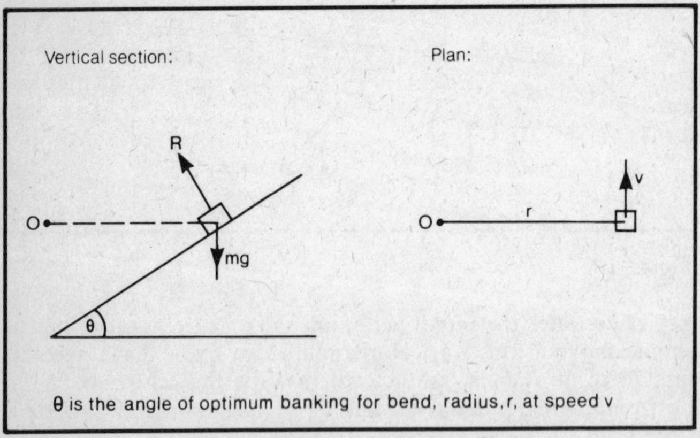

Figure 103.

Refer to figure 103.

Consider the vehicle of mass m, negotiating a bend of radius r at speed v. We shall find the angle θ such that no frictional force or force from the rails is called into play. θ is the angle of **optimum banking.**

Horizontally, there must be a force of magnitude mv^2/r directed towards O. This must come from the component of the normal reaction in the horizontal direction.

Hence, $\boldsymbol{R} \sin \theta = \dfrac{\boldsymbol{mv^2}}{\boldsymbol{r}}$.

Vertically, there must be no motion and therefore no resultant force in this direction.

Hence, $\boldsymbol{R} \cos \theta = \boldsymbol{mg}$.

It follows that the **angle of optimum banking for speed *v*** is given by

$\textbf{arctan}\left(\dfrac{\boldsymbol{v^2}}{\boldsymbol{rg}}\right)$, (which is independent of the mass m).

Alternatively, we may say that a vehicle may travel around a bend of radius r with this angle of optimum banking at speed $\boldsymbol{v} = \sqrt{\boldsymbol{rg} \tan \theta}$ without any tendency to side-slip.

Motion in a circle with variable speed

If a particle is moving on a circular path with variable speed v, then it must be moving under the action of a resultant force with non-zero components in both the radial and tangential directions. The angular speed at any instant is $\dot{\theta}$, so that at that instant, the radial component of force is $m\dot{\theta}^2 r$ and the tangential component is

$$m\frac{\mathrm{d}v}{\mathrm{d}t} = m\ddot{\theta}r.$$

Motion of this type arises when a particle is restricted to movement in a vertical path under the action of its weight. This is the case, for example, when a ring slides on a vertical circular wire; when a particle slides down the outside of a sphere; when a particle is attached to a light inextensible string fixed at the other end, and is given an initial velocity.

We shall now consider vertical circular motion in detail. The motion we shall be concerned with falls into two broad categories: that where the motion is restricted to a circular path, and that where it is not.

We shall often be able to apply the Principle of Conservation of Mechanical Energy to good advantage, since we shall be concerned with systems where the only external force doing work is gravity.

Motion in a vertical circle

We shall first of all consider the motion of a small ring threaded on a smooth vertical circular wire, as being typical of the cases where motion is restricted to the vertical circle.

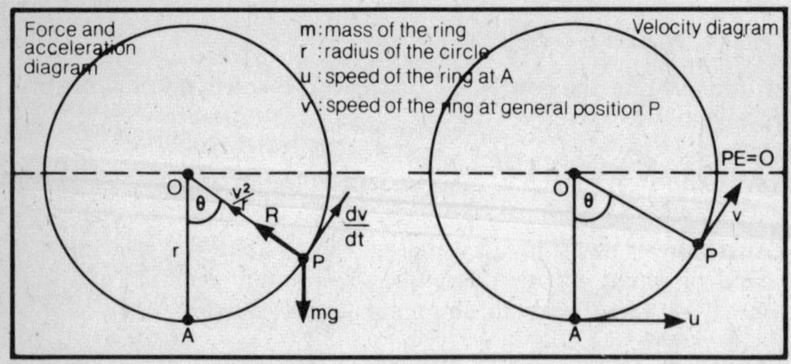

Figure 104.

Analysing the motion radially, we see that the necessary force, mv^2/r, to keep the ring on its circular path must come from the normal reaction R between the ring and the wire, and the component of weight, $mg\cos\theta$, in that direction (see figure 104),

$$\text{i.e., } R - mg\cos\theta = \frac{mv^2}{r}.$$

Tangentially, the only force acting, since the ring is smooth, is the component of weight, $mg\sin\theta$,

$$\text{therefore } -mg\sin\theta = m\frac{dv}{dt}.$$

Applying Conservation of Mechanical Energy to the system, with the horizontal through the centre of the circular wire as the zero potential energy level, $\frac{1}{2}mu^2 - mgr = \frac{1}{2}mv^2 - mgr\cos\theta$,

$$\text{i.e., } \boldsymbol{u^2 - v^2 = 2rg(1 - \cos\theta)}.$$

We see that with this type of motion the ring may either make complete circular motion, or it may come to instantaneous rest before it reaches the top of the vertical circle and then proceed to oscillate, or it may just reach the highest point and come to rest there.

Let us examine the conditions for these cases.

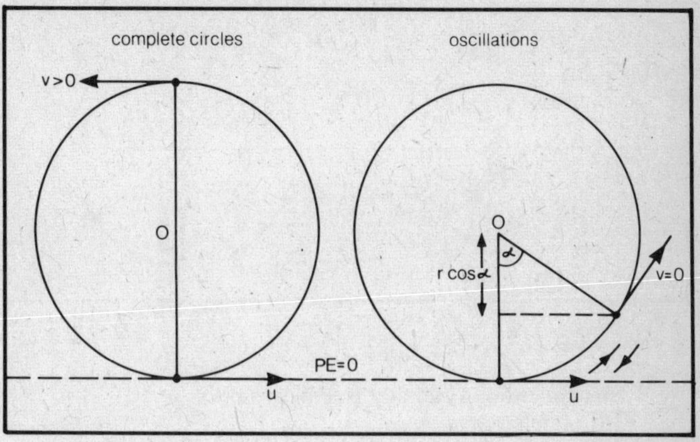

Figure 105.

If the ring is to pass through the highest point of the circle and go on to describe complete circles, its velocity at the highest point must be greater than zero in order to take it over the top point. (If it were zero at the top point, it would stay there.)

At the highest point, from the Conservation of Energy, see figure 105,

$$\tfrac{1}{2}mu^2 = 2mgr + \tfrac{1}{2}mv^2.$$

Therefore $v^2 = u^2 - 4gr$.

If $\boldsymbol{v} > \mathbf{0}$, then $\boldsymbol{u}^2 > \mathbf{4}\boldsymbol{gr}$ **and complete circles** will be made.

If $\boldsymbol{v} = \mathbf{0}$, then $\boldsymbol{u}^2 = \mathbf{4}\boldsymbol{gr}$ and the ring will come to rest at the highest point.

If $\boldsymbol{u}^2 < \mathbf{4}\boldsymbol{gr}$, the ring comes instantaneously to rest when $\theta = \alpha$ and then oscillates. From energy considerations we see that,

$$\frac{1}{2}mu^2 = mgr(1 - \cos\alpha) \text{ and } \boldsymbol{\cos\alpha = 1 - \frac{u^2}{2gr}}.$$

We shall now look at the motion of a particle rotating in a vertical circle where the particle is attached to one end of a light inextensible string, the other end being fixed. This situation is typical of the cases where the motion is not restricted to a vertical circle.

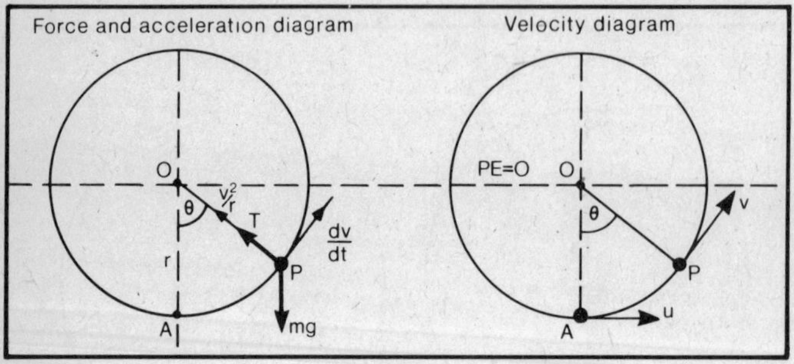

Figure 106.

Analysing the motion radially (see figure 106),

$$T - mg\cos\theta = \frac{mv^2}{r}$$

and tangentially,

$$m\frac{dv}{dt} = -mg\sin\theta.$$

Applying **Conservation of Energy,**

$\frac{1}{2}mu^2 - mgr = \frac{1}{2}mv^2 - mgr\cos\theta$, thus $v^2 = u^2 - 2gr(1 - \cos\theta)$.

Combining these equations,

$$T = mg\cos\theta + \frac{m}{r}(u^2 - 2gr + 2gr\cos\theta)$$

$$= \boldsymbol{m\left(\frac{u^2}{r} - 2g + 3g\cos\theta\right)}.$$

There are three possibilities for the motion of the particle: it may describe complete vertical circles; it may come to instantaneous rest below the level of O and then oscillate, the string always being taut; it may cease travelling on a circular path at some point above the level of O when the string becomes slack and then travel as a projectile until the string becomes taut again.

Let us examine each of these cases and the conditions necessary for them to happen (see figure 107).

If the particle describes **complete vertical circles** then the essential condition is that **the tension should never be zero other than instantaneously at the top of the circular path**. It is not sufficient to ensure that $v > 0$ at the highest point, since the particle may be moving as a projectile inside the circle.

Hence, $T = m\left(\frac{u^2}{r} - 2g + 3g\cos\theta\right) \geq 0$ for all values of θ.

$$\text{i.e.,}\quad \frac{u^2}{r} \geq 2g - 3g\cos\theta.$$

The maximum value of $(2g - 3g\cos\theta)$ occurs when $\theta = 180°$, $\cos\theta = -1$, so that to ensure that T is never zero, other than instantaneously at the highest point, $\boldsymbol{u^2 \geq 5gr}$.

If the particle **oscillates**, then it must come to rest instantaneously at or below the level of O. T will always be greater than zero for $\theta \leq 90°$, $\cos\theta \geq 0$, since

$$T = \frac{mv^2}{r} + mg\cos\theta,$$

so there is no need to test for that.

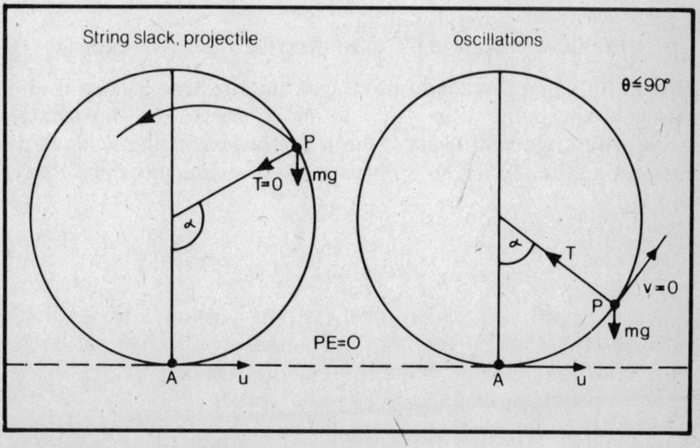

Figure 107.

Hence, for **oscillations**, $v = 0$ **for** $\theta \leq 90°$, i.e., since when $v = 0$, from energy considerations,

$$\cos\theta = 1 - \frac{u^2}{2gr} \quad \text{and} \quad \cos\theta \geq 0, \quad u^2 \leq 2gr.$$

In both of these cases the string is always taut and the particle always moves on a vertical circular path. Hence the values of u for this to be the case are $u \leq \sqrt{2gr}$ or $u \geq \sqrt{5gr}$.

Circular motion ceases at the instant when the string becomes slack,

$$\text{i.e., when } T = 0, \cos\theta = \frac{2gr - u^2}{3gr}.$$

This will be the case for $\sqrt{2gr} < u < \sqrt{5gr}$ from the results above, but also from the fact that for circular motion to cease, $-1 < \cos\theta < 0$, and $T = 0$.

Worked examples

Example 1 A light inextensible string AB of length $7a$, has its ends fixed in space, with A distant $5a$ vertically above B. A bead C of mass m is threaded onto the string and it rotates in a horizontal circle with constant angular velocity ω; $AC = 4a$. Show that

$$\omega^2 = \frac{35g}{12a}$$

and write down the time for C to describe the circle once.

Instead of being threaded onto the string, the bead is fixed to the string at the point were $AC = 4a$ and C moves in a horizontal circle with angular velocity Ω, both parts of the string being taut. Determine the tension in each part of the string and show that

$$\Omega^2 > \frac{5g}{16a}.$$

(S.U.J.B.)

When the bead is threaded on the string, tension is transmitted through the string. If the bead is rotating in a horizontal circle with constant angular velocity, then the tension in the string must be providing the necessary radial force.

Hence $T\cos(90° - \theta) + T\cos\theta = m\omega^2 r$, where r is the radius of the circle.

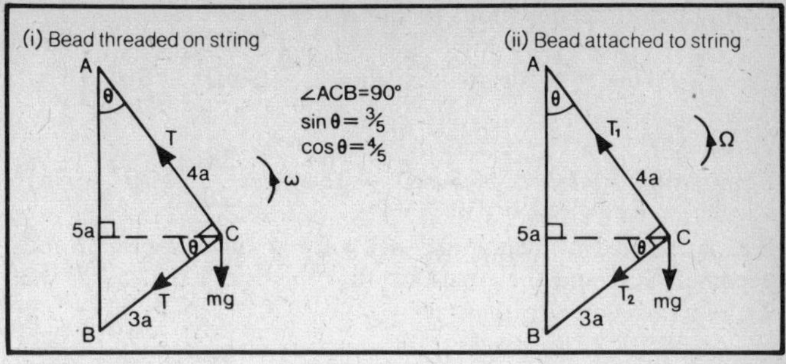

Figure 108.

From figure 108(i), we see that $r = 4a \sin\theta = \frac{12}{5}a$.

Using this value,

$$T = \frac{m\omega^2}{7}\,12a.$$

Vertically there is no motion and hence no resultant force in that direction, therefore:

$T \sin(90° - \theta) = T \sin\theta + mg$. Hence $T = 5mg$.

Combining these results, $5mg = m\omega^2 \dfrac{12a}{7}$, i.e., $\boldsymbol{\omega^2 = \dfrac{35g}{12a}}$.

The angular velocity is ω radians per second, therefore the particle will take $2\pi/\omega$ seconds to complete one revolution,

$$\text{i.e., } \mathbf{2\pi\sqrt{\frac{12a}{35g}}\ \text{secs.}}$$

Refer to figure 108(ii).

If the bead is **fixed** to the string, then tension is not transmitted across it and we shall not necessarily have the same tensions in both parts of the string. Analysing the motion radially,

$$T_1 \cos(90° - \theta) + T_2 \cos\theta = m\Omega^2 \frac{12}{5} a,$$

$$\text{i.e., } 3T_1 + 4T_2 = 12am\Omega^2.$$

Vertically, $T_1 \sin(90° - \theta) = mg + T_2 \sin\theta$,

Therefore, $4T_1 - 3T_2 = 5mg$.

Combining these equations to eliminate T_2,

$$T_1 = \frac{1}{25}(36am\Omega^2 + 20mg), \quad \text{i.e., } \boldsymbol{T_1 = \frac{4m}{25}(9a\Omega^2 + 5g)}.$$

$$\text{Eliminating } T_1,\ \boldsymbol{T_2} = \frac{1}{25}(48am\Omega^2 - 15mg) = \boldsymbol{\frac{3m}{25}(16\Omega^2 - 5g)}.$$

If Ω is reduced in magnitude, AC will remain taut but CB may become slack, and the condition that this is not the case is that $T_2 > 0$,

$$\text{i.e., } 16a\Omega^2 > 5g \quad \text{and} \quad \boldsymbol{\Omega^2 > \frac{5g}{16a}}.$$

Example 2 A particle P is given a horizontal velocity u from the lowest point on the smooth inner surface of a hollow spherical shell of radius r and centre O. Find the range of values for u so that P will not leave the surface of the shell.
If the particle does leave the surface when OP makes an acute angle α with the upward vertical, show that $3gr \cos \alpha = u^2 - 2gr$. In this case find the reaction between the particle and the surface when OP makes an angle α with the downward vertical.

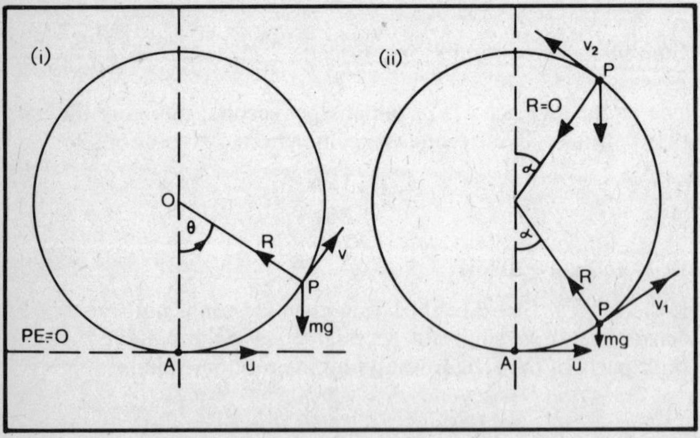

Figure 109.

Since the surface is smooth, the reaction of the surface on the particle will be normal to the surface, i.e., directed towards the centre of the circle.

Radially, for general position P,

$$R - mg\cos\theta = \frac{mv^2}{r}.$$

From Conservation of Energy:

$$\tfrac{1}{2}mu^2 = \tfrac{1}{2}mv^2 + mgr(1 - \cos\theta),$$

$$\text{i.e., } v^2 = u^2 - 2gr(1 - \cos\theta).$$

Combining these equations, $R = m\left(\dfrac{u^2}{r} - 2g + 3g\cos\theta\right)$.

If the particle is not to leave the surface of the shell, it must either be **oscillating** below the level of O or making **complete vertical circles**.

If it is oscillating below O, then $v = 0$ for $\theta \leq 90°$. When $v = 0$,

$$\cos\theta = 1 - \frac{u^2}{2gr} \text{ and since } \cos\theta \geq 0,$$

$$\boldsymbol{u^2 \leq 2gr}.$$

If the particle is making **complete vertical circles**, then when $\theta = 180°$, $\cos\theta = -1$, it is necessary that $R \geq 0$,

i.e., $R = m\left(\dfrac{u^2}{r} - 2g - 3g\right) \geq 0$, so $\boldsymbol{u^2 \geq 5gr}$.

Hence the ranges for u are $\boldsymbol{u \geq \sqrt{5gr}}$ and $\boldsymbol{u \leq \sqrt{2gr}}$.

If the particle leaves the surface when OP makes an acute angle α with the upward vertical, i.e., $\theta = 180° - \alpha$, then $R = 0$ at this point,

$R = m\left(\dfrac{u^2}{r} - 2g - 3g\cos\alpha\right) = 0$ and $\boldsymbol{3gr\cos\alpha = u^2 - 2gr}$.

When $\theta = \alpha$, $R = m\left(\dfrac{u^2}{r} - 2g + 3g\cos\alpha\right)$

and since $\dfrac{u^2}{r} = (3g\cos\alpha + 2g)$ then $\boldsymbol{R = 6mg\cos\alpha}$.

Example 3 A smooth narrow tube is in the form of a circle, centre O, and radius a, fixed in a vertical plane. Two particles,

A and B, of mass $3m$ and m respectively, are connected by a light inextensible string of length $\pi a/2$.

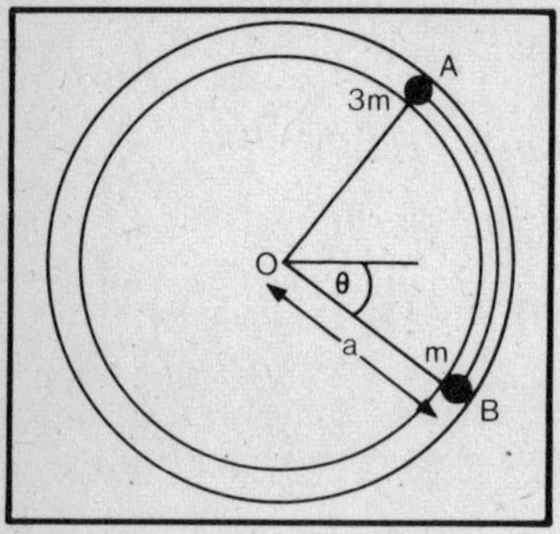

Figure 110.

The system is released from rest with A at the highest point and B at the level of O. If OB makes an angle of θ with the horizontal diameter of the tube at time t after the release of the system (as shown in figure 110), find, by energy considerations, an equation of motion for the system in terms of θ and t, whilst the string remains taut.

Hence, or otherwise, show that

$$a\frac{d^2\theta}{dt^2} = \frac{g}{4}(3\sin\theta + \cos\theta).$$

Find, in terms of m, g, and θ, the tension, T in the string and the reaction between the tube and B. Find also the value of θ when the string becomes slack.

Particle B is at the level of O, when A is at the highest point. As long as the string is taut, both particles will be moving with the same speed, $a\,d\theta/dt$, and angle BOA will be $\pi/2$.

The tube is smooth, therefore we are in a conservative system of forces and we may apply Conservation of Energy to the system.

Taking the zero potential energy line at the level of O,

$$3mga = 3mga\sin(90° - \theta) + \frac{3m}{2}\left(a\frac{d\theta}{dt}\right)^2$$

$$- mga\sin\theta + \frac{m}{2}\left(a\frac{d\theta}{dt}\right)^2$$

thus, $\mathbf{2a\left(\frac{d\theta}{dt}\right)^2 = g(3 - 3\cos\theta + \sin\theta)}$,

which is the required **equation of motion**.

Taking the forces as shown in figure 111, for particle B, radially,

$$S - mg \sin\theta = ma\left(\frac{d\theta}{dt}\right)^2 = \frac{mg}{2}(3 - 3\cos\theta + \sin\theta),$$

therefore, $\boldsymbol{S = \frac{3}{2}mg(1 - \cos\theta + \sin\theta)}$.

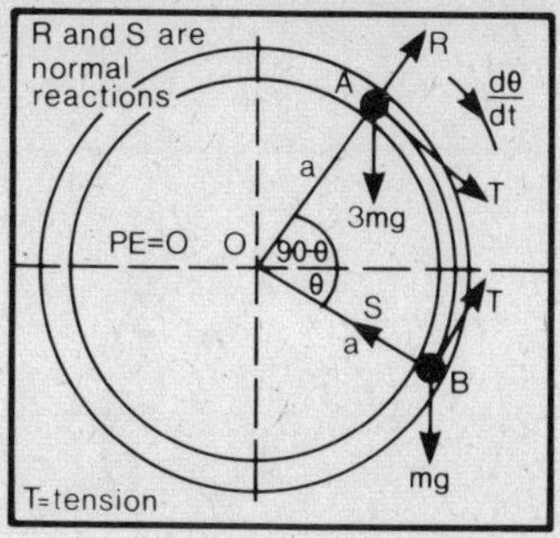

Figure 111.

In order to find $a\dfrac{d^2\theta}{dt^2}$ in terms of g and θ, we shall differentiate the energy equation with respect to time, hence

$$4a\frac{d\theta}{dt}\frac{d^2\theta}{dt^2} = g(3\sin\theta + \cos\theta)\frac{d\theta}{dt},$$

i.e., $$\boldsymbol{a\frac{d^2\theta}{dt^2} = \frac{g}{4}(3\sin\theta + \cos\theta)}.$$

Looking at the tangential forces acting on B, we see

$$mg\cos\theta - T = ma\frac{d^2\theta}{dt^2},$$

therefore, $$\boldsymbol{T} = mg\sin\theta - \frac{mg}{4}(3\sin\theta + \cos\theta)$$

$$= \boldsymbol{\frac{3mg}{4}(\cos\theta - \sin\theta)}$$

The string becomes slack when $T = 0$,

i.e., when $\sin\theta = \cos\theta$, $\boldsymbol{\theta = \frac{\pi}{4}}$.

Example 4 A particle is placed on the top of a fixed smooth vertical hoop, centre O and radius a, and is slightly displaced from rest. It comes away from the hoop at a point P. Calculate:
a) the angle which OP makes with the vertical,
b) the velocity of the particle when it leaves the hoop.
A second particle is placed at the lowest part of the hoop and projected along the inside smooth surface. It also leaves the hoop at the same point P. Calculate:
c) the velocity with which this particle leaves the hoop,
d) the velocity of projection of this second particle.

(S.U.J.B.)

Let us consider the equations of motion for the particle on the outside of the hoop. Radially, the resultant force must be of magnitude mv^2/a and directed towards the centre in order to keep the particle on a circular path (figure 112(i)),

i.e., $mg \cos\theta - R = \dfrac{mv^2}{a}$.

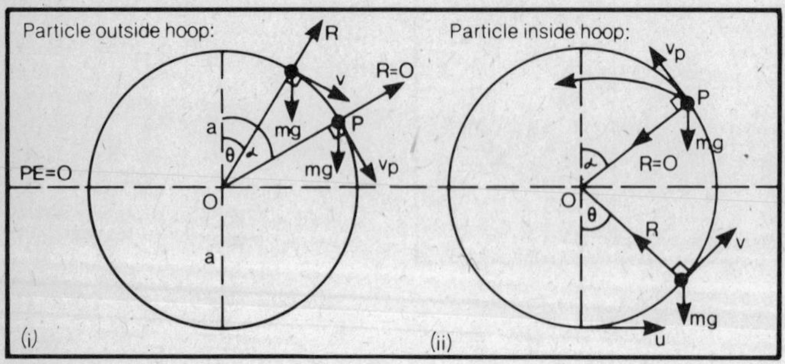

Figure 112.

Thus, $R = mg\cos\theta - \dfrac{mv^2}{a}$. From energy considerations,

$mga = mga\cos\theta + \frac{1}{2}mv^2$ so that $v^2 = 2ga(1 - \cos\theta)$.

Hence, $\boldsymbol{R} = mg\cos\theta - 2gm(1 - \cos\theta) = \boldsymbol{3mg\cos\theta - 2gm}$.

a) When the particle leaves the hoop, $R = 0$ so that $\cos\alpha = (\frac{2}{3})$, i.e., $\boldsymbol{\alpha} = \textbf{arccos}\ (\frac{2}{3})$.

b) The velocity V_p with which the particle leaves the hoop is given by

$$V_p{}^2 = 2ga(1 - \cos\alpha) = \frac{2ga}{3}.$$

Therefore $\boldsymbol{V_p} = \sqrt{\dfrac{\boldsymbol{2ga}}{\boldsymbol{3}}}$

The equation of motion, radially, for the particle moving inside the hoop is (figure 112(ii)),

$$R - mg\cos\theta = \frac{mv^2}{a}$$

and from energy considerations, $v^2 - u^2 = -2ga(1 - \cos\theta)$.

c) The particle leaves the sphere at P, therefore $R = 0$ when $\theta = (180° - \arccos(\frac{2}{3}))$.

Hence the second particle leaves the hoop with velocity

$$\sqrt{\frac{2ga}{3}}. \text{ (The same as the first particle.)}$$

d) The velocity of projection, u, is, from energy considerations, such that

$$-mga + \frac{1}{2}mu^2 = \frac{2}{3}mga + mg\frac{a}{3}.$$

Thus $\boldsymbol{u = 2\sqrt{ga}}$.

Key terms

In order that a particle should move on a circular path, radius r, the **resultant** force acting on it must have a **radial component of magnitude mv^2/r directed towards the centre**, and a **tangential component** of $m\,dv/dt$.

When considering motion in a **vertical** circle we distinguish between cases where **motion is restricted to a circular path**, e.g., a bead threaded onto a vertical circular wire, and where motion is **not restricted to the circular path**, e.g., a particle attached to one end of a string fixed at the other end.

In the case of **motion restricted to the circular path**, the condition that **complete vertical circles** should be described by the particle is that its **velocity is greater than zero at the highest point**. For it to **oscillate**, its **velocity will be zero before it reaches the highest point**.

In the case of **motion not restricted to the circular path**, the condition that **complete vertical circles** should be described is that when it is in a **vertical line with the highest point, the force which has been restricting it to the circular path**, (i.e., the tension, or normal reaction, for example), **is greater than or equal to zero**. For **oscillation**, the **velocity** of the particle must be **instantaneously zero** before it goes above the level of the centre of the circle.

Chapter 11
Simple Harmonic Motion

Simple harmonic motion (S.H.M.) is defined by the equation of motion

$$\frac{d^2s}{dt^2} = -n^2s,$$

where s is some measure of displacement, and n is a constant. As we can see from this equation of motion, **the acceleration of the body moving with S.H.M. is proportional to its displacement from some fixed point and is directed towards the fixed point**. (We use n^2 and not n as the constant, to keep the motion directed towards the fixed point, since $n^2 \geq 0$).

If we were considering motion in the straight line Ox, then the equation of motion for simple harmonic motion would be

$$\frac{d^2x}{dt^2} = -n^2x,$$

where x is the linear displacement from O.

$\dfrac{d^2\theta}{dt^2} = -n^2\theta$ is the equation of motion for angular S.H.M.,

where θ represents the angular displacement from some fixed line.

Let us derive further equations for S.H.M. from the basic one, $d^2s/dt^2 = -n^2s$.

Acceleration $= v\dfrac{dv}{ds} = -n^2s$, so that $\int v\,dv = -n^2 \int s\,ds$

and $v^2 = -n^2s^2 + k$ where k is the constant of integration.

If we have the condition that when $v = 0$, $s = a$, then $k = n^2a^2$, and $v^2 = n^2(a^2 - s^2)$.

This equation tells us that $a^2 \geq s^2$ in order to keep v^2 positive or zero.

When $v = 0$, $s = \pm a$, therefore S.H.M. is oscillatory between $s = +a$ and $s = -a$.

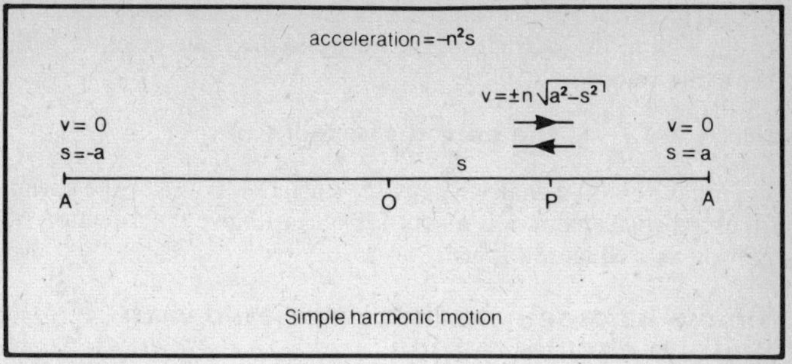

Simple harmonic motion

Figure 113.

We have used linear S.H.M. in figure 113, but of course these are general equations for S.H.M.

We say that a is the **amplitude** of the motion.

From $v^2 = n^2(a^2 - s^2)$ we see that the maximum speed occurs when $s = 0$, therefore the maximum and minimum velocities are na and $-na$ respectively and occur when the particle passes through O.

Since $v = \dfrac{\mathrm{d}s}{\mathrm{d}t} = \pm n\sqrt{a^2 - s^2}$, $\displaystyle\int \frac{1}{\sqrt{a^2 - s^2}}\,\mathrm{d}s = n\int \mathrm{d}t$

and $\arcsin\left(\dfrac{s}{a}\right) = nt + k$.

This constant of integration, k, will depend entirely on the point at which $t = 0$ and the value of s at that time. For example, if the particle starts from O, then when $t = 0$, $s = 0$, therefore $k = 0$ and $s = a \sin nt$. If when $t = 0$, $s = a$, then $s = a \sin(nt + \pi/2) = a \cos nt$.

Generally, $s = \sin(nt + \varepsilon)$ where ε depends upon the instant at which we begin measuring s. We could equally well write $s = a\cos(nt + \varepsilon')$ or $s = A \sin nt + B \cos nt$ from basic trigonometrical results. By differentiating, this general value for s,

i.e., $s = a \sin(nt + \varepsilon)$, $v = \dfrac{\mathrm{d}s}{\mathrm{d}t} = an\cos(nt + \varepsilon)$

and $a = \dfrac{\mathrm{d}v}{\mathrm{d}t} = -an^2 \sin(nt + \varepsilon) = -n^2 s$.

If for t we substitute $t + 2\pi/n$ in the equations for s and v, we obtain identical results. This shows that after successive intervals of time $2\pi/n$, the particle passes through the same position with the same velocity.

Hence, $T = \dfrac{2\pi}{n}$ **is the period of oscillation.**

It should be noted that whereas $v^2 = n^2(a^2 - s^2)$ gives the speed for any displacement, $v = an \cos(nt + \varepsilon)$ will give the direction of motion as well as the speed.

Simple harmonic motion associated with uniform circular motion

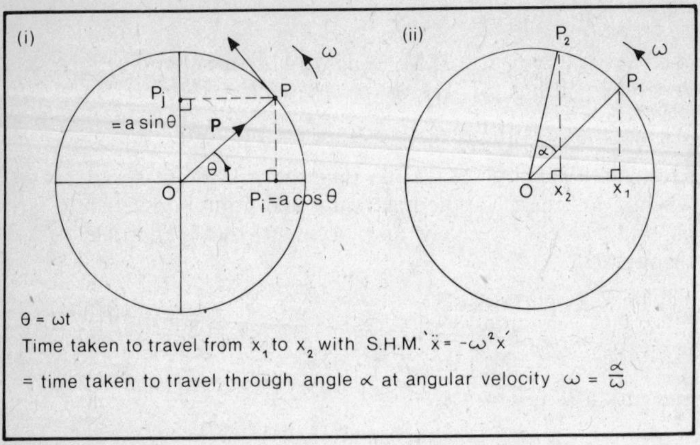

Figure 114.

Referring to figure 114, consider the position vector, **P**, of a point moving in a circle, radius a, centre O.

$$\mathbf{P} = a\cos\theta\,\mathbf{i} + a\sin\theta\,\mathbf{j}$$

If P is a point moving around the circle with constant angular speed ω rad s^{-1}, then $\theta = \omega t$ and

$$\mathbf{P} = a\cos\omega t\,\mathbf{i} + a\sin\omega t\,\mathbf{j}$$

Thus, $\quad \dot{\mathbf{P}} = a\omega\sin\omega t\,\mathbf{i} + a\omega\cos\omega t\,\mathbf{j}$

and $\quad \ddot{\mathbf{P}} = -a\omega^2\cos\omega t\,\mathbf{i} - a\omega^2\sin\omega t\,\mathbf{j}$

The **orthogonal projections** of P onto the **i** and **j** axes, P_i and P_j, respectively, move with **S.H.M.**, since

$$\ddot{\mathbf{P}}_i = -\omega^2\mathbf{P}_i \quad\text{and}\quad \ddot{\mathbf{P}}_j = -\omega^2\mathbf{P}_j.$$

Hence, referring to figure 114(ii), the time, t, taken by P_i to travel between X_1 and X_2 with S.H.M., is equal to the time taken by P to travel over the corresponding arc P_1P_2 with constant angular velocity, ω, i.e.,

$$t = \frac{\alpha}{\omega}$$

The simple pendulum

One of the most frequent examples of angular S.H.M. is that of the simple pendulum. A simple pendulum consists of a bob connected to a fixed point by a light inextensible string, oscillating through a small angle in a vertical plane.

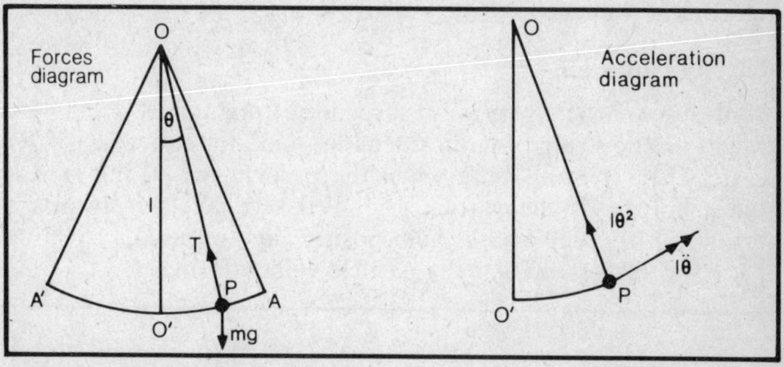

Figure 115.

Refer to figure 115.

The bob, P, has mass m and the string of length l is fixed at O. OO' is vertical and the angle made by the string to the vertical at any time, t, is θ. The angular acceleration, in the direction of increasing θ, is $\ddot{\theta}$, given by $-mg \sin\theta = ml\ddot{\theta}$, from Newton's second law applied tangentially.

As we stated earlier, the bob oscillates through a small angle, therefore

$$\sin\theta \simeq \theta \quad \text{and} \quad \ddot{\theta} = -\frac{g}{l}\theta.$$

Comparing this equation with the basic equation for S.H.M., '$\ddot{s} = -n^2 s$', we see that the bob performs angular S.H.M. with period '$T = 2\pi/n$', which in this case is

$$T = 2\pi\sqrt{\frac{l}{g}}. \qquad \left(n^2 = \frac{g}{l}\right)$$

Forces which produce simple harmonic motion

We have seen that in order for S.H.M. to be performed, the acceleration of a body must be proportional to its displacement from a fixed point and directed towards that point. By Hooke's law, the tension in an elastic string and the tension or thrust in a spring is $\lambda x/l$, with the usual notation, and acts in such a direction as to attempt to restore the string or spring to its natural length. Hence we should expect a particle attached to an elastic string, or to a spring, to perform S.H.M. We shall now illustrate that this is the case.

Refer to figure 116.

Consider a particle of mass m suspended from a fixed point A by a light elastic string of natural length l and modulus of elasticity λ. Let the extension be e when the particle is hanging at rest under gravity. If the particle is pulled vertically downwards a distance a below its equilibrium position and released, providing $a \leq e$, we shall show that the particle will perform S.H.M.

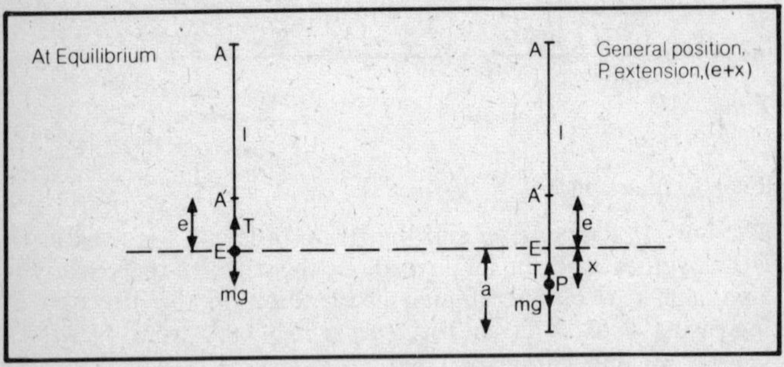

Figure 116.

At equilibrium, $T = mg$.

By Hooke's Law, $T = \dfrac{\lambda e}{l}$, $\dfrac{\lambda e}{l} = mg$.

At the general position, P, in the direction of increasing x, the equation of motion is $mg - T = m\ddot{x}$.

By Hooke's Law,

$$T = \frac{\lambda(e + x)}{l} = mg + \frac{\lambda x}{l}, \quad \text{thus} \quad \ddot{x} = -\frac{\lambda}{ml}x.$$

Comparing this with the basic equation for S.H.M., we can see that the particle will perform S.H.M. about E as the centre, with period

$$\text{'}T = \frac{2\pi}{n}\text{'}, \quad \text{where} \quad n = \sqrt{\frac{\lambda}{lm}}.$$

Hence $T = 2\pi\sqrt{\frac{lm}{\lambda}}$.

However this equation of motion only holds while the string is **stretched** so that the **amplitude** of motion which will be a (since that is the extension when the velocity is zero) must be less than or equal to e for S.H.M.

If $a > e$, then the particle will perform S.H.M. as long as the string is stretched, i.e., until it reaches A' on its upward journey. It will then move upwards under gravity, come to instantaneous rest, fall to A' and then move with S.H.M. again.

We chose to measure x from E because at the equilibrium position the resultant force acting on the particle is zero, therefore the acceleration is zero. We know that this is a property of S.H.M. at the centre of oscillation and so we anticipated, correctly, that E would be the centre of this oscillation. In general, the equilibrium position will be the centre of S.H.M. oscillations.

Consider an elastic string of natural length l and modulus of elasticity λ, attached at one end to point A and at the other to point B. Both points are on a smooth plane and at a distance of $2l$ apart. A particle P is attached to the mid-point, M, of the string. P is pulled towards B through a distance $a \leq l/2$ and then released. The equilibrium position is clearly M. We shall anticipate that this will be the centre of oscillation and measure the displacement x from M. Attaching P to the midpoint of the elastic string has in effect divided AB into two separate elastic strings, AP and PB, each of natural length $l/2$.

Let us consider the equation of motion for P with the direction of increasing x as positive (see figure 117).

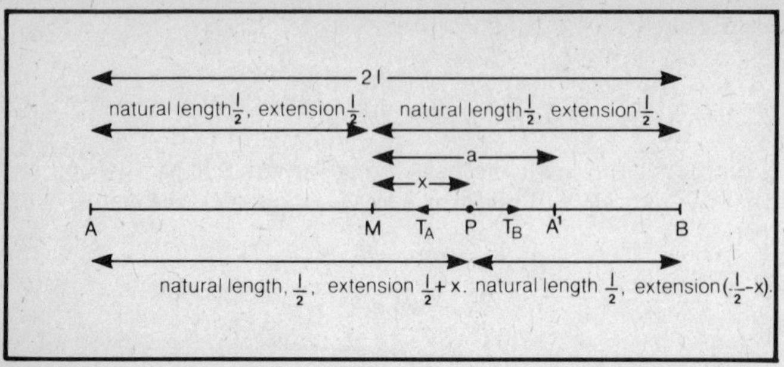

Figure 117.

The forces acting on P are T_A and T_B where, from Hooke's Law,

$$T_A = \frac{\lambda}{l/2}(l/2 + x) \quad \text{and} \quad T_B = \frac{\lambda}{l/2}(l/2 - x)$$

Applying Newton's second law: $T_B - T_A = m\ddot{x}$, therefore

$$\lambda - \frac{2\lambda x}{l} - \lambda - \frac{2\lambda x}{l} = m\ddot{x} \quad \text{and} \quad \ddot{x} = -\frac{4\lambda x}{lm}.$$

Hence the particle will perform S.H.M. with period

$$T = \pi\sqrt{\frac{ml}{\lambda}},$$

centre of oscillation, M, and amplitude a.

If the centre of oscillation is not apparent in a problem, then as long as an equation of motion of the form $\ddot{s} = -n^2(s + k)$ is derived, where k is a constant, there will be S.H.M. about the point where $s + k = 0$.

Substituting $x = s + k$, gives $\ddot{x} = \ddot{s}$, and $\ddot{s} = -n^2(s + k)$ becomes $\ddot{x} = -n^2x$ which is of the standard form.

Worked examples

Example 1 A mass of 10 kg moves with simple harmonic motion. When it is 2 m from the centre of oscillation, the velocity and acceleration of the body are 12 ms^{-1} and 162 ms^{-2} respectively. Calculate:

(i) the number of oscillations per minute;

(ii) the amplitude of the motion;

(iii) the force being applied to the body when it is at the extremities of its motion.

(S.U.J.B.)

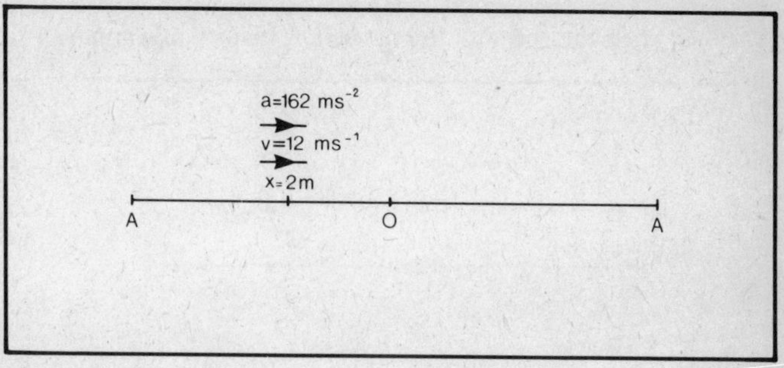

Figure 118.

The only part of the motion when both the acceleration and the velocity are positive is on the journey $A'O$. The displacement, $\mathbf{x}$, is $-\mathbf{2}$ when $\mathbf{v} = \mathbf{12}$ and $\mathbf{a} = \mathbf{162}$.

We are told that the motion is simple harmonic, therefore the equation of motion is $\ddot{x} = -n^2x$.

Substituting $x = -2$ and $\ddot{x} = 162$, $\mathbf{n^2 = 81}$.

(i) The period of oscillation is $\boldsymbol{T} = \dfrac{2\pi}{n} = \dfrac{\mathbf{2\pi}}{\mathbf{9}}$ secs.

Hence the number of oscillations per minute is $\dfrac{\mathbf{270}}{\pi}$.

(ii) A standard equation of S.H.M. is $v^2 = n^2(a^2 - x^2)$ where a is the amplitude. Substituting $v = 12$ when $x = -2$ and $n^2 = 81$, $144 = 81(a^2 - 4)$,

i.e.,
$$\mathbf{a} = \frac{2}{3}\sqrt{\mathbf{13}}\ \mathbf{m}.$$

(iii) When the particle is at the extremities of its motion, its displacement is the amplitude. From $\ddot{x} = -n^2x$,

$$\ddot{x} = \pm 81 \times \tfrac{2}{3}\sqrt{13} = \pm \mathbf{54\sqrt{13}\ ms^{-2}}.$$

From Newton's second law '$f = ma$' we see that the force being applied to the particle at these points is of magnitude $\mathbf{540\sqrt{13}\ N}$ directed towards O.

Example 2 A particle is moving with linear S.H.M. of amplitude 0.5 m and of period 4 s. Find the maximum speed and acceleration of the particle. Find also the speed of the particle when its displacement from the centre of oscillation is 0.25 m. Find the least time which elapses between two consecutive instants when the speed of the particle is half its maximum.

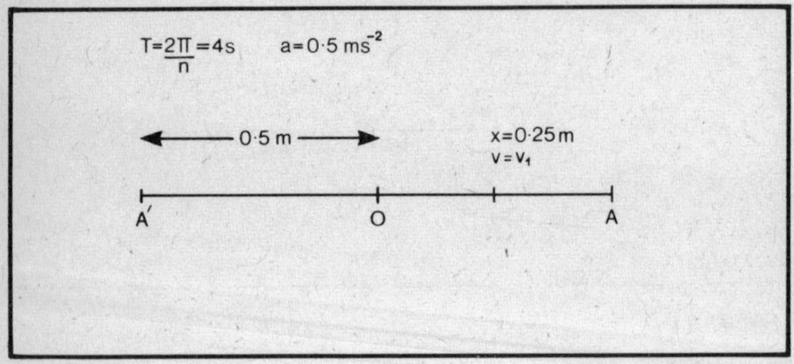

Figure 119.

We are told that the period is 4 s,

i.e., $T = \dfrac{2\pi}{n} = 4$, therefore $n = \dfrac{\pi}{2}$, $\boldsymbol{n^2 = \dfrac{\pi^2}{4}}$.

Hence from the basic equations for S.H.M.:

$$\ddot{x} = -\frac{\pi^2}{4}x \quad \text{and} \quad v^2 = \frac{\pi^2}{4}\left(\frac{1}{4} - x^2\right).$$

The **maximum speed, *V***, will occur when x^2 is a minimum,

i.e., when $x = 0$. Thus $V^2 = \dfrac{\pi^2}{16}$, $\boldsymbol{V = \dfrac{\pi}{4}}$ **ms^{-1}**

The **maximum acceleration** occurs when x is a **minimum**, as we can see from

$$\ddot{x} = -\frac{\pi^2}{4}x.$$

Hence since the amplitude is $\frac{1}{2}$, the maximum acceleration occurs when $x = -\frac{1}{2}$ and is

$$\left(\frac{\pi^2}{8}\right) \text{ms}^{-2}$$

The speed of the particle when its displacement from the centre of oscillation is 0·25 is found by substituting $x = \frac{1}{4}$ into $v^2 = n^2(a^2 - x^2)$, i.e.,

$$v_1^{\,2} = \frac{\pi^2}{4}\left(\frac{1}{4} - \frac{1}{16}\right) \quad \text{and} \quad \mathbf{v_1 = \frac{\pi}{8}\sqrt{3}\ ms^{-1}}$$

If we measure time from when the particle passes through O then $x = a \sin nt$ and $\dot{x} = an \cos nt$.

Hence when the speed is half the maximum, i.e., $\dot{x} = \pm\pi/8$,

$$\text{then } \pm\frac{\pi}{8} = \frac{\pi}{4}\cos\frac{\pi}{2}t, \quad \cos\frac{\pi}{2}t = \pm\frac{1}{2}$$

$$\text{and} \quad \frac{\pi}{2}t = \frac{\pi}{3}, \frac{2\pi}{3}, \frac{4\pi}{3}, \frac{5\pi}{3}, \ldots$$

Therefore the least time which elapses between instants when the speed of the particle is $\pi/8$,

$$\text{is} \quad \mathbf{\frac{2}{3}\,s}$$

Example 3 A particle of mass m is suspended from a fixed point by an elastic string of natural length $4l$ and modulus of elasticity $4mg$. The particle is pulled down a distance $2l$ below its equilibrium position and released from rest. Show that the particle starts to move with S.H.M. and find its period.

Find the time that elapses before the string first becomes slack, and from energy considerations find the greatest height above the equilibrium position that the particle achieves.

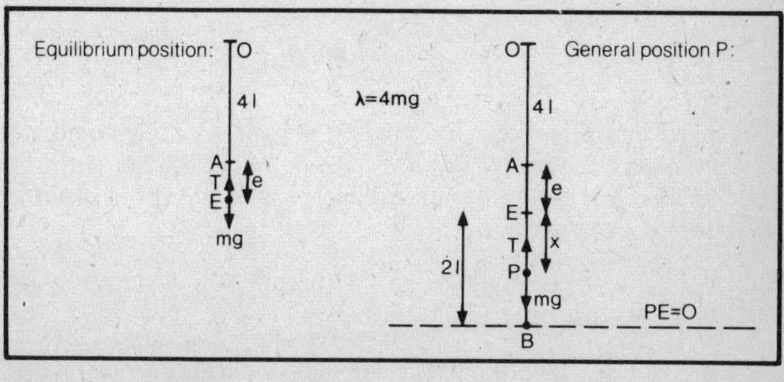

Figure 120.

At equilibrium, $T = mg$ and by Hooke's Law (see figure 120)

$$`T = \frac{\lambda x}{l}', \quad \text{therefore} \quad \frac{4mge}{4l} = mg \quad \text{and} \quad e = l.$$

We shall now find the general equation of motion for the particle, measuring x from E in anticipation of the equilibrium position being the centre of S.H.M.

In the direction of increasing x, $mg - T = m\ddot{x}$. (From '$f = ma$').

$$\text{Since } T = \frac{4mg}{4l}(e + x) \quad \text{then} \quad \ddot{x} = -\frac{gx}{l}$$

which is an **equation of motion for S.H.M. of period**

$$`\boldsymbol{T = \frac{2\pi}{n}}', \quad \text{i.e., } \textbf{of period} \quad \boldsymbol{2\pi\sqrt{\frac{l}{g}}}$$

The **amplitude** of the motion is **$2l$** since that is the value of x when $v = 0$. We see that since $e = l$, the particle will cease to move with S.H.M. at A and move under gravity until it returns to A and then once more move with S.H.M.

If we measure time from when the particle is released at B then,

$$x = a \cos nt = 2l \cos \sqrt{\frac{g}{l}}\, t.$$

$$\text{When } x = -l, \cos \sqrt{\frac{g}{l}}\, t = -\frac{1}{2} \quad \text{and} \quad \sqrt{\frac{g}{l}}\, t = \frac{2}{3}\pi.$$

(we are only concerned with the **first** value). Thus, the time that elapses before the string first becomes slack is,

$$t = \frac{2}{3}\sqrt{\frac{l}{g}}\,\pi \textbf{ secs.}$$

The particle will achieve its greatest height when it comes to instantaneous rest on its upward journey under gravity with the string slack. The total mechanical energy at B is just the **elastic potential energy** of

$$\frac{4mg}{8l}(3l)^2 = \frac{9}{2}mgl.$$

At the point at which the particle comes to instantaneous rest, at height h above E, the total mechanical energy is $mg(h + 2l)$.

Equating these two values, since we are in a conservative system of forces and can apply the Principle of Conservation of Mechanical Energy:

$$mg(h + 2l) = \frac{9}{2}mgl \quad \text{and} \quad \boldsymbol{h = \frac{5l}{2}}$$

Example 4 One end of a light elastic string, of natural length l, is attached to a fixed point. A particle of mass m is attached to the other end and hangs freely, stretching the string by a length a in its equilibrium position. The particle is now pulled down a further distance $\frac{1}{2}a$, and released from rest. Prove that the particle moves with simple harmonic motion, and find the period of its oscillations.

If, instead of the particle being pulled down a further distance $\frac{1}{2}a$, it is pulled down a further $1\frac{1}{2}a$, find how far it will rise before coming to instantaneous rest.

(S.U.J.B.)

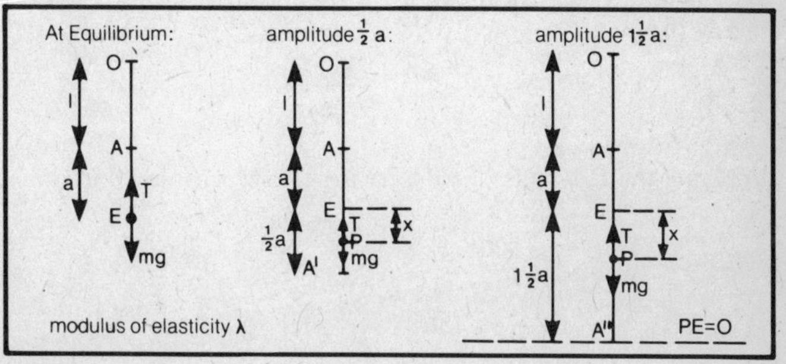

Figure 121.

Refer to figure 121.

At equilibrium, we know that $T = mg$, and by Hooke's Law,

$$T = \frac{\lambda a}{l}, \quad \text{therefore} \quad \lambda = \frac{lmg}{a}.$$

Applying Newton's second law to the particle in general position P, with x measured from E, (in anticipation of that being the centre of S.H.M. oscillations): $mg - T = m\ddot{x}$. We know that

$$T = \frac{lmg}{al}(a + x). \quad \text{Hence,} \quad \ddot{x} = -\frac{g}{a}x,$$

which shows that **the particle moves with S.H.M. of period**

$$\boldsymbol{T = \frac{2\pi}{n}}, \quad \text{i.e.,} \quad \boldsymbol{T = 2\pi\sqrt{\frac{a}{g}}}.$$

The amplitude of this S.H.M. will be $\frac{1}{2}a$, since that is the value of x when $v = 0$. Hence the string will not go slack since it will never reach A.

If the particle is released from a distance $1\frac{1}{2}a$ from the equilibrium position, the particle will have the same general equation and so move with S.H.M. as long as the string is taut. In this case the amplitude will be $1\frac{1}{2}a$. The particle will rise to A with S.H.M. and then move under gravity, come to instantaneous rest, fall to A, and then move again with S.H.M. When it comes to rest at height h above the equilibrium position the only mechanical energy it will have will be gravitational potential energy of $mg(h + 1\frac{1}{2}a)$.

At A'' it had mechanical energy in the form of elastic potential energy of

$$\frac{lmg(2\frac{1}{2}a)^2}{a2l} = \frac{25mga}{8}.$$

Applying the Principle of Conservation of Mechanical Energy,

$$mg(h + 1\tfrac{1}{2}a) = \frac{25mga}{8},$$

$$\boldsymbol{h = \frac{13a}{8}}.$$

Key terms

Simple harmonic motion is defined by the equation of motion

$$\frac{\mathrm{d}^2 s}{\mathrm{d}t^2} = -n^2 s,$$

where s is a displacement and n is a constant.

Forces which produce S.H.M. are proportional to the displacement from a fixed point and directed towards that fixed point. The centre of S.H.M. oscillation will be the equilibrium position for the body.

The following equations of motion for S.H.M. are quotable:

$v^2 = n^2(a^2 - x^2)$; $x = a \sin(nt + \varepsilon)$ where ε depends upon the value of x when $t = 0$; $\dot{x} = an \cos(nt + \varepsilon)$. The amplitude, a, of the oscillation is the maximum distance x and occurs when $v = 0$. The time for a complete oscillation, **the period of the motion is $2\pi/n$.**

The **maximum magnitude of acceleration** occurs when $x = \pm a$, and is $\boldsymbol{n^2 a}$. The **maximum speed** occurs when $x = 0$ and is $\boldsymbol{na}$.

The tension in an elastic string, or the thrust or tension in a spring, produces S.H.M. about the equilibrium position as the centre of oscillation. The distance from the equilibrium position of the point from which the particle is originally released, is the amplitude of the motion.

Chapter 12
Damped Oscillations

If a particle is moving in a **straight line under the action of a force directed towards a fixed point on that line and proportional to its displacement from that fixed point, then the particle moves with S.H.M.**, with equation of motion

$$\frac{d^2x}{dt^2} + n^2x = 0.$$

(Refer to figure 122(i)).

If, in addition to this force, it is also subject to a **resisting force proportional to its speed** then its equation of motion becomes

$$\frac{d^2x}{dt^2} + k\frac{dx}{dt} + n^2x = 0.$$

(Refer to figure 122(ii)).

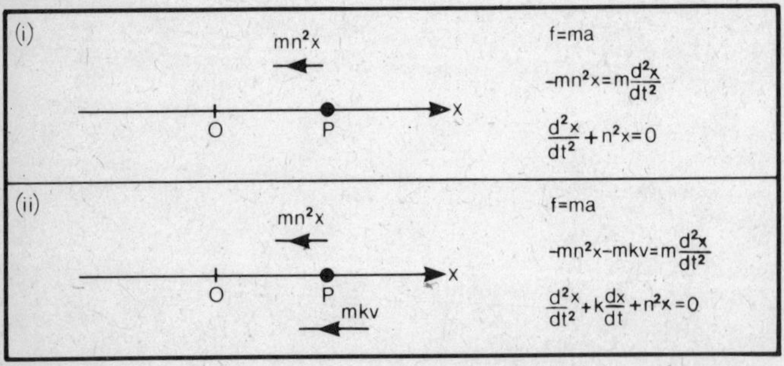

Figure 122.

This equation is a **linear differential equation of the second order with constant coefficients**, which we need to solve in order to investigate the ensuing motion fully. A full explanation of the techniques for solution of this type of equation will be found in most standard Pure Mathematics books but a brief outline will be given here.

Solution of second order linear differential equations with constant coefficients

Consider the general equation

$$\frac{\mathrm{d}^2x}{\mathrm{d}t} + 2k\frac{\mathrm{d}x}{\mathrm{d}t} + n^2x = 0$$

('k' has become '$2k$' for ease of algebraic manipulation).

We can see by differentiation and substitution that a solution of the form $\boldsymbol{x = Ae^{mt}}$, where A and m are constants, would satisfy the differential equation provided that m satisfies the quadratic equation $m^2 + 2km + n^2 = 0$, called **the auxiliary equation**.

There are two roots of this equation, say m_1 and m_2. The general solution of the differential equation depends upon the nature of m_1 and m_2.

If m_1 and m_2 are real and distinct, the general solution is $\boldsymbol{x = Ae^{m_1t} + Be^{m_2t}}$, where A and B are constants.

If $m_1 = m_2 = m$, then the general solution is $\boldsymbol{x = Ae^{mt} + Bte^{mt}}$.

If the roots of the quadratic equation are **complex**, then they will be $m_1 = -k + i\sqrt{n^2 - k^2}$ and $m_2 = -k - i\sqrt{n^2 - k^2}$, $(i = \sqrt{-1})$ and the general solution may be written as

$$x = A'\,e^{(-k+i\sqrt{n^2-k^2})t} + B'\,e^{(-k-i\sqrt{n^2-k^2})t}$$

where A' and B' are complex constants.

However, using the general results that

$$\text{'}e^{-i\theta} = \cos\theta - i\sin\theta\text{'} \quad \text{and} \quad \text{'}e^{i\theta} = \cos\theta + i\sin\theta\text{'}$$

we can rewrite the general solution as

$$\boldsymbol{x = e^{-kt}(A\cos\sqrt{n^2 - k^2}\,t + B\sin\sqrt{n^2 - k^2}\,t)}$$

where $A = A' + B'$ and $B = (A' - B')i$.

A more convenient form of this general solution, is

$$\boldsymbol{x = Ce^{-kt}\cos(\sqrt{n^2 - k^2}\,t + \varepsilon)}$$

We shall now investigate the physical significance of these solutions when x represents displacement and t time.

Let the equation of motion of a particle be $\ddot{x} + 2k\dot{x} + n^2x = 0$.

As we have seen, this equation of motion occurs when a particle is moving in a straight line subject to two forces, one proportional to its displacement from a fixed point on that line and directed towards that fixed point, and the other a resisting force proportional to its speed.

The auxiliary equation for the general solution of the equation of motion is $m^2 + 2km + n^2 = 0$.

If the auxiliary equation has real, distinct roots m_1 and m_2 then the general solution is

$$x = Ae^{m_1 t} + Be^{m_2 t}.$$

This motion will **not be oscillatory** since there is no value of t for which $x = 0$, or $\dot{x} = 0$. This situation occurs when the damping force $2km\dot{x}$ dominates the oscillatory force mn^2x, and we refer to this situation as **heavy damping**.

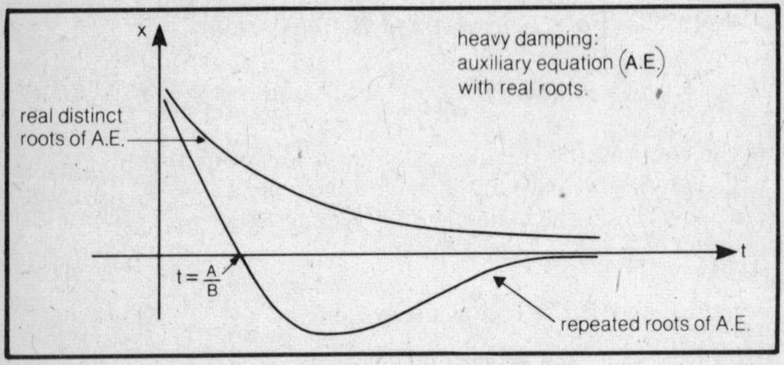

Figure 123.

If the **auxiliary equation** has **repeated** roots $m_1 = m_2 = m$, then the general solution is

$$x = (A + Bt)e^{mt}$$

Hence, there is one finite value of t, for which $x = 0$, $t = A/B$ and one finite value of t, for which

$$\dot{x} = 0, t = -\frac{(B + mA)}{mB}.$$

Again, this situation is one of **heavy damping**.

If the auxiliary equation has complex roots, then the general solution may be written as

$$\boldsymbol{x = Ce^{-kt} \cos(\sqrt{n^2 - k^2}\, t + \varepsilon)}.$$

Since this equation contains a cosine function, then clearly, this motion is oscillatory and is called **damped harmonic motion**, or **light damping**.

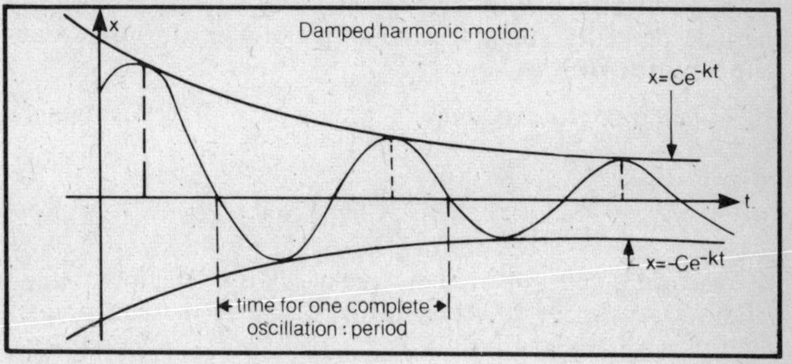

Figure 124.

Since $-1 \leq \cos\theta \leq 1$, $\boldsymbol{x = Ce^{-kt}}$ and $\boldsymbol{x = -Ce^{-kt}}$ form **upper** and **lower boundaries** for the curve. $\boldsymbol{e^{-kt}}$ is called the **damping factor**. The period of oscillation is the time for one complete cycle.

We see that $x = 0$ when $\cos(\sqrt{n^2 - k^2}\, t + \varepsilon) = 0$,

i.e., when $\sqrt{n^2 - k^2}\, t + \varepsilon = \dfrac{\pi}{2}, \dfrac{3\pi}{2}, \dfrac{5\pi}{2}, \dfrac{7\pi}{2}, \ldots$.

Thus $x = 0$ at times $t = \dfrac{1}{\sqrt{n^2 - k^2}}\left(\dfrac{\pi}{2} - \varepsilon\right)$,

$$\frac{1}{\sqrt{n^2 - k^2}}\left(\frac{3\pi}{2} - \varepsilon\right), \quad \frac{1}{\sqrt{n^2 - k^2}}\left(\frac{5\pi}{2} - \varepsilon\right), \ldots.$$

The difference between the successive values of t is constant. Hence, the **period of oscillation is constant and equal to**

$$\boldsymbol{\frac{2\pi}{\sqrt{n^2 - k^2}}}.$$

The **amplitude** of oscillation is the maximum value of x attained in that oscillation. The stationary values of x occur when

$\dot{x} = 0$. The maximum and minimum values may be found from this condition, and it is the case that **the amplitudes of successive oscillations decrease with time in geometric progression**.

In general, if $k^2 \geq n^2$, **the particle approaches the position $x = 0$ asymptotically**. In the case of $k^2 = n^2$, the particle passes through the origin once before returning to **approach the origin asymptotically**. If $k^2 < n^2$ then the particle oscillates about the origin, performing **damped harmonic motion**.

If, in addition to the forces represented by the equation of motion

$$\frac{d^2x}{dt^2} + 2k\frac{dx}{dt} + n^2x = 0,$$

the particle is acted upon by a force which is a function of time, $mf(t)$, acting in the direction of increasing x, the equation of motion is

$$\frac{d^2x}{dt^2} + 2k\frac{dx}{dt} + n^2x = f(t).$$

The solution of this type of equation, and again a full explanation of this, will be found in a standard Pure Mathematics book, and involves finding a particular solution of the equation, called the **particular integral**, and adding it to the solution of

$$\frac{d^2x}{dt} + 2k\frac{dx}{dt} + n^2x = 0,$$

called the **complementary function**.

We have just dealt with the techniques for finding the **complementary function**, and finding the **particular integral** involves trying a solution which is of the general form of $f(t)$. For example, if $f(t) = 3t^2$ we would try $x = at^2 + bt + c$. If $f(t) = 4\cos 3t$, we would try $f(t) = A\cos 3t + B\sin 3t$ or $f(t) = C\cos(3t + \varepsilon)$, whichever is the more convenient. If $f(t)$ is already in the complementary function, then a particular integral of t times the general type of $f(t)$ should be tried. Generally, either a particular solution will be obvious or straightforward, or it will be given.

Consider when $mf(t)$ is itself an oscillatory force, e.g., $ma\cos bt$.

The equation of motion is $\frac{d^2x}{dt^2} + 2k\frac{dx}{dt} + n^2x = a \cos bt$.

The particular integral of this equation will be of the form $x = A \cos bt + B \sin bt$ and the constants A and B would be found by substitution into the differential equation. This particular integral is of **simple harmonic form** and continues **indefinitely**.

The general solution of the differential equation is the sum of the complementary function and particular integral.

We have seen already that the complementary function is the equation of a damping motion which diminishes in time whether heavy or light. Hence in this case, the motion will tend towards that of the particular integral and the particular integral represents that part of the motion called the **forced oscillation**.

Worked examples

Example 1 A particle P of mass m moves along a straight line under the action of a force kx directed towards O, where x is the displacement of P from O. The particle is also subject to a resisting force equal to $2\sqrt{mk}\,v$, where v is the speed of the particle, and a constant force F in the direction of Ox. Initially the particle is at rest at O. Deduce an equation of motion for the particle.

Find the displacement of the particle from O at time t.

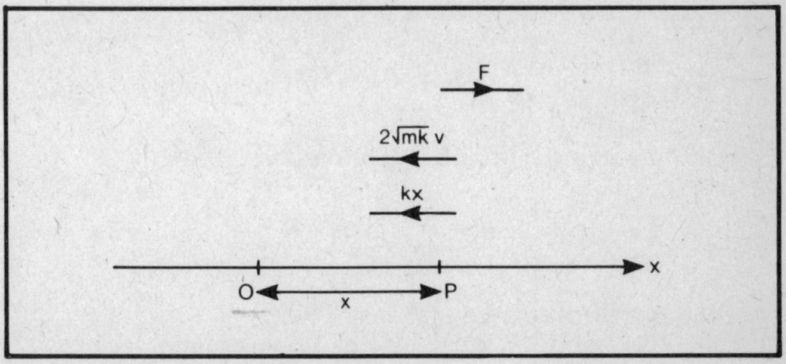

Figure 125.

Applying Newton's second law 'f = *ma*', the equation of motion for the particle is (see figure 125)

$$F - 2\sqrt{mk}\,v - kx = m\frac{d^2x}{dt^2},$$

which rearranges to

$$\frac{d^2x}{dt^2} + 2\sqrt{\frac{k}{m}}\frac{dx}{dt} + \frac{k}{m}x = \frac{F}{m}.$$

In order to find the displacement x at time t we need to solve this second order linear differential equation with constant coefficients.

We shall find the **complementary function** by solving,

$$\frac{d^2x}{dt^2} + 2\sqrt{\frac{k}{m}}\frac{dx}{dt} + \frac{k}{m}x = 0$$

The **auxiliary equation** is $p^2 + 2\sqrt{\frac{k}{m}}\,p + \frac{k}{m} = 0.$

Therefore $p = -\sqrt{\frac{k}{m}}$ twice.

Hence, the complementary function is

$$\boldsymbol{x = (A + Bt)\exp\left(-\sqrt{\frac{k}{m}}\,t\right)}.$$

We need to find the **particular integral**, and since the right hand side of the equation is a constant, we shall try the particular solution of $x = K$, where K is a constant.

If $x = K$, $\frac{dx}{dt} = 0$, $\frac{d^2x}{dt^2} = 0.$

Substituting into the original equations,

$$\frac{k}{m}.K = \frac{F}{m}, \quad \text{thus} \quad K = \frac{F}{k}$$

and the **particular integral** is $x = \frac{F}{k}$.

Hence the complete solution, which is the sum of the complementary function and the particular integral, is

$$\boldsymbol{x = \left[(A + Bt)\exp\left(-\sqrt{\frac{k}{m}}\,t\right)\right] + \frac{F}{k}}.$$

The constants A and B need to be found, and to do this we use the information that initially the particle is at rest at O,

i.e., when $t = 0$, $x = 0$ and $\dfrac{dx}{dt} = 0$.

If $x = 0$ when $t = 0$, then since

$$x = \left[(A + Bt)\exp\left(-\sqrt{\frac{k}{m}}t\right)\right] + \frac{F}{k},$$

$0 = A + \dfrac{F}{k}$ so that $A = -\dfrac{F}{k}$.

By differentiating the complete solution,

$$\frac{dx}{dt} = \left[B\exp\left(-\sqrt{\frac{k}{m}}t\right)\right] - \sqrt{\frac{k}{m}}(A + Bt)\exp\left(-\sqrt{\frac{k}{m}}t\right).$$

Since $\dfrac{dx}{dt} = 0$ when $t = 0$,

$$\text{then } 0 = B - \sqrt{\frac{k}{m}}A \text{ and } B = -\sqrt{\frac{k}{m}}\frac{F}{k}.$$

Hence the complete solution is

$$\begin{aligned} x &= \left[\left(-\frac{F}{k} - \sqrt{\frac{k}{m}}\frac{F}{k}t\right)\exp\left(-\sqrt{\frac{k}{m}}t\right)\right] + \frac{F}{k} \\ &= \frac{F}{k}\left[1 - \left(1 + \sqrt{\frac{k}{m}}t\right)\exp\left(-\sqrt{\frac{k}{m}}t\right)\right] \end{aligned}$$

Example 2 A light spring of natural length l and modulus of elasticity $8mn^2l$ is lying at rest in a straight line on a smooth horizontal table. A particle of mass m is attached to one end of the spring and the other end is made to move with constant speed V in the direction which extends the spring. The particle is subject to a resistance of magnitude $4mnv$ where v is the speed of the particle at time t.

If x is the extension of the spring and t is the time for which the system has been in motion, find the equation of motion of the system in terms of x and t. Solve this equation and hence show that the extension of the spring is approximately constant at $v/2n$ for large t.

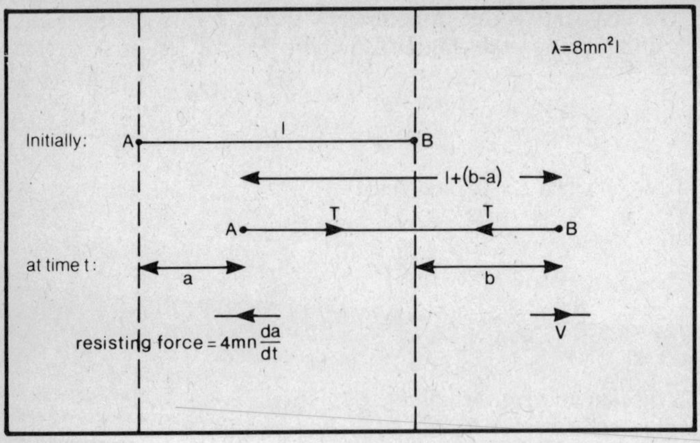

Figure 126.

The extension, x, at time t, of the spring is $x = (b - a)$. The tension T in the spring is, by Hooke's Law (see figure 126),

$$T = 8mn^2 \frac{lx}{l} = 8mn^2 x.$$

The end B moves with constant speed V in the direction AB i.e., $\boldsymbol{b = Vt}$. Hence $x = (Vt - a)$.

The equation of motion for A is, from '$f = ma$,'

$$T - 4mn\frac{da}{dt} = m\frac{d^2a}{dt^2},$$

Therefore $8mn^2x - 4mn\dfrac{da}{dt} - m\dfrac{d^2a}{dt^2} = 0.$

Since $x = Vt - a$, then $a = Vt - x$,

$$\frac{da}{dt} = V - \frac{dx}{dt} \quad \text{and} \quad \frac{d^2a}{dt^2} = -\frac{d^2x}{dt^2}.$$

Substituting these values into the equation of motion for A,

$$8mn^2x - 4mn\left(V - \frac{dx}{dt}\right) + m\frac{d^2x}{dt^2} = 0.$$

Thus $$\boldsymbol{\frac{d^2x}{dt^2} + 4n\frac{dx}{dt} + 8n^2x = 4nV.}$$

To find x in terms of t, we need to solve this second order linear differential equation with constant coefficients. We shall find the complementary function first by solving

$$\frac{d^2x}{dt^2} + 4n\frac{dx}{dt} + 8n^2x = 0.$$

The auxiliary equation is $p^2 + 4np + 8n^2 = 0$,

therefore $p = -2n \pm 2ni$.

We have complex roots of the auxiliary equation. Hence the complementary function is an equation of damped harmonic motion and is $x = Ce^{-2nt}\cos(2nt + \varepsilon)$, where C and ε are constants.

We must now find the particular integral. Since the R.H.S. of the equation is a constant, we shall try a constant for the particular solution, i.e., $x = K$.

If $x = K$, $\dfrac{dx}{dt} = 0$, and $\dfrac{d^2x}{dt^2} = 0$.

Substituting into the original equation,

$$8n^2K = 4nV,\ K = \frac{V}{2n}.$$

Hence, the solution of the differential equation is,

$$x = Ce^{-2nt}\cos(2nt + \varepsilon) + \frac{V}{2n}.$$

We must find the constants C and ε.

We know that the spring is originally unstretched,

i.e., when $t = 0$, $x = 0$. Thus, $C\cos\varepsilon + \dfrac{V}{2n} = 0$.

At $t = 0$, B begins to move with velocity V, i.e., $\dfrac{dx}{dt} = V$.

Since $\dfrac{dx}{dt} = -2ne^{-2nt}C\cos(2nt + \varepsilon) - C2ne^{-2nt}\sin(2nt + \varepsilon)$,

then $V = -2n\left(-\dfrac{V}{2n}\right) - 2nC\sin\varepsilon$.

Therefore $\varepsilon = 0$ and $C = -\dfrac{V}{2n}$.

Hence, $$x = -\frac{V}{2n} e^{-2nt} \cos 2nt + \frac{V}{2n}.$$

As $t \to \infty$, $-\frac{V}{n} e^{-2nt} \cos 2nt \to 0$ so that $x \to \frac{V}{2n}$.

Hence, **for large t**, x is approximately constant at $\frac{V}{2n}$.

Example 3 A particle P of mass m is moving in the straight line OX such that at time t its displacement from O is x. It is subject to: a resistance of magnitude $4mv$ where v is its speed at time t; a force $F \cos t$ where F is constant and in the same direction as the displacement; and a force of magnitude $3m$ times its distance from O, directed towards O.

Find the equation of motion of the particle in terms of x and t and solve it, given that when $t = 0$, $x = a$ and $x = 0$. Show that for large t, the motion is approximately simple harmonic.

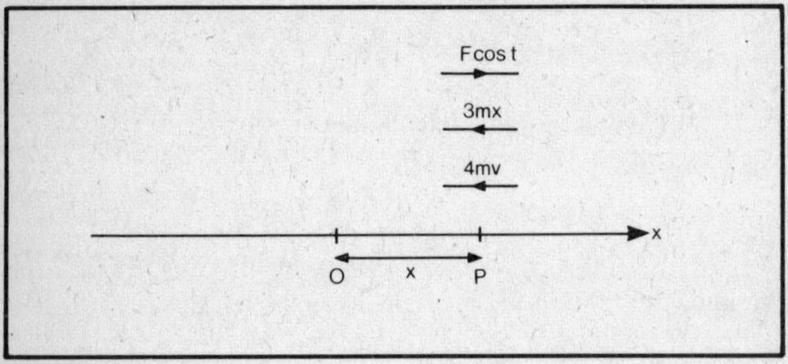

Figure 127.

Applying Newton's second law to the particle (see figure 127),

$$F \cos t - 4mv - 3mx = m\frac{d^2x}{dt^2}$$

i.e., $$\frac{d^2x}{dt^2} + 4\frac{dx}{dt} + 3x = \frac{F}{m} \cos t.$$

which is the required **equation of motion**.

We shall solve $\frac{d^2x}{dt^2} + 4\frac{dx}{dt} + 3x = 0$

in order to get the complementary function.

The auxiliary equation is $n^2 + 4n + 3 = 0$, therefore $n = -3$ or $n = -1$, and the complementary function is $x = Ae^{-3t} + Be^{-t}$.

In order to find the particular integral we shall try a solution of the form $x = L\cos t + M\sin t$.

Hence,
$$\frac{dx}{dt} = -L\sin t + M\cos t$$

and
$$\frac{d^2x}{dt^2} = -L\cos t - M\sin t.$$

Substituting into the differential equation:

$$-L\cos t - M\sin t - 4L\sin t + 4M\cos t + 3L\cos t + 3M\sin t = \frac{F}{m}\cos t.$$

Equating coefficients of cos t,

$$-L + 4M + 3L = \frac{F}{m} \quad \text{and} \quad 2L + 4M = \frac{F}{m}.$$

Equating coefficients of sin t,

$$-M - 4L + 3M = 0,\ M = 2L.$$

Therefore,
$$\boldsymbol{M} = \frac{\boldsymbol{F}}{\boldsymbol{5m}} \quad \text{and} \quad \boldsymbol{L} = \frac{\boldsymbol{F}}{\boldsymbol{10m}}.$$

Hence, the particular integral is $\frac{F}{10m}(\cos t + 2\sin t)$

and the complete solution is

$$x = Ae^{-3t} + Be^{-t} + \frac{F}{10m}(\cos t + 2\sin t).$$

We need to find constants A and B, and to do this we use the information that when $t = 0$, $dx/dt = 0$ and $x = a$.

Since $\frac{dx}{dt} = -3Ae^{-3t} - Be^{-t} + \frac{F}{10m}(-\sin t + 2\cos t)$,

then $0 = -3A - B + \frac{F}{5m}$.

Since $x = a$ when $t = 0$, then $a = A + B + \dfrac{F}{10m}$.

These two equations give $A = -\dfrac{a}{2} + \dfrac{3F}{20m}$, $B = \dfrac{3a}{2} - \dfrac{F}{4m}$.

$$\text{Thus, } \mathbf{x} = \left(-\frac{a}{2} + \frac{3F}{20m}\right)e^{-3t} + \left(\frac{3a}{2} - \frac{F}{4m}\right)e^{-t} + \frac{F}{10m}(\cos t + 2\sin t).$$

When **t is large**, the complementary function,

$$\left(-\frac{a}{2} + \frac{3F}{20m}\right)e^{-3t} + \left(\frac{3a}{2} - \frac{F}{4m}\right)e^{-t}$$

is very small, and the particular integral **dominates**, hence x is approximately

$$\frac{F}{10m}(\cos t + 2\sin t),$$

which is of **simple harmonic form**.

Example 4 A pendulum consists of a small weight of mass m suspended from a fixed point by a light string of length l. The pendulum executes small oscillations in a vertical plane about the position of equilibrium. The oscillations are damped by a resisting force equal to k times the speed of the weight. Write down an equation of motion for the weight, in θ, where θ is the small angle of inclination of the string to the vertical at any instant. Given that

$$k < 2m\sqrt{\frac{g}{l}}, \quad \text{and} \quad \theta = \theta_0 \quad \text{and} \quad \dot{\theta} = 0 \quad \text{when} \quad t = 0,$$

find θ at time t.

Analysing the linear motion of the weight, we look for the forces acting tangential to the string and apply Newton's second law in this direction (figure 128).

Hence, $-kv - mg\sin\theta = m\dfrac{dv}{dt}$.

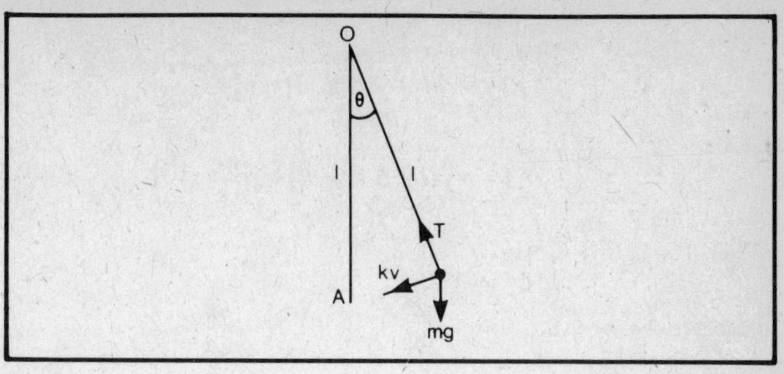

Figure 128.

Since $v = l\dot{\theta}$ and $\dfrac{dv}{dt} = l\ddot{\theta}$, then $ml\ddot{\theta} + kl\dot{\theta} + mg\sin\theta = 0$.

We are told that θ is small, therefore $\sin\theta \simeq \theta$, and the equation of motion for the weight is $\boldsymbol{ml\ddot{\theta} + kl\dot{\theta} + mg\theta = 0}$.

We shall now solve this second order linear differential equation.

The auxiliary equation is $mlp^2 + klp + mg = 0$, therefore

$$p = \frac{-kl \pm \sqrt{k^2l^2 - 4m^2lg}}{2ml}.$$

We are told that $k < 2m\sqrt{\dfrac{g}{l}}$,

and since both sides of this inequality are positive, $k^2l^2 < 4m^2lg$. Hence the roots of the auxiliary equation are **complex**.

We now know that we have an equation of motion for **damped harmonic motion** of the general form

$$\theta = \left(A\cos\sqrt{\frac{g}{l} - \frac{k^2}{4m^2}}\,t + B\sin\sqrt{\frac{g}{l} - \frac{k^2}{4m^2}}\,t\right)\exp\left(-\frac{kt}{2m}\right).$$

For ease of algebraic manipulation we shall let $h = \sqrt{\dfrac{g}{l} - \dfrac{k^2}{4m^2}}$.

Thus, $\theta = (A\cos ht + B\sin ht)\exp\left(-\dfrac{kt}{2m}\right)$

and

$$\dot{\theta} = -\frac{k}{2m}(A\cos ht + B\sin ht)\exp\left(-\frac{kt}{2m}\right)$$

$$+ h(-A\sin ht + B\cos ht)\exp\left(-\frac{kt}{2m}\right).$$

When $t = 0$, $\theta = \theta_0$ and $\dot{\theta} = 0$,

therefore, $\theta_0 = A$ and $0 = -\frac{k}{2m}\theta_0 + hB$, $B = \frac{k\theta_0}{2mh}$.

Thus, the general solution is

$$\boldsymbol{\theta = \theta_0\left(\cos ht + \frac{k}{2mh}\sin ht\right)\exp\left(-\frac{kt}{2m}\right)}$$

where $\boldsymbol{h = \sqrt{\frac{g}{l} - \frac{k^2}{4m^2}}}$.

Key terms

The equation of motion of a particle moving under the action of a force proportional to its distance from a fixed point on its line of motion, and directed towards that point, and a resisting force proportional to its speed, is

$$\frac{d^2x}{dt^2} + 2k\frac{dx}{dt} + n^2x = 0.$$

The ensuing motion depends on the nature of the roots of the **auxiliary equation**, $p^2 + 2kp + n^2 = 0$.

If the roots are **real**, then the motion is **heavily damped** and the particle approaches the origin (fixed point) asymptotically. The particle passes through $x = 0$ no more than once.

If the roots are **complex**, then we have **damped harmonic motion**: an oscillatory motion which diminishes in time.

This motion has **constant period** $\frac{2\pi}{\sqrt{n^2 - k^2}}$

If, in addition to the forces mentioned, a further force $F(t)$ which is a function of time acts on the particle, the equation of motion is

$$\frac{d^2x}{dt} + 2k\frac{dx}{dt} + n^2x = \frac{F}{m}.$$

The complementary function of this equation is the equation of damped motion. Hence, the particular integral for this equation dominates the motion as $t \to \infty$ provided it does not also approach zero as $t \to \infty$.

Chapter 13
Forces Acting at a Point

Force is a **line localised** vector. Vector techniques may be applied to a set of forces to find the **magnitude** and **direction** of the **resultant** force. Since we are discussing **concurrent** forces in this chapter, we shall postpone discussion of the location of the line of action of the resultant force until the next chapter. Clearly, forces meeting at a point have their resultant force passing through that point. Concurrent forces have no turning effect on a body.

When considering bodies other than particles we take their weight to be concentrated at their centre of gravity. In the case of uniform bodies, this is at their geometric centre of length, area or volume.

Concurrent forces in equilibrium

A particle is in equilibrium if it is at rest or moving with uniform velocity, i.e., it has zero acceleration. If a particle is at rest under the action of a set of forces $\mathbf{F}_1, \mathbf{F}_2, \ldots$, then $\mathbf{F}_1 + \mathbf{F}_2 + \ldots = \mathbf{0}$ since the resultant force must be zero.

If each force is resolved into its components in three non-coplanar directions $\hat{\mathbf{a}}, \hat{\mathbf{b}}, \hat{\mathbf{c}}$, i.e., $\mathbf{F} = F_a \hat{\mathbf{a}} + F_b \hat{\mathbf{b}} + F_c \hat{\mathbf{c}}$

then $\sum F = \sum F_a \mathbf{a} + \sum F_b \mathbf{b} + \sum F_c \mathbf{c} = \mathbf{0}$. Hence $\sum F_a = 0$, $\sum F_b = 0, \sum F_c = 0$.

For a particle to be in equilibrium under the action of a set of forces, the sum of the components in three non-coplanar directions must be zero. Conversely, if the sum of the components of a system of forces acting on a particle in each of three non-coplanar directions is zero, then that system is in equilibrium. In practice we usually consider three mutually perpendicular directions. In the case of coplanar forces we need consider two non-parallel directions only. (The sum of the components in the direction perpendicular to the plane of the forces is zero).

The magnitude and direction of the resultant of a system of forces may be found from the vector polygon. A system of forces in equilibrium may be represented by a **closed** vector polygon since the magnitude of their resultant must be zero. Conversely, it is the case that, if a system of concurrent forces may be represented by a closed vector polygon, then they are in equilibrium.

Three concurrent forces in equilibrium

The vector polygon for three concurrent forces in equilibrium is a triangle. If three concurrent forces may be represented by the sides of a triangle taken in order, then those forces are in equilibrium. This is the **Triangle of Forces** property. The application of trigonometry is particularly easy for a triangle.

Consider three **concurrent** forces **P**, **Q**, **R**, represented in magnitude and direction by the sides **AB**, **BC**, **CA**, of the triangle *ABC*. $\mathbf{AB} + \mathbf{BC} = \mathbf{AC}$, $\mathbf{AC} + \mathbf{CA} = \mathbf{0}$. Hence their vector sum is zero and since they are concurrent, the system is in equilibrium.

Applying the **sine rule** to triangle *ABC*,

$$\frac{AB}{\sin(180° - \beta)} = \frac{BC}{\sin(180° - \gamma)} = \frac{CA}{\sin(180° - \alpha)}$$

Thus $\dfrac{P}{\sin\beta} = \dfrac{Q}{\sin\gamma} = \dfrac{R}{\sin\alpha}$.

This result is known as **Lami's theorem**.

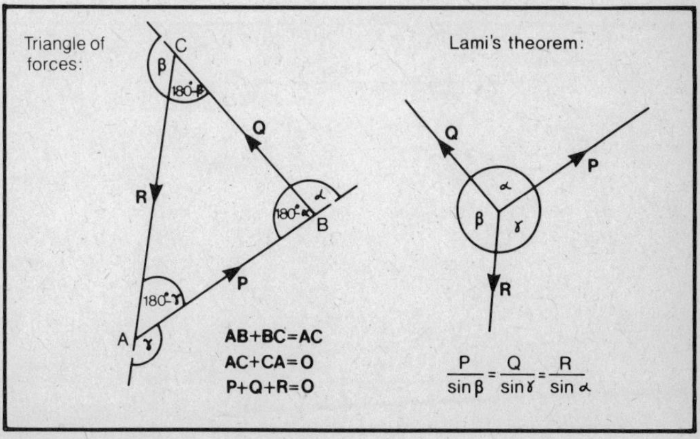

Figure 129.

Lami's theorem states that if three forces acting at a point are in equilibrium, then the ratio of the magnitude of each force to the sine of the angle between the other two forces is constant.

If a rigid body is in equilibrium under the action of three non-parallel forces, then those forces are concurrent. Any two of the forces will have their lines of action meeting at a point. Their resultant will also pass through this point. If the system is in equilibrium then the third force must be of equal magnitude, opposite direction, and act along the same line of action as the resultant of the first two. Hence the three forces must be concurrent. (Parallel forces will be discussed in the next chapter.)

Worked examples

Example 1 i) Forces **AB**, **BC**, **CD**, **ED**, **EF**, **AF**, of magnitude 2, 4, 6, $2p$, $2q$, 18, newtons respectively act along the sides of a regular hexagon $ABCDEF$. If the system is in equilibrium find the values of p and q.

ii) The ends A and E of a light inextensible string are attached to two fixed points on the same horizontal level. $AB = DE$ and $BC = CD$. Particles of weight W, $3W$, W are attached to B, C, D, respectively and hang in equilibrium. AB and BC make acute angles α and β respectively with the horizontal. Find, in terms of α, β and W the tensions in the parts of the string AB and BC and find α in terms of β.

(i)

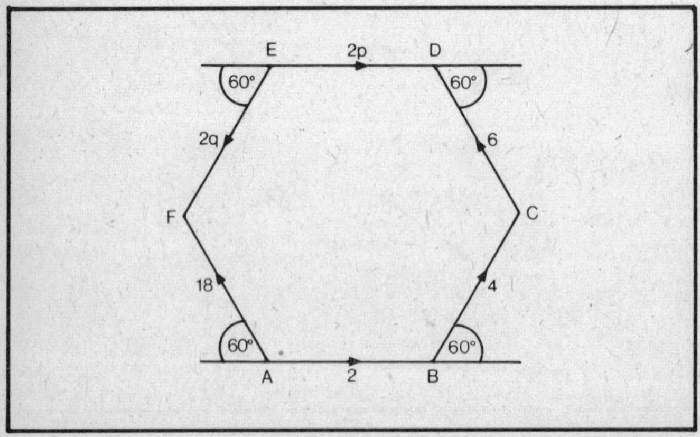

Figure 130.

If the forces are in equilibrium, then their resultant force must be zero, so that the sum of the horizontal components, X, must be zero, and the sum of the vertical components, Y, must be zero.

Resolving in the horizontal direction (see figure 130),

$X = 2 + 4\cos 60° - 6\cos 60° + 2p - 2q\cos 60° - 18\cos 60° = 0,$

therefore $q = 2p - 8$.

Resolving in the vertical direction,

$$Y = 4\sin 60° + 6\sin 60° - 2q\sin 60° + 18\sin 60° = 0,$$

therefore $\boldsymbol{q = 14}$ **and** $\boldsymbol{p = 11}$.

(ii)

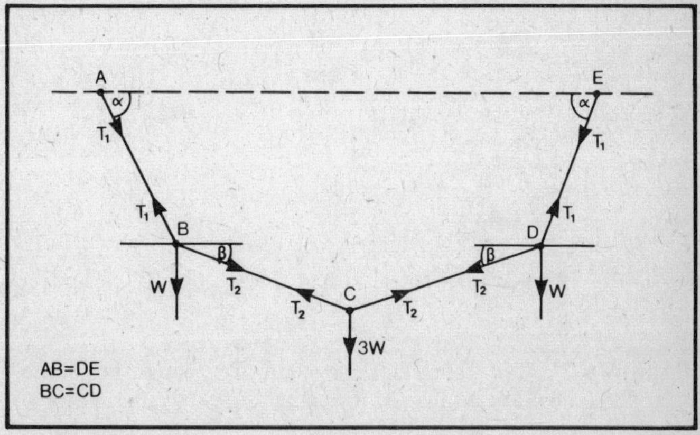

Figure 131.

Refer to figure 131.

We are told that A and E are on the same horizontal level, $AB = DE$ and $BC = CD$. Also we are told that weight W is attached to B and D and $3W$ to C. Clearly, we have a system which is symmetrical about the vertical through C. Hence the angles and tensions are as shown in the diagram. (If we had not found symmetry we should have to assume four different tensions for each part of the string and four different angles that the strings would make with the horizontal. A check for symmetry at the beginning of a problem can save a lot of work.)

If the system is in equilibrium, then each part of the system is also in equilibrium (or that part would move). Hence we can analyse the forces acting on the particles at B, C, D independently to find the conditions for equilibrium there. We shall actually look only at the particles at B and C since, by symmetry, we shall gain no new information from particle D.

Resolving the forces at B horizontally, $T_1 \cos\alpha - T_2 \cos\beta = 0$.

And vertically, $T_1 \sin\alpha - W - T_2 \sin\beta = 0$.

Resolving the forces at C horizontally, $-T_2 \cos\beta + T_2 \cos\beta = 0$ (which is what we would expect).

And vertically, $2T_2 \sin\beta = 3W$, $\boldsymbol{T_2 = \dfrac{3W}{2\sin\beta}}$.

Substituting this value into $T_1 \sin\alpha - W - T_2 \sin\beta = 0$,

$$\boldsymbol{T_1 = \frac{5W}{2\sin\alpha}}.$$

Substituting these values for T_1 and T_2 into $T_1 \cos\alpha - T_2 \cos\beta = 0$,

$$\frac{5W}{2\sin\alpha}\cos\alpha - \frac{3W}{2\sin\beta}\cos\beta = 0.$$

Therefore $\boldsymbol{\tan\alpha = \frac{5}{3}\tan\beta}$.

Example 2 Two light rings can slide on a rough horizontal rod. The rings are connected by a light inextensible string of length l. A weight W is attached to the midpoint of the string. Find the greatest distance between the rings possible for equilibrium. μ is the coefficient of friction between each ring and the rod.

Again symmetry helps us with this problem. We are told that the weight is attached to the midpoint of the string and that the coefficient of friction is the same for both rings and the rod. Hence there is symmetry about the vertical through W, for this system, and the forces and angles will be as shown in figure 132.

When the distance between the rings is greatest, the frictional force will be limiting and acting in the direction to prevent the rings sliding towards each other.

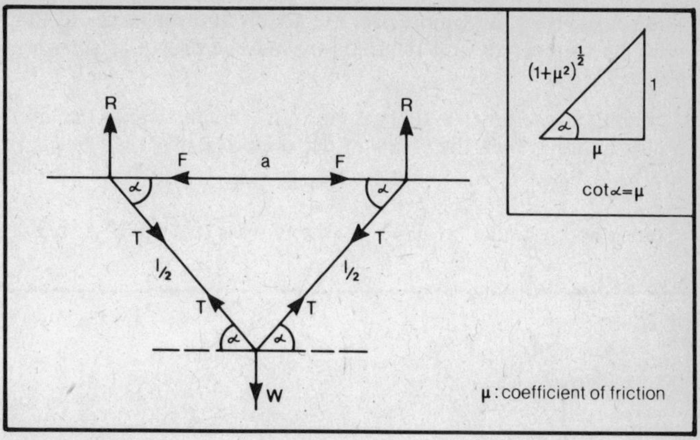

Figure 132.

Hence, resolving for either of the rings,

horizontally $F = T \cos \alpha$, and vertically $R = T \sin \alpha$.

Since friction is limiting, $F = \mu R$ thus $\cot \alpha = \mu$. We could also have deduced this result from the fact that the resultant reaction, i.e., F and R combined, will make an angle λ to the normal reaction where λ is the angle of friction and $\tan \lambda = \mu$, $(\lambda = 90° - \alpha)$.

The greatest distance, a, between the rings is

$$\frac{2l}{2} \cos \alpha \quad \text{and since} \quad \cot \alpha = \mu,$$

$$\cos \alpha = \frac{\mu}{(1 + \mu^2)^{1/2}} \quad \text{so that} \quad \boldsymbol{a} = \frac{\boldsymbol{l\mu}}{(\mathbf{1} + \boldsymbol{\mu}^2)^{1/2}}$$

Example 3 One end of a light string is fixed and the other end is attached to the end B of a uniform rod AB of weight W. The rod is at rest, at an angle α to the horizontal, under the action of a horizontal force P applied to the end A. Calculate the magnitude of P and show that θ, the angle the string makes with the horizontal is given by $\tan \theta = 2 \tan \alpha$.
Calculate the magnitude of the least force through A which will maintain the rod in equilibrium, inclined at an angle α to the horizontal (the least force is at right angles to the rod.) If, in this case, the string is inclined at an angle ϕ to the horizontal, show that $\tan \phi = 2 \tan \alpha + \cot \alpha$.

(A.E.B.)

The rod is in equilibrium under the action of three forces so they must be concurrent and the diagram may be drawn as shown in figure 133(i).

Applying trigonometry to triangle ABC, in particular the mid-point theorem, then the ratios of the sides are as shown in figure 133(ii). Thus

$$\tan\alpha = \frac{c}{b}, \quad \text{and} \quad \tan\theta = \frac{2c}{b} \quad \text{therefore} \quad \mathbf{\tan\theta = 2\tan\alpha.}$$

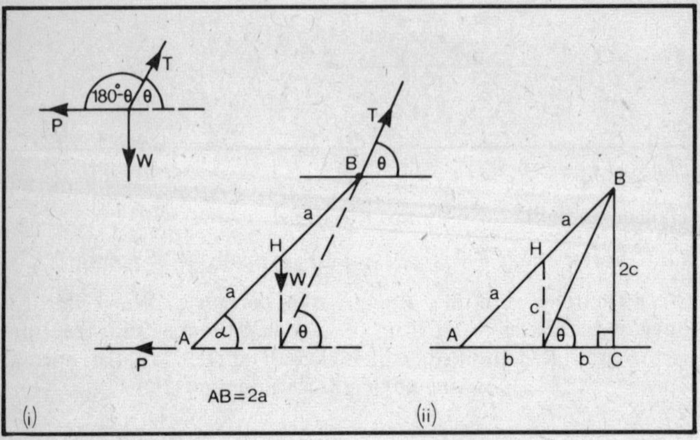

Figure 133.

Applying Lami's theorem to the forces,

$$\frac{P}{\sin(90° + \theta)} = \frac{W}{\sin(180° - \theta)} = \frac{T}{\sin 90°},$$

Therefore, $P = W\dfrac{\cos\theta}{\sin\theta} = W\cot\theta$. Thus, $\mathbf{P = \dfrac{W}{2}\cot\alpha.}$

Again, when we are considering P at right angles to the rod, we have a three force problem, therefore the forces are concurrent. Applying Lami's Theorem (see figure 134),

$$\frac{P}{\sin(90° + \phi)} = \frac{T}{\sin(180° - \alpha)} = \frac{W}{\sin(90° + \alpha - \phi)},$$

Therefore, $\dfrac{P}{\cos\phi} = \dfrac{T}{\sin\alpha} = \dfrac{W}{\cos(\alpha - \phi)}$.

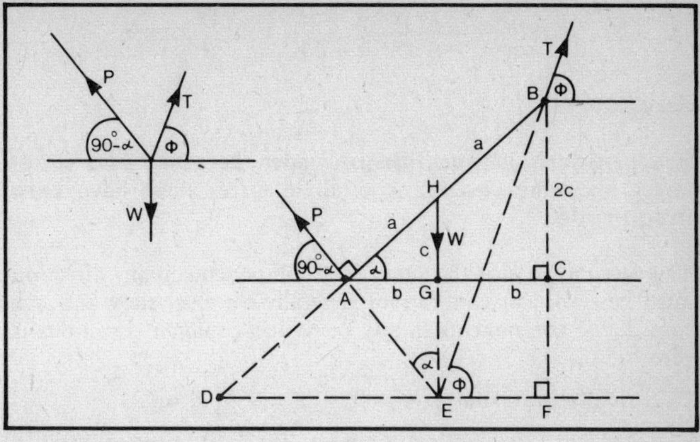

Figure 134.

Again applying trigonometry to the system, in particular the mid-point theorem, $AG = GC = EF = b$ and $BC = 2GH = 2c$.

Since $\tan\alpha = \dfrac{HG}{GA} = \dfrac{c}{b}$, and also

$$\tan\alpha = \frac{AG}{EG} = \frac{b}{EG}, \quad \text{then} \quad EG = \frac{b^2}{c}.$$

$$\text{Since } \tan\phi = \frac{BC + CF}{EF} \quad \text{and} \quad CF = GE = \frac{b^2}{c},$$

$$\tan\phi = \frac{2c + \dfrac{b^2}{c}}{b} = 2\frac{c}{b} + \frac{b}{c}.$$

However, $\dfrac{c}{b} = \tan\alpha$ and $\dfrac{b}{c} = \cot\alpha$,

therefore, $\mathbf{\tan\phi = 2\tan\alpha + \cot\alpha}$.

We can now use this result to find the magnitude of P:

$$\boldsymbol{P} = \frac{W\cos\phi}{\cos(\alpha - \phi)} = \frac{W}{\cos\alpha + \tan\phi\sin\alpha}$$

$$= \frac{W}{\cos\alpha + \sin\alpha\,(2\tan\alpha + \cot\alpha)}$$

$$= \frac{W \cos \alpha}{\cos^2 \alpha + 2 \sin^2 \alpha + \cos^2 \alpha} = \frac{W}{2} \cos \alpha.$$

Key terms

If a particle is in **equilibrium** under the action of a set of forces, then the **resultant** of these forces must have **zero magnitude**.

This fact means that the sum of the components in any direction must be zero. For equilibrium generally we must show that the sum of the components in any three non-coplanar directions is zero.

(We usually show that $\sum X = 0, \sum Y = 0, \sum Z = 0.$

For coplanar forces acting at a point, for equilibrium we need to show that $\sum X = 0, \sum Y = 0.$)

Any system of forces in **equilibrium** may be represented by a **closed vector polygon**.

Lam's theorem states that if a body is in equilibrium under the action of three concurrent forces, then the ratio of the magnitude of each force to the sine of the angle between the other two, is constant.

If a rigid body is in equilibrium under the action of **three non-parallel forces**, then those forces are **concurrent**.

The **triangle of forces** property is that three forces which are in equilibrium may be represented in magnitude and direction by the sides of a triangle taken in order.

Chapter 14
Non-Concurrent Forces: Turning Effect; Parallel Forces; Couples

Resultants of Coplanar Forces: Equilibrium and Equivalent Systems

We have seen that if **three non-parallel forces** act on a rigid body and that body is in equilibrium, then those forces must be **concurrent**. We have also seen that if the sum of the components of a set of forces acting on a particle in three non coplanar directions is zero, then that system is in equilibrium. We observed that there was no possibility of a turning effect on a particle.

Clearly we must now investigate turning effects on rigid bodies and the effect of parallel forces on a rigid body.

Turning effect

Consider a lamina which is free to rotate in its own plane about an axis through a fixed point in its plane, perpendicular to the plane. In general, a force applied in the plane of the lamina will turn the lamina about that axis. The turning effect of the forces of the same magnitude and direction will vary according to the distance of their lines of action from that axis. Experience tells us that the further away from the axis of rotation we apply a force, the greater the turning effect.

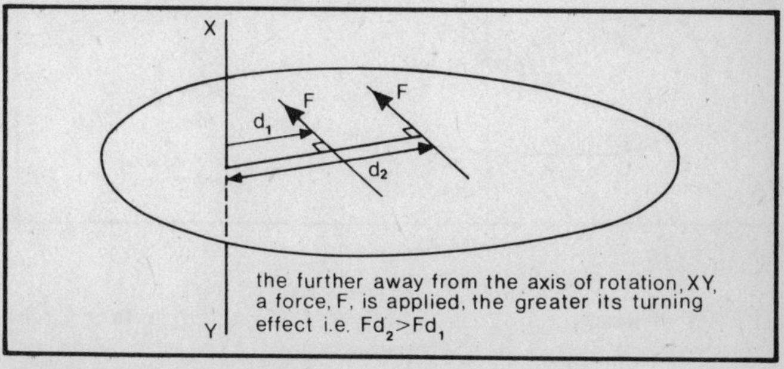

X
F
F
d1
d2
Y
the further away from the axis of rotation, XY, a force, F, is applied, the greater its turning effect i.e. Fd2>Fd1

Figure 135.

The turning effect *C* of a force *F* about a particular axis is measured by multiplying the magnitude of the force with the perpendicular distance of the line of action of the force from the axis, i.e., $C = Fd$.

C is called the **moment of *F* about that axis** or the **torque about the axis.**

Torque is measured in newton metres since it is a product of force and distance. The direction of torque is that of the rotation it would produce. Usually **anticlockwise moments** are taken as **positive**, and **clockwise moments** are taken as **negative**. The opposite convention is perfectly acceptable but unusual. It is wise, when solving problems, to specify the positive sense of torque.

We often refer to the moment of a force about a **point**, where what is actually meant, is the moment of the force about an axis through that point perpendicular to the plane containing the point and the force.

The **resultant torque** of a set of forces about a given axis is the algebraic sum of their individual torques. In figure 136, the resultant turning effect of forces F_1, F_2, F_3, F_4 is given by $d_1F_1 - d_2F_2 + d_3F_3 - d_4F_4$ and the resulting sign of the answer will indicate whether the resultant turning effect is anti-clockwise or clockwise.

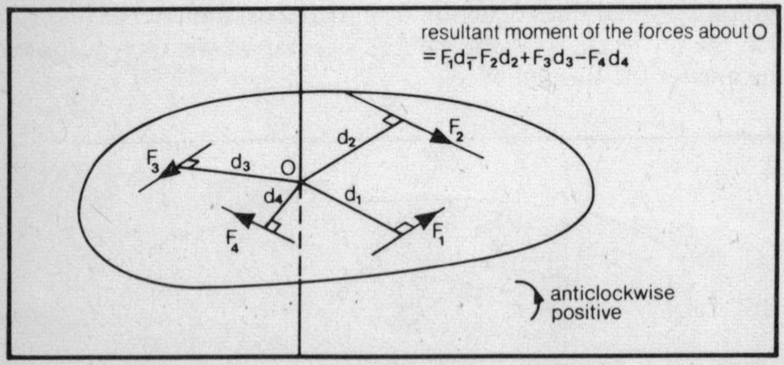

Figure 136.

If a set of forces acting on a body has a single resultant force (this may not always be the case, as we shall see later when we consider couples) then we would expect the moment of that resultant force to be the same as the resultant moment of that set of forces. This is indeed the case as stated in the **Principle of Moments**. The **Principle of Moments states that the algebraic sum of the moments of a set of forces about any axis is equal to the moment, about the same axis, of the resultant force.**

We may **represent the moment of a force graphically**. Consider a force which is represented in magnitude, direction and line of action by the line segment AB. The **moment of that force about O, Fd,** is represented by $ABd =$ **2 (Area of triangle OAB)** (see figure 137).

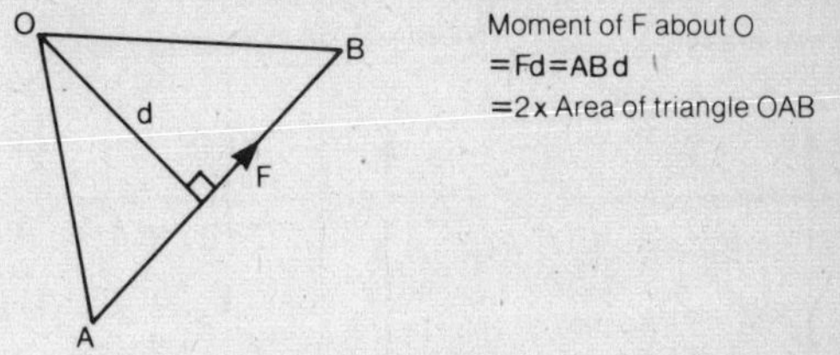

Figure 137.

If a non zero force has zero moment, then its line of action must pass through the point about which moments are being taken.

Parallel forces

If we are to replace two parallel forces P, Q by a single resultant force R, then clearly it will be parallel to these forces and of magnitude equal to the algebraic sum of the forces. From the Principle of Moments, the moment of R about any point must equal the resultant moment of those forces about the same point.

Like parallel forces P, Q, acting through points A, B, such that AB is perpendicular to their lines of action, have resultant force R of magnitude $P + Q$, parallel to P and Q. If R acts through C on AB, then, taking moments about A, $(AC)R = (AB)Q$ and $AC : CB = Q : P$.

Hence **like parallel forces** have a resultant force with magnitude equal to the sum of the magnitudes of the forces, which acts parallel to them, cutting any line from one line of action to the other (from similar triangles, see figure 138) in the inverse ratio of the forces.

Unlike parallel forces P, Q, acting through the same points A, B, have resultant $R = (P - Q)$, parallel to P and Q. Taking moments about A, $(AC)R = -(AB)Q$, so that $\boldsymbol{AC : CB = -Q : P}$.

Hence **unlike parallel forces** have a resultant equal in magnitude to the difference of their magnitudes, which acts parallel to them, in the sense of the larger magnitude force, cutting any line from one line of action to the other, externally in the inverse ratio of the magnitudes of the forces (see figure 138(ii)).

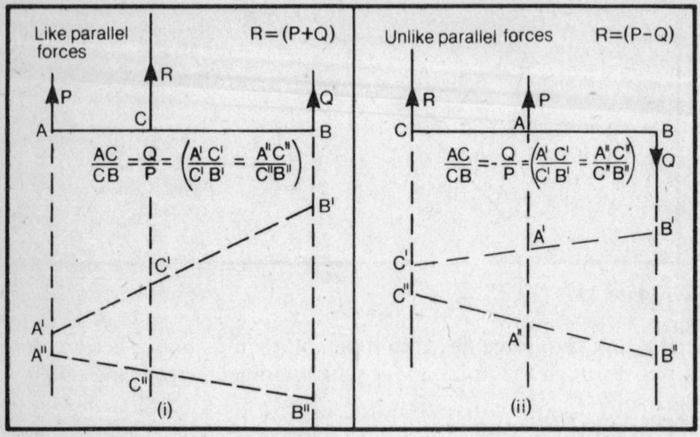

Figure 138.

Couples

A special case arises when we try to find the **resultant of two equal unlike parallel forces**. We see that the resultant force would be of magnitude zero. However, clearly there is a turning effect of dP about A, refering to figure 139. We must conclude that, in fact, we cannot replace two equal unlike parallel forces with a single force, since they represent a **pure turning effect**. They cause no change in the linear motion of a body since their linear resultant is zero. We call such a pair of forces a **couple** and we shall now investigate its turning effect.

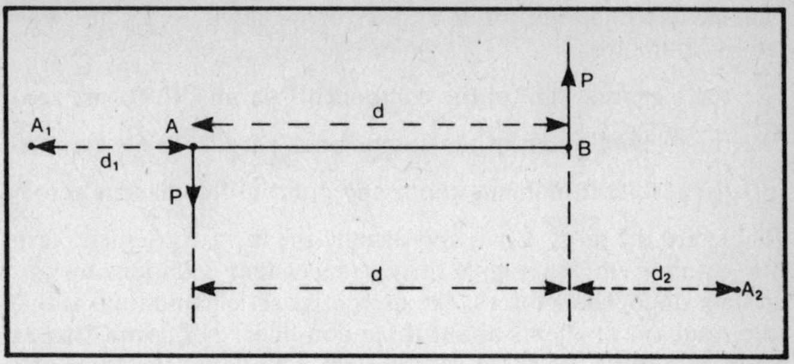

Figure 139.

Taking moments about A gives the anticlockwise torque as dP. Taking moments about B gives the anticlockwise torque as dP. Taking moments about A_1 gives the anticlockwise torque as $-d_1 P + (d_1 + d)P = dP$. Taking moments about A_2 gives the anticlockwise torque as $(d + d_2)P - d_2 P = dP$.

Hence we deduce a very important and useful result, known as the **constant moment property of a couple**, which is that the moment of a couple about any axis perpendicular to its plane, is constant for that couple and equal to the product of the magnitude of one of the forces with the perpendicular distance between the forces. Hence, the **characteristics of a couple are that its linear resultant is zero and it has a constant moment about any point in its plane.**

Resultants of coplanar forces

We shall now apply these results to systems of coplanar forces. We shall first examine the conditions for a system of coplanar forces to be in equilibrium.

Coplanar forces in equilibrium

If a system of coplanar forces is in equilibrium, there must be no resultant force acting, otherwise the system would have an acceleration. We must also be sure that the system has no resultant turning effect, i.e., it does not reduce to a couple, which would satisfy the condition that there was no resultant force acting, but would not leave a system in equilibrium. To check for this we need only take moments about one point in the plane because of the constant moment property of couples. Hence sufficient conditions for a system of coplanar forces to be in equilibrium are:

i) the algebraic sum of the components parallel to Ox are zero,

ii) the algebraic sum of the components parallel to Oy are zero,

iii) the resultant moment about any point in the plane is zero.

These are the most usual, and usually the most convenient, tests for equilibrium. They give three independent equations for any system of coplanar forces. An alternative set of conditions is that the resultant moments about three non-linear points are zero. In each case there are three independent equations, and to resolve further, or to take moments about other points, will give no new information. This is because coplanar forces may produce motion in two dimensions and/or rotation in that plane, i.e., the system has only three degrees of freedom and so only three independent equations may be formed.

Equivalent systems of coplanar forces

When we reduce a system of coplanar forces to an **equivalent** system, we must be clear about the meaning of an **equivalent** system. A system of forces is measured by the effect it produces.

A set of coplanar forces may produce motion in the two dimensions of the plane and/or it may produce rotation in that plane. Hence to replace a system of coplanar forces with an equivalent system we must check that it produces exactly the same linear motion and turning effect. We may do this by ensuring that the components of force in both the Ox and the Oy directions are the same in both systems, and that the turning effect about one point in the plane is the same in both systems. We could also show equivalence if the resultant moment about three non collinear points is the same for both systems.

The **resultant** of a system of forces is the simplest equivalent system we can find.

The **resultant** of a system of **coplanar** forces will be either a **single resultant** force, or a **couple**. To reduce a system of forces to a **resultant force** we need to know its magnitude, direction and line of action. The magnitude and direction of the resultant force may easily be found by finding the sum of the components of the forces in each of two perpendicular directions, usually Ox and Oy. The line of action of the resultant force is found by comparing its moment with the resultant moment of the set of forces about a point, usually the origin.

To reduce the system to a **couple**, we need to check that the linear resultant of the system is zero, and that the moment of the couple is not zero. To do this we need only take moments about one point in the original system, because of the **constant moment** property of couples.

Any system of coplanar forces may be replaced with an equivalent system consisting of a force passing through a particular point together with a couple. If the system reduces to a single force, F, passing through A, and we replace this system with an equivalent system of a force, passing through a particular point B, and a couple, then since the linear resultant of a couple is zero, clearly the new force must be of the same magnitude and direction as the original. Hence, we must find the appropriate couple which would have the effect of reducing this new force to the original force. If this couple is of moment Fd, where d is the perpendicular distance between the lines of action of the two forces, and in the appropriate sense, as shown in figure 140, then one 'arm' of the couple applied to the new force will cancel it out and leave the other 'arm' as the original system.

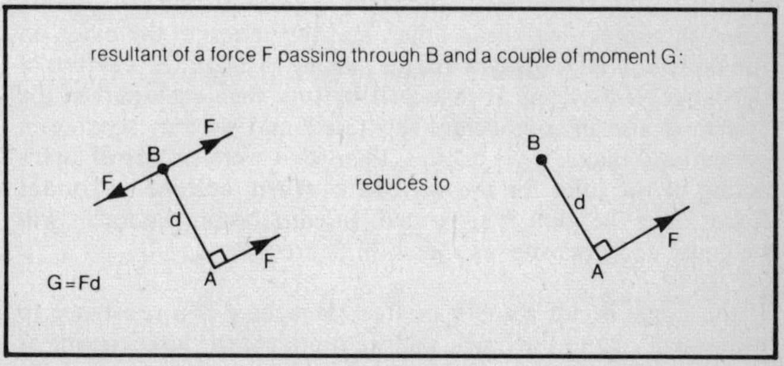

Figure 140.

If the system actually reduces to a couple, then trivially, it may be reduced to a force of zero magnitude, passing through the required point, and a couple of the same moment as the original.

We have illustrated a **general property of couples** above, namely that **when a couple is applied to a force, the effect is to shift that force parallel to itself**.

Hinged bodies

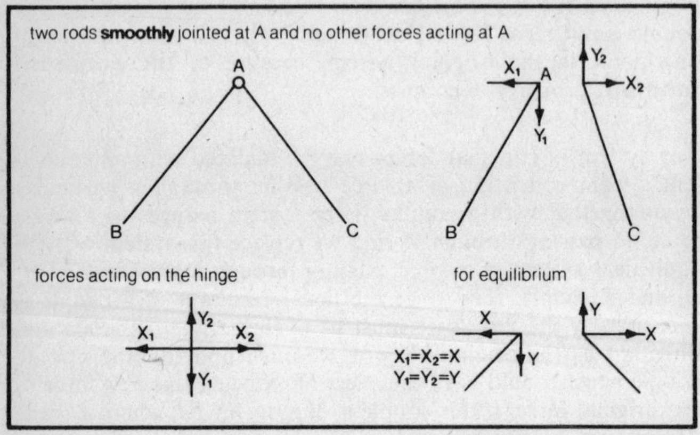

Figure 141.

We shall be dealing with problems involving one body hinged to another. If these bodies are freely or smoothly hinged or jointed, then the only reactions at the joint are due to the two bodies being in contact with each other, and the action of the hinge on the body will be a **single force** passing through the **centre** of the hinge. If a system is in **equilibrium**, then each part of the system is also in equilibrium separately and we may separate a system at a hinge. If we do this, then what were **internal** forces acting at the joint for the **whole system**, become **external** forces once the joint is separated. In equilibrium the forces will be equal and opposite, as shown in figure 141.

If the hinge is not smooth or free, then there is a resistance to movement due to the hinge, and we represent the forces acting at the hinge by a **single force** acting through the centre of the hinge together with a couple. Again, if no other forces are acting on the hinge, then we may separate the system at the hinge, and the action of the hinge on the bodies may be represented by equal and opposite forces and equal and opposite couples as shown. When other forces act on the hinge, the actions of the hinge on the two bodies connected by it are not necessarily equal and opposite and the necessary conditions for equilibrium of the hinge are required, in order to complete the analysis of the total equilibria (see Example 11).

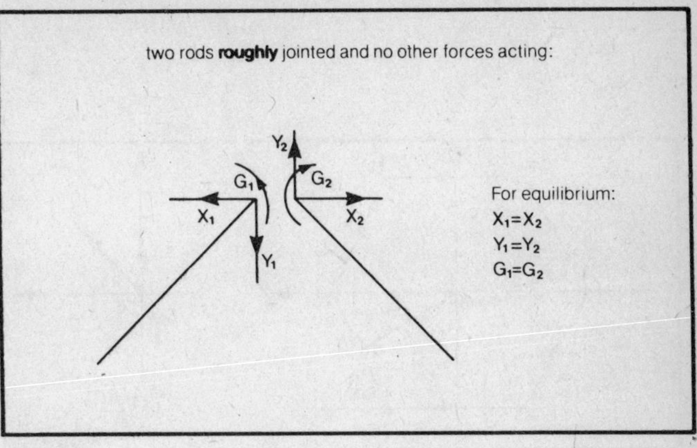

Figure 142.

Worked examples

Example 1 *ABCDEF* if a regular hexagon of side 1 m. Forces of magnitude 4, 6, 10, 8, *P* and 2 newtons act along *AB*, *BC*, *CD*, *ED*, *FE* and *AF* respectively, directions being indicated by the order of the letters.

a) If the resultant passes through the centre of the hexagon, find the value of *P* and the magnitude and direction of the resultant.

b) The force in *FE* is replaced by another so that the new system reduces to a couple. Find the magnitude of the force and the moment of the couple indicating its sense.

(S.U.J.B.)

a) We are told that the system reduces to a resultant force passing through *O*. Let this resultant force be *R* with horizontal component *X* and vertical component *Y*. We know that the sums of the horizontal and vertical components must be the same in both systems. Hence,

$$\boldsymbol{X} = 4 + 6\cos 60° - 10\cos 60° + 8 + P\cos 60° - 2\cos 60°$$

$$= \mathbf{9} + \frac{\boldsymbol{P}}{\mathbf{2}}, \text{ and}$$

$$\boldsymbol{Y} = 6\sin 60° + 10\sin 60° + P\sin 60° + 2\sin 60°$$

$$= \frac{\sqrt{\mathbf{3}}}{\mathbf{2}}(\mathbf{18} + \boldsymbol{P}).$$

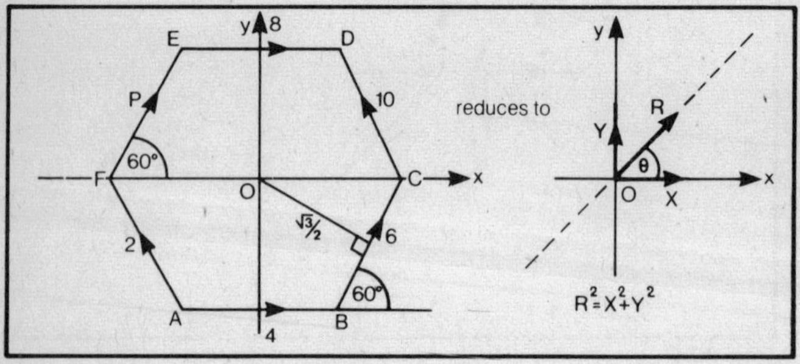

Figure 143.

We know that, since the resultant passes through O, the resultant moment about O in both systems should be zero.

Hence, taking moments about O, with anticlockwise as positive,

$$0 = \frac{\sqrt{3}}{2}(4 + 6 + 10 - 8 - P - 2), \text{ so that } \boldsymbol{P} = \mathbf{10}.$$

Using this value of P gives $\boldsymbol{X} = \mathbf{14}$ and $\boldsymbol{Y} = \mathbf{14\sqrt{3}}$.

The magnitude of R is given by $R = \sqrt{X^2 + Y^2}$.

Therefore $\boldsymbol{R} =$ **28 newtons in a direction**

$$\arctan\frac{Y}{X} = \arctan\sqrt{3} = \mathbf{60°\ to\ the\ horizontal}.$$

b) The force in FE is replaced by another, say, Q, in the direction FE, and the system now reduces to a couple. Let the moment of this couple be G. We know that the characteristics of a couple are that $\sum X = 0$, $\sum Y = 0$ and the moment about all points is the same. Hence,

$$\sum X = 0 = 4 + 6\cos 60° - 10\cos 60° + 8 + Q\cos 60° - 2\cos 60°$$

$$= \mathbf{9 + \frac{Q}{2}} \text{ and}$$

$$\sum Y = 0 = \frac{\sqrt{3}}{2}(18 + Q).$$

Both of these give the magnitude of Q as 18 newtons acting in the direction EF. The moment of the couple will be given by the resultant moment about any point in the system, but we shall choose O as the point. Taking moments about O, with anticlockwise as positive,

$$G = \frac{\sqrt{3}}{2}(4 + 6 + 10 + 18 - 2 - 8) = \mathbf{14\sqrt{3}\ Nm}.$$

Example 2 A rectangular lamina $ABCD$ with $AB : BC = 4 : 3$, has forces in the ratio 1 : 2 : 3 : 10 acting along AD, DC, CB, CA, respectively. Find the ratio of the resultant force to the force acting along CA.

Taking the x-axis along AB and the y-axis along AD, find the gradient of the line of action of the resultant force. Find also the point where the line of action of the resultant cuts the x-axis.

Find the magnitude and sense of the couple which, when added to the system, reduces the system to a resultant force of the same magnitude and direction as originally, but with its line of action now passing through the mid-point of CD.

When the couple is not applied, find forces acting along AB, BC and AD which reduce the system to equilibrium.

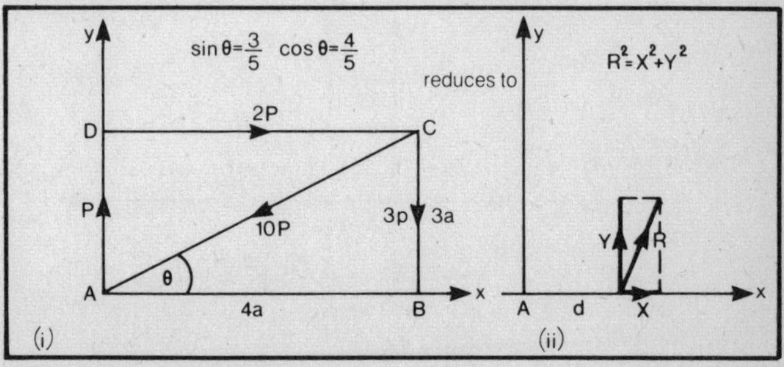

Figure 144.

Refer to figure 144 and let the given forces be P, $2P$, $3P$, $10P$, respectively, and $AB = DC = 4a$, and $AD = BC = 3a$, ('P' and 'a' are constants of proportionality).

The resultant, R, of the system will have horizontal component,

$$X = 2P - 10P\cos\theta = -6P$$

and vertical component,

$$Y = P - 3P - 10P\sin\theta = -8P.$$

Hence, the **magnitude of the resultant force is**

$$\mathbf{R} = \sqrt{X^2 + Y^2} = P\sqrt{36 + 64} = \mathbf{10P}.$$

Therefore $\qquad R : 10P = \mathbf{1 : 1}.$

The gradient of the line of action of the resultant is given by $Y/X = \mathbf{4/3}$.

If R replaces the system of forces in every way, then its moment about any point in the plane will be the same as the resultant moment about that point for the set of forces. Hence, if we let R cut the x-axis at (d,0) then, taking moments about A, $dY = -12aP - 6aP$, and $\mathbf{d} = \mathbf{9a/4}$.

(In this case, when taking the moment about A of R, it is easier to resolve R into its components X and Y, since we already know them and the line of action of X passes through O and therefore has zero moment about that point.)

Hence the resultant R is as shown in figure 145.

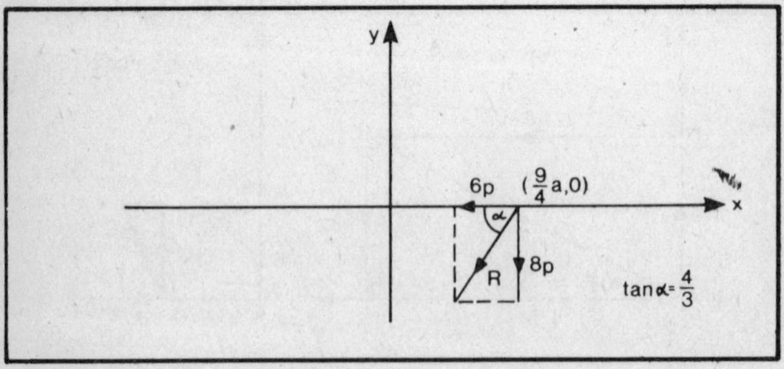

Figure 145.

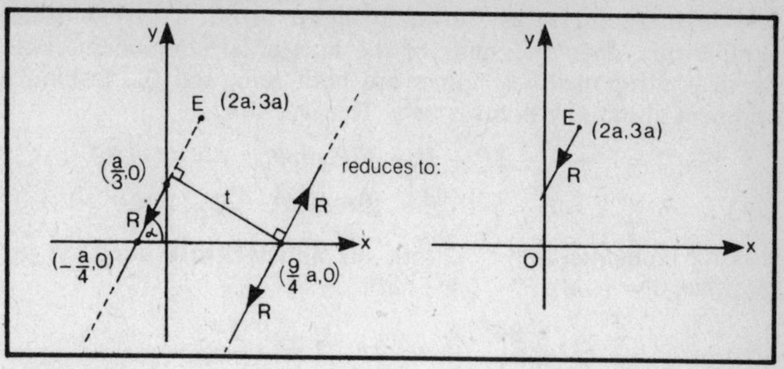

Figure 146.

In order to shift R parallel to itself so that its line of action passes through E, the mid-point of CD, we must apply a couple of anticlockwise moment Rt where t is the perpendicular distance between the two lines of action, as shown in figure 146. We must calculate t.

We know that the gradient of both lines of action is $\frac{4}{3}$, that the original one passes through $(\frac{9}{4}a, 0)$ and that the new one passes through $(2a, 3a)$. Hence their equations are respectively

$$y = \frac{4}{3}x - 3a \quad \text{and} \quad y = \frac{4}{3}x + \frac{a}{3}.$$

From figure 146, we see that $t = \dfrac{10}{4} a \sin \alpha = 2a$.

Hence, the **couple is of magnitude $20aP$ acting in the anticlockwise sense.**

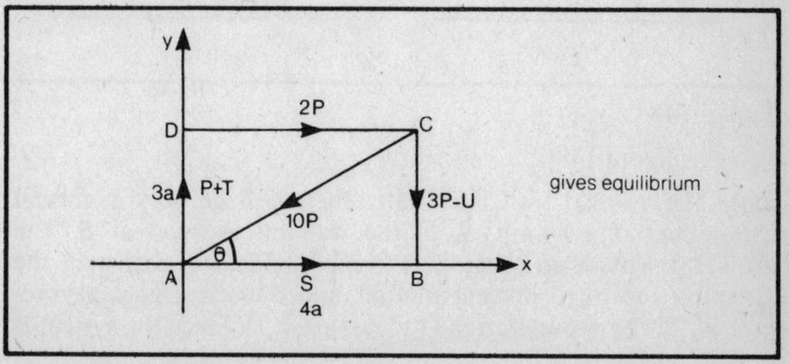

Figure 147.

We take the forces as shown in figure 147. If a system is in equilibrium, then the sums of the horizontal components and vertical components of forces are both zero, and the resultant moment about any point is zero. Hence,

$$\sum Y = 0 = P + T - 3P + U - 10P \sin \theta, \quad T + U = 8P$$

and $\quad \sum X = 0 = S + 2P - 10P \cos \theta, \quad S = -.6P.$

Taking moments about A, with the anticlockwise direction as positive, $0 = -4a(3P - U) - 6aP$,

$$U = \frac{9P}{2} \quad \text{therefore} \quad T = -\frac{7P}{2}.$$

Example 3 A uniform rod AB, of weight W and length $2a$, rests in equilibrium with the end A on rough horizontal ground and the end B in contact with a smooth vertical wall, which is perpendicular to the plane containing the rod. If AB makes an angle arctan $(\frac{4}{3})$ with the horizontal, find the least value of the coefficient of friction between the rod and the ground for equilibrium to be preserved. If the coefficient of friction is $\frac{1}{2}$, find the greatest distance along the rod that a weight W may be attached and equilibrium preserved.

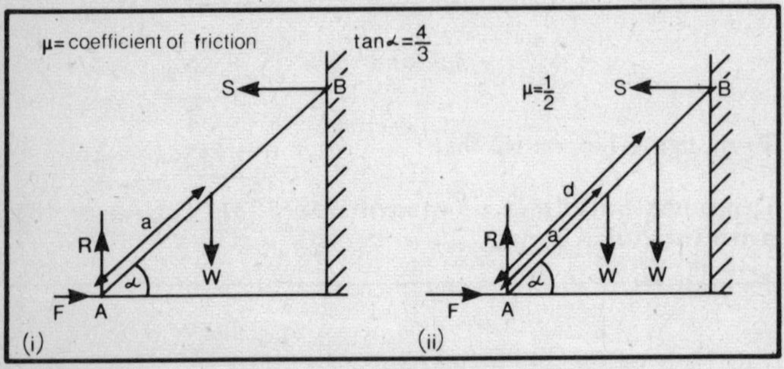

Figure 148.

Refer to figure 148(i).

Since the vertical wall is smooth, there will be only a normal component of reaction, S, of the wall on the rod at B. The ground is rough so there is a frictional force F acting in the direction shown to prevent motion, and there is a normal reaction, R, of the ground on the rod, acting at A. Since the system is in equilibrium:

$$\sum X = 0 = F - S, F = S; \quad \text{and} \quad \sum Y = 0 = R - W, R = W.$$

Taking moments about A, since the resultant moment must be zero, then $0 = 2a \sin \alpha\, S - a \cos \alpha\, W$, $S = \frac{1}{2} \cot \alpha\, W$.

We know that $S = F$, thus $F = \frac{1}{2} \cot \alpha\, W$.

The frictional force F must have this magnitude for equilibrium to be possible. The least possible value for the coefficient of friction for equilibrium to be possible will be when this frictional force is the limiting value, i.e., $F = \mu R$. Therefore $\mu = \frac{1}{2} \cot \alpha = \frac{3}{8}$. **Hence $\frac{3}{8}$ is the least possible value of μ for equilibrium to be possible.**

If $\mu = \frac{1}{2}$ and an extra weight W is added to the rod at the greatest distance d up the rod possible for equilibrium to be preserved, then the system is in **limiting equilibrium**. Thus $\boldsymbol{F = \mu R = \frac{1}{2}R}$. (See figure 148(ii).)

Examining the conditions for equilibrium, we see that

$$\sum X = 0 = F - S, \quad \boldsymbol{S = \tfrac{1}{2}R}$$

and

$$\sum Y = 0 = R - 2W, \quad \boldsymbol{R = 2W}.$$

Hence, $\boldsymbol{S = W}$.

Taking moments about A, for equilibrium,

$$0 = a \cos \alpha\, W + d \cos \alpha\, W - 2a \sin \alpha\, S.$$

Therefore $\boldsymbol{d} = 2a \tan \alpha - a = \boldsymbol{5a/3}$.

Example 4 Figure 149 shows a uniform rod AB of length $6a$ and weight W, inclined at 45° to the vertical. The rod rests in a vertical plane perpendicular to a rough vertical wall, with its foot A against the wall and supported on a smooth peg C.

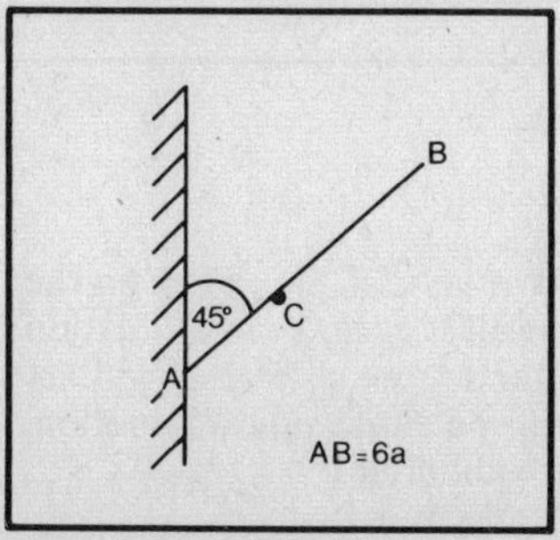

Figure 149.

If $AC = 2a$ and the rod is on the point of slipping downwards, find the coefficient of friction between the rod and the wall.

Find the least vertical force, which would cause the rod to be on the point of sliding upwards, when applied to the end A of the rod.

If the rod is on the point of sliding downwards, the system will be in **limiting equilibrium**. The frictional force F will be acting upwards, to prevent motion. There will be a normal reaction R of the wall against the rod at A. At C, since the peg is smooth, it will exert only the normal reaction S. The weight W will act from the centre of gravity of the rod, distance $3a$ from A. (See figure 150(i).)

Applying the conditions for equilibrium:

horizontally, $R = S \cos 45° = \dfrac{S}{\sqrt{2}}$,

and vertically, $F + S \sin 45° = W$.

Since $F = \mu R$, $\mu = \dfrac{W}{R} - 1$.

Taking moments about A, for equilibrium, $2aS = \dfrac{3aW}{\sqrt{2}}$,

therefore $R = \dfrac{3W}{4}$, and $\mu = \dfrac{1}{3}$.

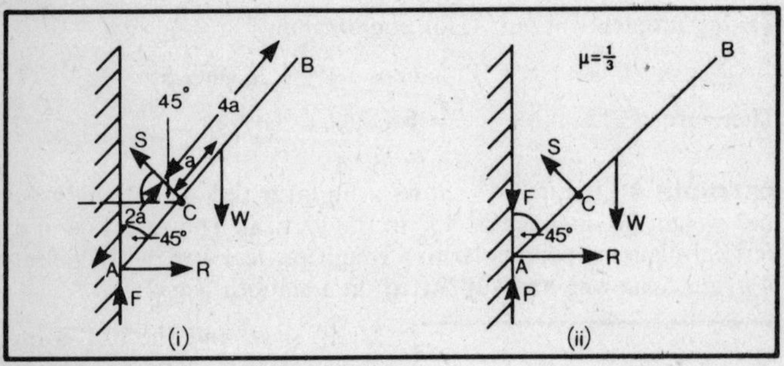

Figure 150.

Refer to figure 150(ii).

If a force P is applied vertically at A so that the rod is on the point of sliding upwards, the system must again be in a situation of limiting equilibrium. This time the frictional force F will act downwards in order to prevent the potential upward motion. Hence, applying the conditions for equilibrium:

horizontally, $R = \dfrac{S}{\sqrt{2}}$, and vertically, $P + \dfrac{S}{\sqrt{2}} = W + F$.

Taking moments about A, for equilibrium $2aS = \frac{3aW}{\sqrt{2}}$.

Rearranging these equations, and using the fact that

$$F = \mu R \quad \text{and} \quad \mu = \frac{1}{3}, \quad \mathbf{P} = \frac{\mathbf{W}}{\mathbf{2}}.$$

Example 5 Forces represented completely by λ**OA** and μ**OB** act at O. Show that if $\lambda + \mu \neq 0$, their resultant is $(\lambda + \mu)$**OC** where C is the point of AB such that $\lambda AC = \mu CB$. By considering the resultant of forces represented completely by **OA** and **BO**, deduce the resultant when $\lambda + \mu = 0$.

$ABCD$ is a quadrilateral, and forces represented completely by 6**BA**, 2**BC**, **DC** and 3**DA** act along its sides. Their resultant is represented completely by k**PQ**, P and Q being points on BD and AC respectively. Find k and the ratios into which P and Q divide BD and AC respectively.

If an extra force 3**AC** is introduced at A and the resultant intersects AP at R, find $AR : RP$.

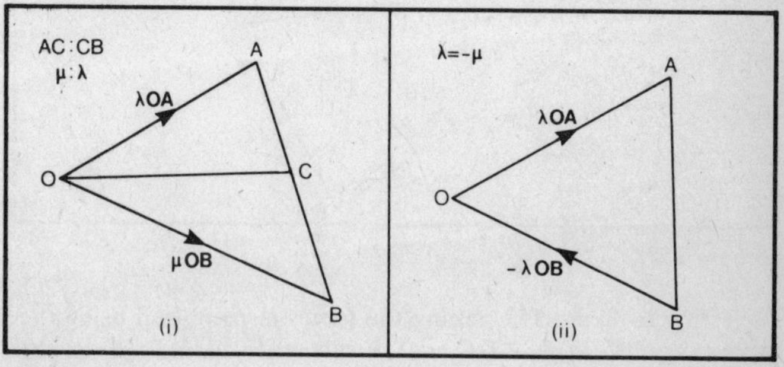

Figure 151.

Referring to figure 151(i)

$$\lambda\mathbf{OA} = \lambda\mathbf{OC} + \lambda\mathbf{CA}, \text{ and } \mu\mathbf{OB} = \mu\mathbf{OC} + \mu\mathbf{CB}.$$

Hence, $\lambda\mathbf{OA} + \mu\mathbf{OB} = (\lambda + \mu)\mathbf{OC} + \lambda\mathbf{CA} + \mu\mathbf{CB}$.

Since $\lambda\mathbf{AC} = \mu\mathbf{CB}$, then $\lambda\mathbf{CA} + \mu\mathbf{CB} = \mathbf{0}$.

Thus, $\lambda\mathbf{OA} + \mu\mathbf{OB} = (\lambda + \mu)\mathbf{OC}$.

Hence, the magnitude and direction of the resultant of forces λ**OA** and μ**OB** will be represented by $(\lambda + \mu)$**OC**.

Since the lines of action of λ**OA** and μ**OB** intersect at O, their resultant will also pass through O. Hence the **resultant of forces represented completely by λOA and μOB** is **represented completely by $(\lambda + \mu)$OC where C** is **on AB, such that λAC = μCB.**

When $(\lambda + \mu) = 0$ then λ**OA** + μ**OB** = λ**OA** − λ**OB** = λ**OA** + λ**BO**.

Hence, referring to figure 151(ii), from the triangle of vectors, the magnitude and direction of the resultant of λ**OA** and λ**BO** will be λ**BA** and its line of action must pass through O since the lines of action of λ**OA** and λ**BO** intersect at O.

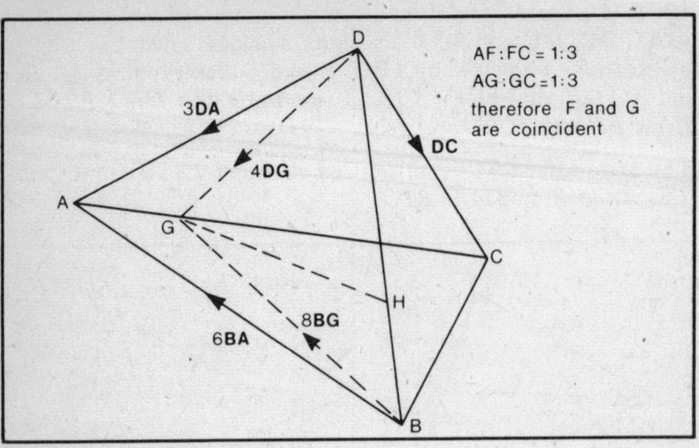

Figure 152.

Referring to figure 152, taking the forces in pairs, and using the above results, 3**DA** + **DC** = 4**DF** where F divides AC in the ratio 1 : 3, i.e., $3AF = FC$.

Similarly, 6**BA** + 2**BC** = 8**BG** where G divides AC in the ratio 2 : 6, i.e., $3AG = GC$.

Hence, F and G are coincident. We have reduced the system to forces 4**DG** and 8**BG**.

Also, 4**DG** + 8**BG** = 12**HG** where H divides DB in the ratio 8 : 4, i.e., $2BH = HD$. Therefore, the resultant force is represented completely by 12**HG**.

We are told that the resultant force is represented completely by k**PQ** where P and Q are points on BD and AC respectively.

Hence $\quad k = 12$, H is P and $\mathbf{BP} : \mathbf{PD} = 1 : 2$;
G is Q and $\mathbf{AQ} : \mathbf{QC} = 1 : 3$.

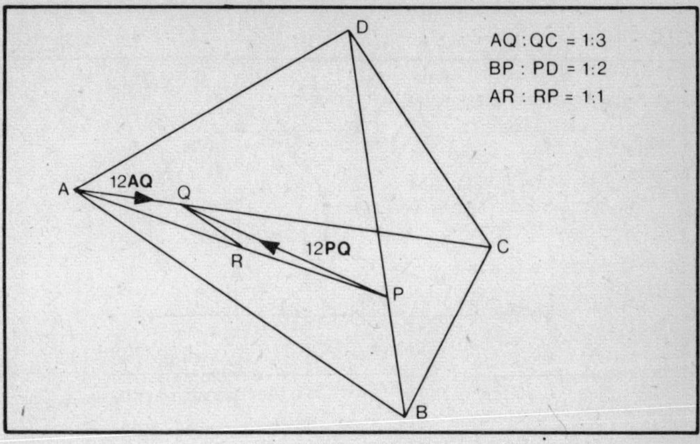

Figure 153.

Refer to figure 153.

An extra force represented completely by $3\mathbf{AC}$ is now introduced. Since $3\mathbf{AC} = 12\mathbf{AQ}$, then $3\mathbf{AC} + 12\mathbf{PQ} = 12\mathbf{AQ} + 12\mathbf{PQ} = 24\mathbf{RQ}$, where R divides AP in the ratio $12 : 12$, i.e., $\mathbf{AR} = \mathbf{RP}$.

Example 6 Figure 154 shows a uniform rod AB of length $2a$ and weight W. The rod is smoothly pivoted to the rough horizontal ground at A and smoothly pivoted at B to the centre of a light disc which is touching the ground at C.

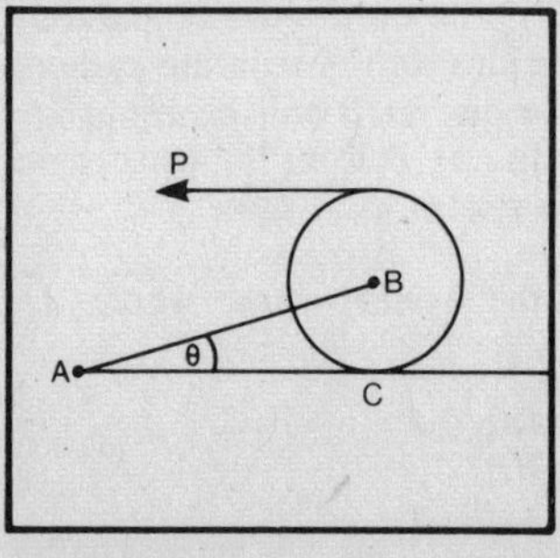

Figure 154.

The coefficient of friction between the disc and the ground is μ. The disc and the rod are free to rotate in the same vertical plane. The system is in equilibrium under the action of the horizontal force P, as shown, applied to the highest point of the disc.

Find the horizontal and vertical components of the reaction at C. Find also, the maximum value of P possible for equilibrium, in terms of μ, W and θ.

Find the magnitude of the reaction of the pivot A and find the angle that its line of action makes with the horizontal. Find the horizontal and vertical components of the reaction of the rod on the disc.

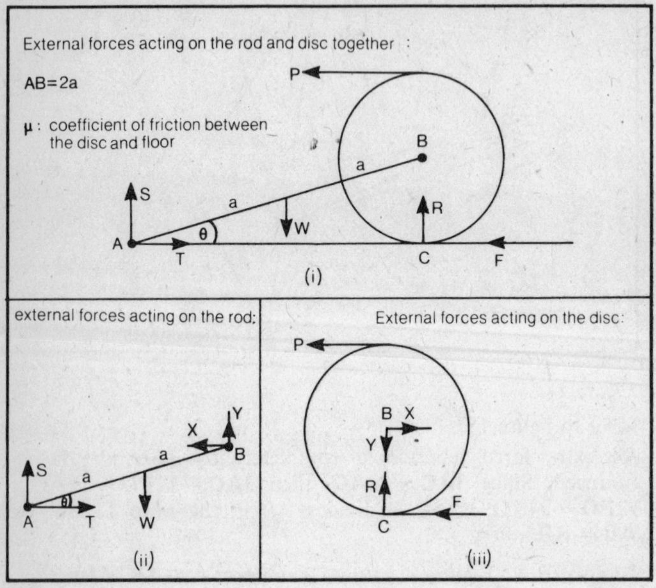

Figure 155.

While the system is in equilibrium, each part of the system is also in equilibrium. Hence, as shown above, we can consider the forces acting on the **rod** separately, and on the **disc** separately, and impose conditions for equilibrium on the separate systems. The reaction on the rod from the disc at B will be equal and opposite to the reaction on the disc at B from the rod. Hence they will be internal forces for the system as a whole.

Taking moments about A for the system as a whole, for equilibrium, $0 = -a \cos \theta\, W + 2a \cos \theta\, R + 4a \sin \theta\, P$.

Thus
$$\boldsymbol{R} = \frac{\boldsymbol{W} - 4 \tan \theta\, \boldsymbol{P}}{2}.$$

Resolving horizontally and vertically for the disc as a separate system, for equilibrium, $\boldsymbol{Y} = \boldsymbol{R}$ and $P + F = X$.

Taking moments about B in the same system, for equilibrium, $F = P$. Therefore, $\mathbf{X = 2P}$.

Hence the **horizontal component of reaction at C is P and the vertical component is**

$$\frac{W - 4\tan\theta\, P}{2}.$$

In equilibrium, $F = P$. Therefore equilibrium will be broken if P is increased to a magnitude that F cannot equal. Hence for equilibrium, P must be of magnitude less than or equal to the limiting value of friction, i.e., $P \le \mu R$,

$$P \le \frac{\mu W - 4\mu\tan\theta\, P}{2} \quad \text{and} \quad P \le \frac{\mu W}{2 + 4\mu\tan\theta}.$$

Resolving horizontally and vertically for the rod as a separate system, for equilibrium, $S + Y = W$ and $X = T$.

Therefore $\mathbf{S = W - R} = \dfrac{W + 4\tan\theta\, P}{2}$ and $\mathbf{T = 2P}$.

Hence the magnitude of the force exerted by the pivot A on the rod, $\sqrt{S^2 + T^2}$, is

$$\sqrt{\left(\frac{W + 4\tan\theta\, P}{2}\right)^2 + 4P^2}$$

$$= \frac{\sqrt{W^2 + 8WP\tan\theta + 16P^2\sec^2\theta}}{2}$$

This force makes an angle of $\arctan\left(\dfrac{S}{T}\right)$ to the horizontal.

If this angle is ϕ, then $\tan\phi = \dfrac{W}{4P} + \tan\theta$.

The horizontal component of the force exerted by the rod on the disc is $2P$ and the vertical component is

$$\frac{W - 4\tan\theta\, P}{2},$$

as shown in figure 155(iii).

Example 7 State one set of conditions sufficient to ensure that a system of coplanar forces is in equilibrium.

The points A and B have coordinates $(a, 0)$, (a, a) referred to perpendicular axes Ox, Oy. A system of forces acts in the plane xOy and has a clockwise moment $4aP$ about O and anticlockwise moments of $6aP$ and $2aP$ about A and B respectively. Find the magnitude of the resultant of the system and find the equation of its line of action.

It is required to reduce the system to one force acting at the point $(-a, 0)$ and a couple. Find the magnitude of the force, the equation of its line of action and the moment of the couple.

(W.J.E.C.)

A sufficient set of conditions for a system of coplanar forces to be in equilibrium is that the sum of the horizontal components of force should be zero, the sum of the vertical components of force should be zero, and the resultant moment about one point in the plane should be zero.

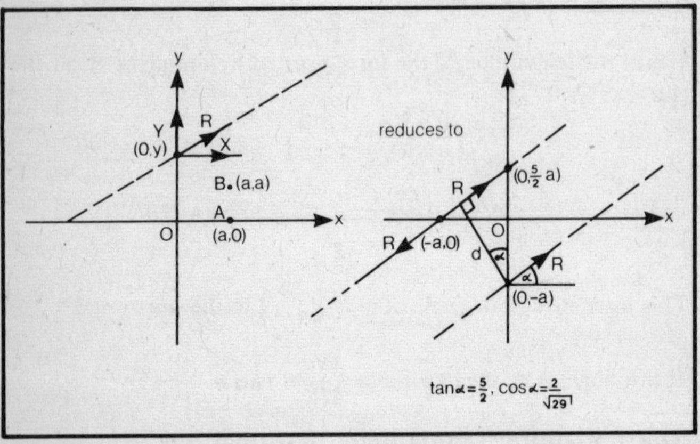

Figure 156.

We are told that the resultant moments about three different points are different. Hence the system does not reduce to a couple and it can be reduced to a single resultant force. Let this resultant force be R, with horizontal component X and vertical component Y. Without loss of generality, since force is transmissible along its line of action, we shall let it act from the point $(0, y)$. (See figure 156.)

We shall take anticlockwise moments as positive. Taking moments about O, $-4aP = -yX$. Taking moments about A, $6aP = -yX - aY$, and about B, $2aP = -aY - (y-a)X$.

Hence, $Y = -10P$, $X = -4P$, and $y = -a$, and the magnitude of the resultant, $\sqrt{X^2 + Y^2}$, is $2\sqrt{29}P$.

The gradient of the line of action of the resultant is $Y/X = 5/2$ and the line cuts the y-axis at $(0, -a)$. Hence **the equation of the line of action of the resultant is**

$$y = \frac{5}{2}x - a.$$

If we reduce the system to one force acting at the point $(-a, 0)$ and a couple, then that force must be of the same magnitude and direction as the resultant force R, since the linear resultant of a couple is zero, and the two systems must be equivalent. The couple must be of moment such that the resultant of that couple and the force, of magnitude and direction R, passing through $(-a, 0)$ is the single resultant force R, passing through $(0, -a)$. Hence the magnitude of the couple is Rd, as shown in figure 156, where

$$d = \frac{7}{2}a \cos \alpha = \frac{7}{\sqrt{29}}a,$$

in the anticlockwise direction. **The system therefore reduces to a single force of magnitude $2\sqrt{29}P$, line of action**

$$y = \frac{5}{2}x + \frac{5}{2}a,$$

together with an anticlockwise couple of moment $14aP$.

Example 8 A uniform ladder of weight W and length $2a$ rests with its lower end on rough horizontal ground and its upper end leaning against a smooth vertical wall. The coefficient of friction between the ladder and the ground is $\frac{1}{2}$. If the system is in limiting equilibrium, find the angle the ladder makes with the horizontal, and find the reaction at the wall.
A man of weight W climbs the ladder, find how far up he can climb before the ladder slips.
A weight W is placed at the foot of the ladder, find how far the man can climb before the ladder slips.

Since the vertical wall is smooth the only component of reaction at the wall is the normal reaction. At the lower end of the ladder there will be the normal reaction and the frictional force acting in the direction shown in the figures, in order to prevent motion.

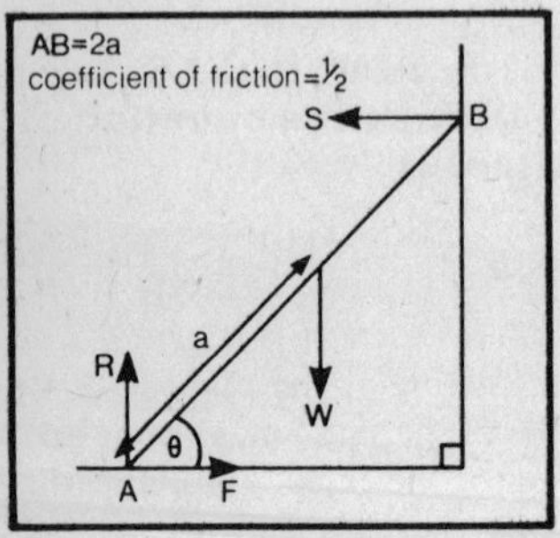

Figure 157.

Referring to figure 157 and imposing conditions for equilibrium:

Resolving horizontally $F = S$.

Resolving vertically $R = W$.

Taking moments about A,

$a \cos \theta \, W = 2a \sin \theta \, S.$

We are told that friction is limiting, therefore,

$$F = \mu R = \frac{R}{2}.$$

Combining these equations, $\boldsymbol{\theta = \arctan 1 = \frac{\pi}{4}}$ and $\boldsymbol{S = \frac{W}{2}}$.

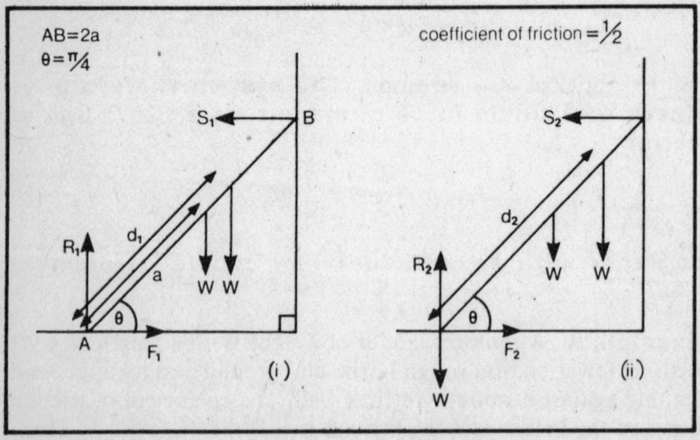

Figure 158.

When a man of weight W climbs the ladder, he will increase the normal reaction at the ground and so raise the magnitude of the limiting value of the frictional force.

Referring to figure 158(i) and imposing conditions for equilibrium. Resolving vertically, $R_1 = 2W$; resolving horizontally, $S_1 = F$.

Taking moments about A, $aW \cos\theta + d_1 W \cos\theta = 2a \sin\theta\, S_1$.

If d_1 is the greatest distance up the ladder that the man can climb, then the frictional force is limiting, i.e., $F_1 = \mu R_1 = W$. Since $\theta = \pi/4$, when we combine these equations, $\boldsymbol{d_1 = a}$. Hence **the man can climb halfway up the ladder before the ladder begins to slip.**

When a weight W is placed at the foot of the ladder, again the normal reaction (and hence the limiting value of friction) will be raised. Referring to figure 158(ii) and imposing the conditions for equilibrium:

Resolving horizontally, $S_2 = F_2$; resolving vertically, $R_2 = 3W$.

Taking moments about A, $a \cos\theta\, W + d_2 \cos\theta\, W = S_2 2a \sin\theta$.

If d_2 is the maximum distance up the ladder that the man can climb before the ladder slips, then friction is limiting.

Thus $F_2 = \dfrac{3W}{2}$. Combining these equation, $d_2 = 2a$.

Hence the man can now climb to the top of the ladder without it slipping

Example 9 (i) A non-uniform heavy beam $ABCD$ rests horizontally on supports at B and C, where $AB = BC = CD = a$. When a load of weight W is hung from A the beam is on the point of rotating about B. An additional load of weight $5W$ is hung from D and the beam is then on the point of rotating about C. Show that the weight of the beam is $4W$ and find the distance of the centre of gravity from B.

(ii) Uniform rods AB and BC, each of weight W and length $2a$ are freely jointed at B and rest in a horizontal straight line ABC on two supports, one at A and the other at a point D of BC. Find the length BD and find the force exerted on each support.

(i) Since the beam is **non-uniform**, the weight cannot be taken to be acting from its mid-point. If we let the centre of gravity of the rod be at G, where $BG = d$ as shown in the diagrams, then we can consider its weight, say w, to be acting from there.

When the rod is on the point of rotating about one support, then the reaction at the other support will just have become zero. Hence in figure 159(i), where the rod is **on the point of rotating about *B*, $S_1 = 0$**, and in figure 159(ii), where the rod is **on the point of rotating about *C*, $R_2 = 0$.**

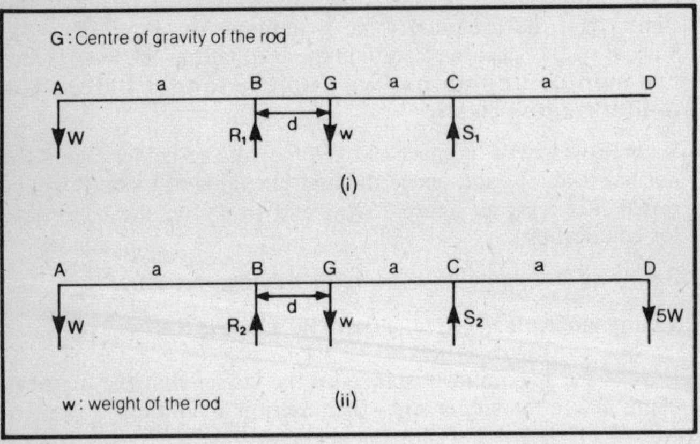

Figure 159.

In figure 159(i), where the rod is on the point of rotating about *B*, it is wise to take moments about *B* (that is from where we measure *d*, and we are not interested in the value of R_1).

Thus $$aW = dw.$$

In figure 159(ii), where the rod is on the point of rotating about *C*, taking moments about *C*, $a5W = (a - d)w + 2aW$.

Combining these two equations,

$$\mathbf{w = 4W} \quad \text{and} \quad \mathbf{d = \frac{a}{4}}.$$

(ii) We are told that rods *AB* and *BC* are freely jointed at *B* and that the system is in equilibrium with *ABC* a horizontal straight line, supported as shown in figure 160. If the system is in equilibrium, then each part of the system is also in equilibrium.

When we consider the system as a whole (diagram (i)) then the reactions at the joint are internal forces which cancel each other out. When we consider rods *AB* or *BC* separately, then the reaction from one rod onto the other at the joint is an external force and must be taken into account. As we have already said, these reactions are equal and opposite. There is no horizontal component of reaction. (Taking one rod as the system, there are no other horizontal forces).

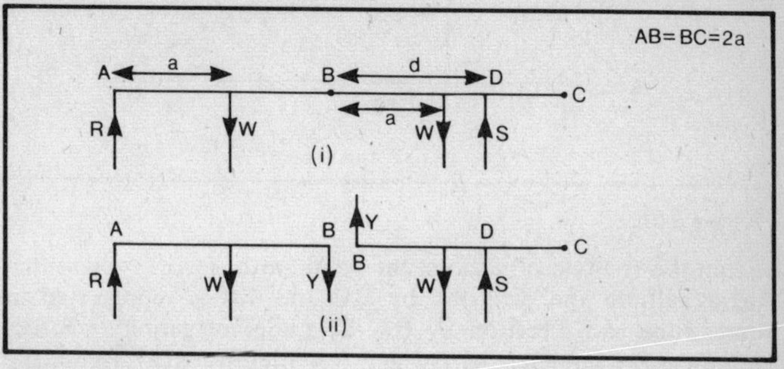

Figure 160.

Taking moments about B in diagram (i) for equilibrium,

$$aW - 2aR = aW - dS.$$

Resolving vertically for the system as a whole, $2W = R + S$.

Taking moments about B for rod AB in diagram (ii) for equilibrium, $aW = 2aR$.

Combining these equations, $R = \frac{W}{2}$, $S = \frac{3W}{2}$ and $d = \frac{2a}{3}$.

Example 10 (i) Show that the resultant of three forces represented completely by the sides of a triangle taken in order is a couple of moment represented by twice the area of the triangle. (ii) Forces $4P$, $4P$, $6P$ act along the sides AB, BC, CA respectively of an equilateral triangle of side a. The line of action of the resultant, R, of these forces meets BC produced at D. Find the magnitude and direction of R and distance BD.

(i) If the forces are represented **completely** by the sides of a triangle taken in order, then they are represented, not only in magnitude and direction, but also in line of action. Let us consider the resultant of forces represented completely by the sides AB, BC, CA of triangle ABC.

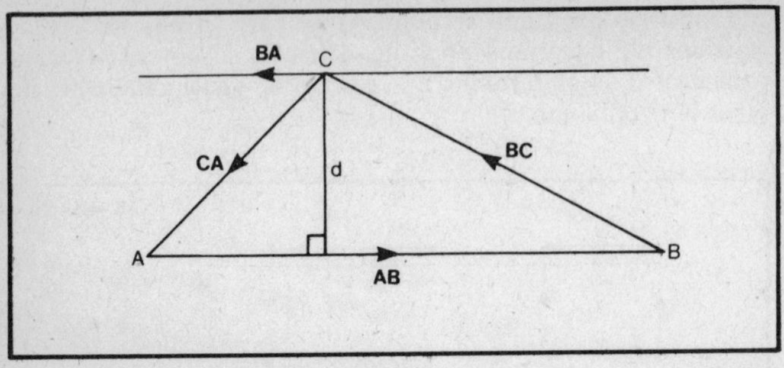

Figure 161.

From the triangle of vectors, the resultant of forces represented in magnitude and direction by **BC** and **CA** is represented in magnitude and direction by **BA**. The lines of action of forces represented completely by **BC** and **CA** interesect at *C*. Hence the line of action of their resultant passes through *C*. Therefore the system reduces to **BA** acting through *C* and **AB** acting along *AB*, i.e., equal, unlike parallel forces. Clearly we have a couple of moment *ABd*, acting in the sense *ABC* where *d* is the perpendicular distance between the two forces.

From figure 161, we see that *d* is the perpendicular height of the triangle. The triangle is of area $\frac{1}{2}ABd$ and so the system reduces to a couple of moment represented by **twice the area of the triangle**.

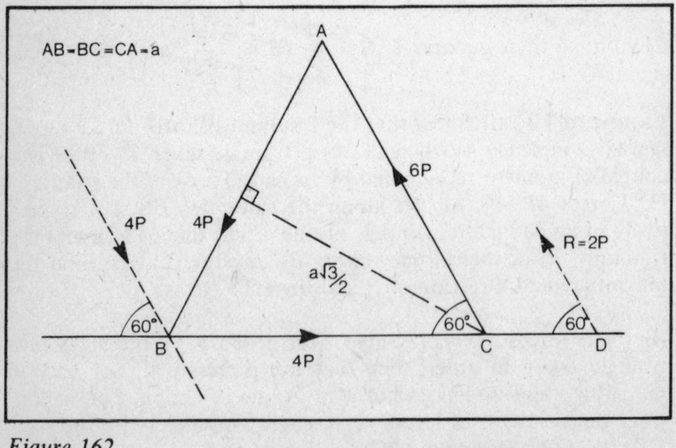

Figure 162.

(ii) Taking the forces as shown in figure 162. Force 4**P** acting along AB, and force 4**P** acting along BC, have as their resultant a force of magnitude 4**P** acting through B, parallel to AC and in the direction of AC.

Hence, the resultant force **R** will be of magnitude 6**P** – 4**P** = 2**P**, and acts in the direction parallel to CA. We find its line of action by taking moments about C in both systems.

In the original system, the resultant moment about C was $2a\sqrt{3}P$ in the anticlockwise direction. When the system is reduced to a single force of magnitude $2P$ acting parallel to, and in the same sense as CA, it must have the same moment about A as the original system, i.e., $2a\sqrt{3}\,P = CD \sin 60°\, 2P$, so that $\boldsymbol{CD = 2a}$ and $\boldsymbol{BD = 3a}$.

Example 11 Four equal rods each of weight W are smoothly jointed together. A light string connects the joints at A and C. The system is suspended from A and hangs in equilibrium. If angle $DAB = 2\theta$, find the horizontal and vertical components of the force exerted by AB on BC. Hence or otherwise find the tension in the string.

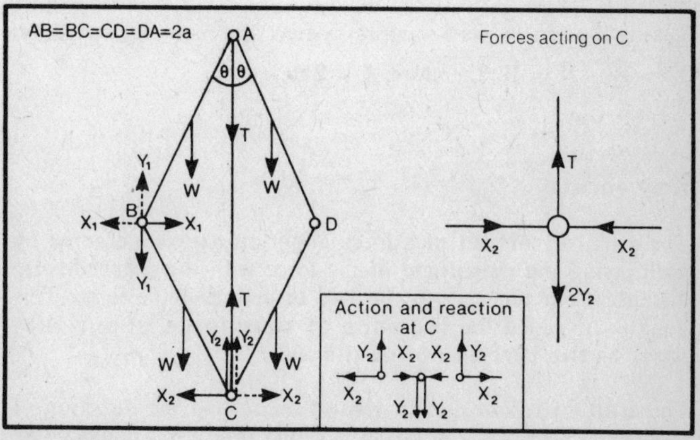

Figure 163.

The system is clearly a rhombus and is symmetrical about AC. The forces exerted by the hinge at C on the rods CB and CD and the forces exerted by the hinge at B on rods BC and BA are as shown in figure 163.

Since the system is symmetrical about AC, the vertical forces acting at C are both in the upward direction. (This is not contrary to 'action and reaction' since the forces shown are those exerted **by** the hinge at C (see insert, figure 163)).

We are asked to find the horizontal and vertical components of the force exerted by AB on BC. We must first find X_1 and Y_1.

Taking the rod BC as a separate system in equilibrium, and taking moments about C,

$$-2aX_1 \cos\theta + 2aY_1 \sin\theta + aW \sin\theta = 0.$$

Taking the rod AB as a separate system and taking moments about A,

$$-2aX_1 \cos\theta - 2aY_1 \sin\theta + aW \sin\theta = 0.$$

Hence, $\boldsymbol{X_1 = \frac{W}{2}\tan\theta}$ and $\boldsymbol{Y_1 = 0}$.

We have to find the tension in the string. Let us consider the forces acting on the hinge at C. They must be in equilibrium, so that $T = 2Y_2$.

Taking the rod BC as a separate system, and resolving vertically,

$Y_2 = Y_1 + W = W$, therefore, $\boldsymbol{T = 2W}$.

Key terms

The **turning effect** of a force about an axis is measured by multiplying the magnitude of the force with the perpendicular distance from the axis to the line of action of the force. This quantity is called the **moment of that force about that axis, or the torque about the axis.**

The **unit** of torque is the newton metre and the direction of moment is that of the rotation it would produce. Anticlockwise and clockwise moments are given positive and negative signs to differentiate between them. Anticlockwise is usually positive.

The **resultant moment** of a system about an axis is the algebraic sum of the moments of the individual forces about the same axis, the sign of the resultant indicating the direction.

The Principle of Moments states that the resultant moment of a system of forces is equal to the moment of the resultant force about the same axis (if the system reduces to a single force).

The resultant of two like parallel forces, P and Q, is of magnitude $(P + Q)$, of direction parallel to P and Q. Its line of action divides any line from the line of action of P to the line of action of Q in the ratio $Q : P$.

The resultant of two unlike parallel forces is of magnitude $(P - Q)$ and acts in the direction and sense of the force of larger magnitude. Its line of action divides any line from the line of action of P to the line of action of Q in the ratio $-Q : P$, i.e., it divides any transversal externally in the ratio $Q : P$.

A couple is the name given to two equal unlike parallel forces. Its linear resultant is zero and it represents a pure turning effect. A couple has moment equal to the product of the magnitude of one of the forces with the perpendicular distance between the lines of action of the parallel forces, the sign representing the direction of rotation. A couple has the same moment about any point in its plane, this is known as the **constant moment property of couples**.

A system of coplanar forces may produce motion in two dimensions and/or a rotation about an axis perpendicular to the plane. Hence it has three degrees of freedom and we can find three independent equations for a system of coplanar forces. We may obtain independent information by resolving the forces in two non-parallel directions and finding their sums, together with taking moments about one point in the plane, or by taking moments about three non-collinear points in the plane. The most usual technique is to resolve in two perpendicular directions, Ox, Oy, and take moments about O.

If a system of coplanar forces is in equilibrium, then there must be no resultant force and no resultant turning effect. We must test that this is the case by using one of the above sets of criteria. We usually test that the sum of the components in two perpendicular directions is zero, and that the resultant moment about the origin is zero.

Equivalent systems of coplanar forces are systems that produce exactly the same effect, both in terms of their resultant linear motion, and their resultant rotational motion.

The resultant of a system of coplanar forces is the simplest equivalent system. A system of coplanar forces reduces either to a single force, or to a couple.

Any system of coplanar forces may be reduced to a single force passing through a particular point together with a couple. A system of coplanar forces is equivalent to another if the sum of the components of force in each of two perpendicular directions and the resultant moment about the same axis, are identical for each system.

Chapter 15
Centres of Mass, Centres of Gravity

We have already defined the **centre of gravity** of a body as being that point from which the weight of the body may be considered to be acting.

Before we investigate this idea further we need to prove a general result for **parallel** forces, namely that **the resultant of a set of parallel forces, acting from fixed points, passes through a fixed point whatever the orientation of the system.** (See figure 164.)

Consider the parallel forces $\mathbf{P}_1$, $\mathbf{P}_2$, $\mathbf{P}_3$ acting through the fixed points A, B, C respectively.

The resultant of $\mathbf{P}_1$ and $\mathbf{P}_2$ is $(\mathbf{P}_1 + \mathbf{P}_2)$ acting through the point Q on AB such that

$$\frac{AQ}{QB} = \frac{P_2}{P_1}.$$

The resultant of $(\mathbf{P}_1 + \mathbf{P}_2)$ acting at Q, and $\mathbf{P}_3$ acting through C, is $(\mathbf{P}_1 + \mathbf{P}_2 + \mathbf{P}_3)$ acting through a point R on QC such that

$$\frac{QR}{RC} = \frac{P_3}{(P_1 + P_2)}.$$

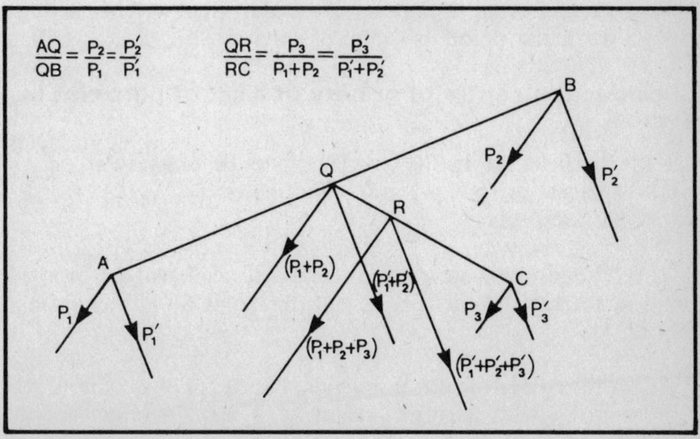

Figure 164.

Let us now consider parallel forces $\mathbf{P}'_1$, $\mathbf{P}'_2$, $\mathbf{P}'_3$ acting through the same fixed points A, B, C and of the same magnitudes as $\mathbf{P}_1$, $\mathbf{P}_2$, $\mathbf{P}_3$ respectively, but with a different direction.

We have seen, in Chapter 13, that the resultant of parallel forces cuts any transversal in a constant ratio. Therefore the resultant of $\mathbf{P}'_1$, $\mathbf{P}'_2$, $\mathbf{P}'_3$ is $(\mathbf{P}'_1 + \mathbf{P}'_2 + \mathbf{P}'_3)$ **acting through** $\boldsymbol{R}$. The magnitude of $(\mathbf{P}'_1 + \mathbf{P}'_2 + \mathbf{P}'_3)$ is the same as that of $(\mathbf{P}_1 + \mathbf{P}_2 + \mathbf{P}_3)$.

This argument may be applied successively to sets of any number of parallel forces, like or unlike, and the general result is proved.

The **weights of a set of particles form a set of parallel forces**, (in a uniform gravitational field). Therefore their resultant weight passes through a fixed point (the centre of gravity) whatever the orientation of the system.

The **weight of a rigid body** is the algebraic sum of the weights of its constituent parts and acts vertically downwards through a fixed point in the body, called the centre of gravity of the body.

To locate the centre of gravity of a body or set of particles, we use the fact that the sum of the moments of the weights of the constituent particles about an axis, will be equal to the moment of the resultant weight about that axis. We are, in effect, replacing the body or system, consisting of its constituent parts with their individual weights, with an equivalent system, consisting of a single weight equal to the sum of the individual weights, acting from a particular point, its centre of gravity.

Consider the centre of gravity of a set of particles in a plane.

Let the particles be in the xy-plane, and be of mass m_1, m_2, $m_3, \dots$ placed at points with coordinates (x_1, y_1), (x_2, y_2), $(x_3, y_3), \dots$ respectively.

Clearly, the resultant weight of these masses will also act from a point in the same plane. Let us call this point G, with coordinates $(\bar{x}, \bar{y})$.

We wish to replace the set of individual particles of weight $m_1 g$, $m_2 g$, $m_3 g, \dots$ with a single weight of $\sum_p m_p g$, acting from a particular point, G.

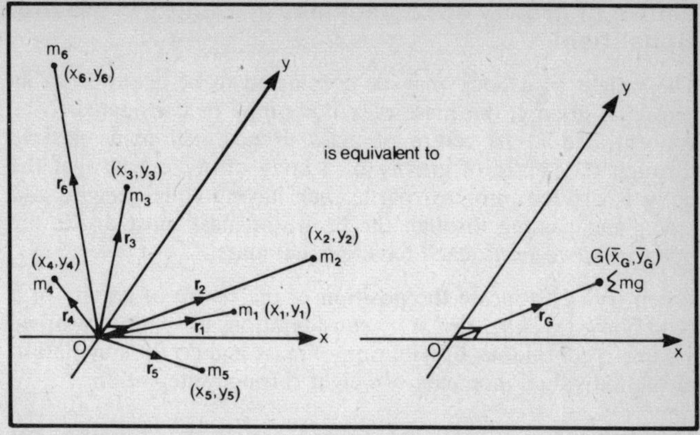

Figure 165.

Without loss of generality, we may let the plane be horizontal. Taking moments about the y axis for both systems,

$$m_1 x_1 g + m_2 x_2 g + m_3 x_3 g + \cdots = \bar{x} \sum_p m_p g.$$

Taking moments about the x axis for both systems,

$$m_1 g y_1 + m_2 g y_2 + m_3 g y_3 + \cdots = \bar{y} \sum_p m_p g.$$

Hence, $$\bar{x} = \frac{\sum_p m_p x_p}{\sum_p m_p} \quad \text{and} \quad \bar{y} = \frac{\sum_p m_p y_p}{\sum_p m_p}.$$

Thus, the position vector, $\mathbf{r}_G$, of the centre of gravity is

$$\mathbf{r}_G = \bar{x}\mathbf{i} + \bar{y}\mathbf{j}.$$

We have made the assumption that the acceleration due to gravity is constant. This is a perfectly valid assumption at a terrestrial level and on the scale we are discussing.

The centre of mass of a system is defined as the point with position vector

$$\mathbf{r}_m = \frac{\sum_p m_p \mathbf{r}_p}{\sum_p m_p}.$$

It is the point about which the mass of the system is equally distributed. As we can see, **the centre of mass and the**

centre of gravity are coincident in a uniform gravitational field.

The weight of a body may be considered to be acting from its centre of gravity, the mass of a body may be considered to be concentrated at its centre of mass. Hence, any plane passing through the centre of gravity of a body must be such that the body is divided into two parts, each having equal weight, and any plane passing through the centre of mass must divide the body into two parts, each having equal mass.

When trying to locate the position of the centre of gravity of a rigid body, we either find it by considerations of symmetry, or we assume a continuous distribution of mass and do the summation of the individual moments of weight through integration.

Centre of gravity of some uniform bodies by symmetry

A uniform body has its mass uniformly distributed throughout the body. Hence, a uniform lamina will have its mass distributed equally about any line of symmetry, and a uniform solid body, about any plane of symmetry. It follows that the centres of mass and gravity of a body lie on every line or plane of symmetry of the body, and may be located at the intersection of any two lines of symmetry, or three planes of symmetry intersecting at a point. Hence, the centres of gravity of: a uniform rod; a circular lamina; a rectangular lamina; a sphere; a rectangular prism; lie at their geometrical centres: the mid-point of the rod; the centre of the circle; the intersection of the diagonals of the rectangle; the centre of the sphere; the intersection of the lines joining the mid-points of the opposite faces of a rectangular prism.

We can extend these ideas to help us find the centres of gravity of uniform bodies which may be considered to be made up of uniform bodies of known centres of gravity.

Consider a **uniform triangular lamina**. It may be considered to be made up of uniform rods parallel to one side, as shown in figure 166.

Clearly, the centre of gravity must lie on the line joining all the mid-points of these rods, i.e., the median. We could repeat this argument for rods parallel to sides *AB*, *AC*. Hence, the **centre of gravity of a uniform triangular lamina is at the intersection of its medians: the centroid.**

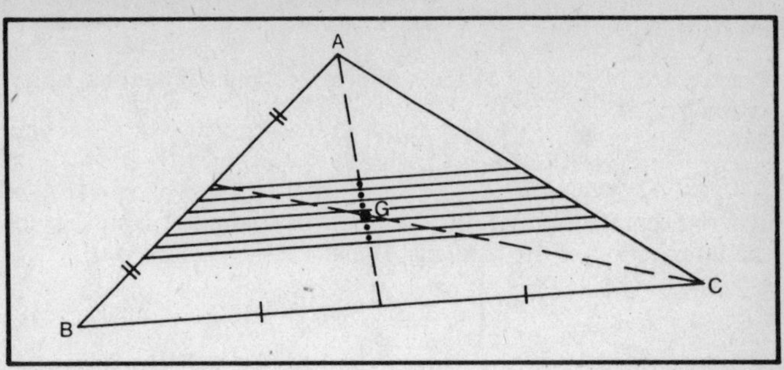

Figure 166.

Here we see that the **centre of gravity**, the **centre of mass** and the **centroid of a uniform triangular lamina** coincide. The **centroid** of a body is its geometric centre: the point about which its area, in the case of a lamina, or volume, in the case of a solid, is equally distributed. Clearly, when the mass of a body is uniformly distributed, and the body is in a uniform gravitational field, these three measures of centre will coincide.

When a body is made up of two or more parts, whose weights and centres of gravity are known, then we can use the **principle of moments** to find the centre of gravity of the **composite body**.

Consider, for example, a uniform lamina *ABCDE*, made up of the rectangle *ABDE*, with $AB = DE = 3a$, and $AE = BD = 2a$, and an isosceles triangle *BCD*, where $BC = CD = 2a$.

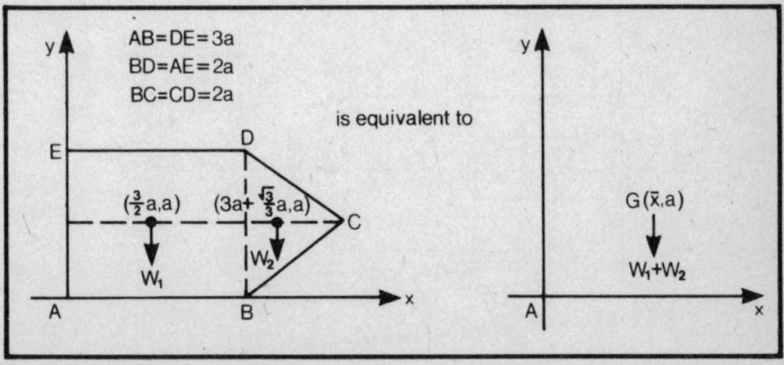

Figure 167.

Refer to figure 167.

The centre of gravity of the uniform rectangular lamina is, by symmetry, at

$$\left(\frac{3a}{2}, a\right)$$

and the centre of gravity of the uniform triangular lamina is at the intersection of the medians, i.e., at

$$\left(3a + \frac{\sqrt{3}\,a}{3}, a\right).$$

By symmetry, the centre of gravity of the compound body lies on $y = a$.

Since the lamina is uniform, it will have a constant weight per unit area. Let this be w.

The weight of $ABCD = W_1 = 2a \times 3a \times w = \mathbf{6a^2w}$, and

the weight of $BCD = W_2 = \frac{1}{2} \times 2a \times \sqrt{3}\,a \times w = \mathbf{\sqrt{3}\,a^2w}$.

Taking the xy-plane to be horizontal, and taking moments about the y-axis in both systems,

$$\frac{3}{2}a \times 6a^2w + \left(3a + \frac{\sqrt{3}\,a}{3}\right)\sqrt{3}\,a^2w = (6a^2w + \sqrt{3}\,a^2w)\bar{x}.$$

Hence, $$\bar{x} = \left(\frac{10 + 3\sqrt{3}}{6 + \sqrt{3}}\right)a, \quad \text{and} \quad \bar{y} = a.$$

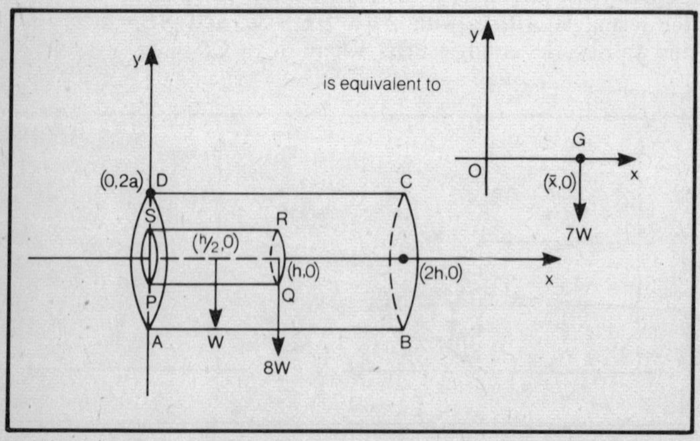

Figure 168.

Consider now the location of the centre of gravity of a uniform solid, consisting of a cylinder of radius $2a$ and height $2h$, with a cylindrical hole of radius a and height h, removed centrally from one plane end.

We always choose our axes so that the calculations are easiest, and hence the axes in figure 168, with the origin at the centre of the plane end with the hole in it.

The removal of part of the cylinder is equivalent to imposing a moment equal in magnitude to the moment of the removed part, but in the opposite sense, i.e., 'supporting' that part of the complete cylinder. The body is uniform so the weight per unit volume is constant. The removed part, $PQRS$, and the complete cylinder $ABCD$ are similar solids, with their lengths in the ratio 1 : 2 and hence their volumes in the ratio 1 : 8.

If $8W$ is the weight of the complete cylinder, then W is the weight of the removed part. By symmetry, the centre of gravity of a right cylinder lies halfway up its axis. Hence, the centre of gravity of the remaining solid will also lie on the axis of these cylinders.

Taking moments about the y-axis,

$$-\frac{h}{2}W + h8W = 7W\bar{x}, \quad \text{therefore} \quad \bar{\boldsymbol{x}} = \frac{\mathbf{15}}{\mathbf{14}}\boldsymbol{h}.$$

Alternatively, we may consider the whole cylinder to be made up of the part to be removed and the remainder. Hence, taking moments about the same axis, the moment of the whole cylinder is equal to the moment of the part to be removed plus the moment of the remaining part.

Taking moments about the y-axis,

$$h8W = h\frac{W}{2} + 7W\bar{x}, \quad \text{and} \quad \bar{x} = \frac{15}{14}h, \text{ as above.}$$

If there is no line of symmetry for the composite body, then we have to take moments about both axes in order to get both coordinates of the centre of gravity.

When a body cannot be divided into a small number of constituent parts whose weights and centres of gravity are known, then we divide the body into an infinite number of infinitely small elements and use calculus to sum their individual moments.

We shall now derive some **standard results** for centres of gravity of common solids using **integration** to do so. Without loss of generality, **we shall take the *xy*-plane to be horizontal.**

A uniform solid right circular cone of radius *r* and height *h*:

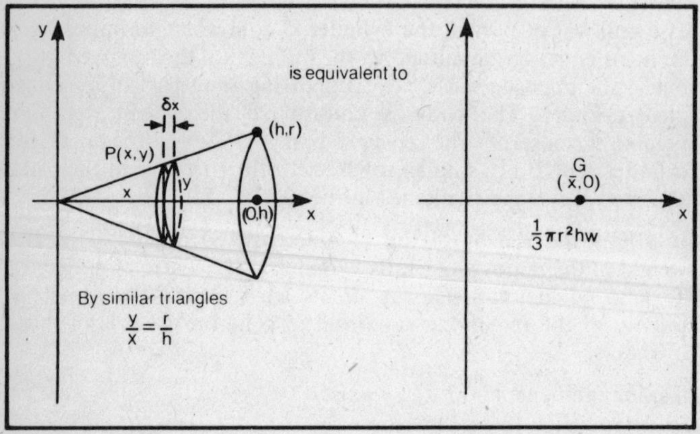

Figure 169.

By symmetry, the centre of gravity of the cone must lie on its axis. If we divide the solid into sections parallel to its plane face, of thickness δx, each section is approximately a thin cylinder with its centre of gravity distance x from the y-axis. As δx gets smaller, the better the approximation. (See figure 169.)

Let w be the weight per unit volume. The moment of a typical element of volume at point $P(x, y)$ about the y-axis, is $x\pi y^2\ \delta x\ w$.

The total weight of these elements is the weight of the cone, $\frac{1}{3}\pi r^2 hw$.

Hence, taking moments about the y-axis for both systems,

$$\sum_{x=0}^{h} x\pi y^2\ \delta x\ w = \bar{x}\,\frac{1}{3}\,\pi r^2 hw.$$

Taking the limit as $\delta x \to 0$, $\int_0^h \pi y^2 wx\ \mathrm{d}x = \bar{x}\,\frac{\pi}{3}\,r^2 hw.$

Since $y = \frac{xr}{h}$, then $\frac{1}{h^3}\int_0^h \mathrm{d}x = \frac{\bar{x}}{3}$,

$$\bar{x} = \frac{3}{4}h.$$

Uniform semicircular lamina of radius *a*:

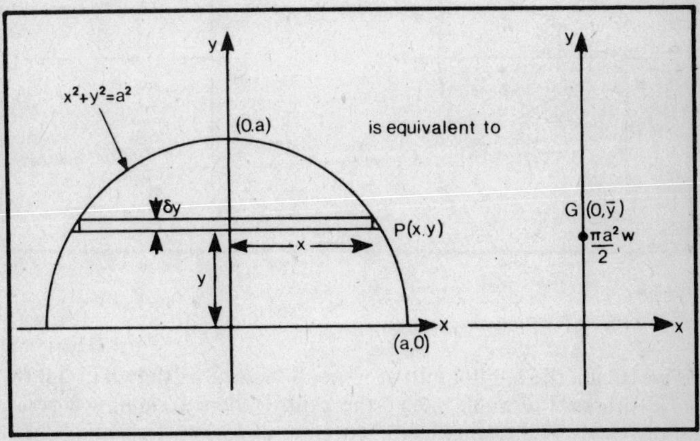

Figure 170.

By symmetry, $\bar{x} = 0$. We divide the lamina into elements parallel to the x-axis of weight $2x\,\delta y\,w$, where w is the constant weight per unit area, as shown in figure 170.

Taking moments about the x-axis in both systems,

$$\int_0^a wy2x\,\mathrm{d}y = \bar{y}\frac{a^2\pi}{2}w. \quad \text{Since} \quad x^2 + y^2 = a^2,$$

$$\bar{y} = \frac{4}{\pi a^2}\int_0^a y(a^2 - y^2)^{1/2}\,\mathrm{d}y$$

$$= \frac{4}{\pi a^2}\left[-\frac{1}{3}(a^2 - y^2)^{3/2}\right]_0^a = \frac{4a}{3\pi}.$$

Sector of a circle of radius *a*, subtending an angle of 2α at the centre:

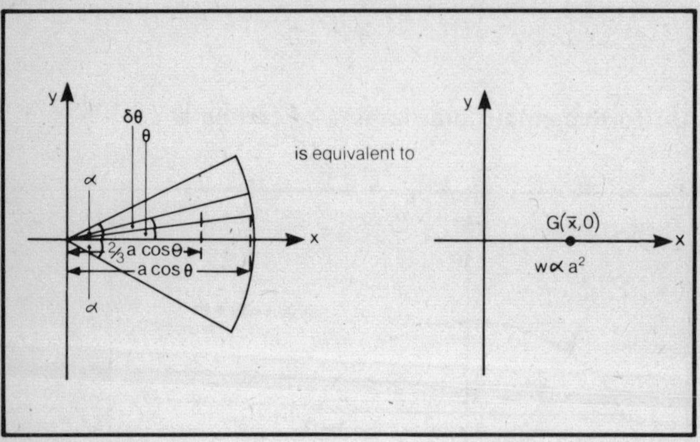

Figure 171.

If we divide the lamina into very small sectors, as shown in figure 171, subtending angle $\delta\theta$ at the centre, then a small sector is approximately a triangle with its centre of gravity $\frac{2}{3}a$ from O, and hence distance $\frac{2}{3}a\cos\theta$ from the y-axis.

Taking the xy-plane as horizontal and w as the constant weight per unit area, the moment about the y-axis of a small sector is

$$\frac{2}{3}a\cos\theta\left(\frac{1}{2}a^2\,\delta\theta\,w\right).$$

Hence, summing all the individual moments and comparing with the equivalent system,

$$\int_{-\alpha}^{\alpha}\frac{2}{3}a\cos\theta\frac{1}{2}a^2w\,d\theta = \frac{2\alpha}{2\pi}\pi a^2 w\bar{x}.$$

Therefore, $\frac{a}{3\alpha}\Big[\sin\theta\Big]_{-\alpha}^{\alpha} = \bar{x} = \boldsymbol{\frac{2a\sin\alpha}{3\alpha}}$.

By symmetry, $\bar{y} = \mathbf{0}$.

Arc of a circle of radius *a*, subtending angle 2α at the centre:

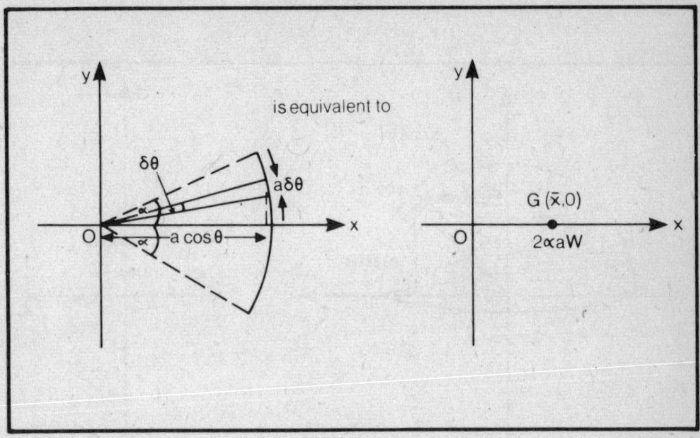

Figure 172.

Dividing the arc into infinitely small arcs, subtending angles $\delta\theta$ at the centre, the moment of a small arc about the y-axis is $a\ \delta\theta\ a \cos\theta\ W$, where W is the constant weight per unit length.

Hence summing for all the small sectors and comparing with the equivalent system,

$$\int_{-\alpha}^{\alpha} a^2 \cos\theta\ W\, d\theta = \frac{2\alpha}{2\pi} 2\pi a W \bar{x}.$$

Thus, $\bar{x} = \dfrac{a}{2\alpha}\Big[\sin\theta\Big]_{-\alpha}^{\alpha} = \dfrac{\mathbf{a \sin\alpha}}{\boldsymbol{\alpha}}$.

By symmetry, $\bar{y} = \mathbf{0}$.

Uniform hemispherical shell of radius *a*:

By symmetry, taking the axes as shown in figure 173, $\bar{x} = 0$. We must divide the shell into bands which approximate in surface area to cylinders of height $a\ \delta\theta$ and surface area $2\pi(a \cos\theta)a\ \delta\theta$. (It is tempting to use the Cartesian equation of a circle and approximate an element to a cylinder of radius x and height δy.

The margin of error here is unacceptable because we are dealing with a shell; if it were a solid hemisphere the percentage error in the evaluation of the volume, using the height as δy, would be negligible, but this is not so with a shell.)

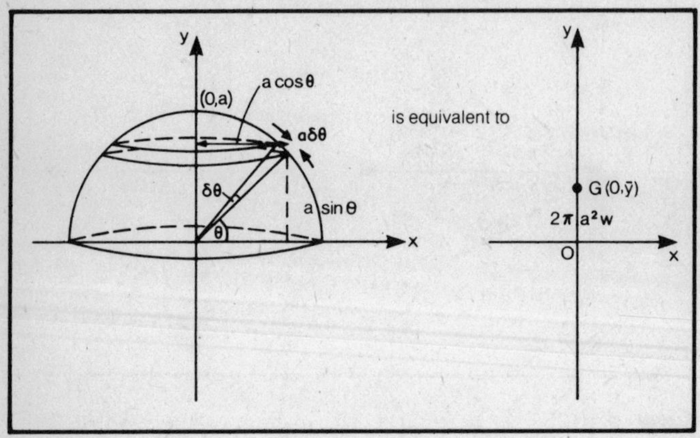

Figure 173.

The moment of an element about the x-axis, taking the xy-plane as horizontal, is $a \sin \theta \; 2\pi(a \cos \theta) \; a \; \delta\theta$.

Summing the moments about the x-axis and comparing with the equivalent system,

$$w \int_0^{\pi/2} 2\pi a^3 \sin \theta \cos \theta \, d\theta = 2\pi a^2 w \bar{y},$$

where w is the constant weight per unit area.

Hence, $\frac{a}{2}\left[-\frac{1}{2} \cos 2\theta\right]_0^{\pi/2} = \bar{y} = \frac{a}{2}$.

Uniform solid tetrahedron of base area *K* and height *h*:

If we divide the tetrahedron into plane sections parallel to the base, we see that we have a series of triangular laminae, each with their centre of gravity at the intersection of their medians. These triangular sections are all similar to the base ABC.

Thus, the centre of gravity of the tetrahedron will lie along the line joining its vertex to the centroid of its plane base. Hence, we choose this line as one of the axes, Ox, in figure 174.

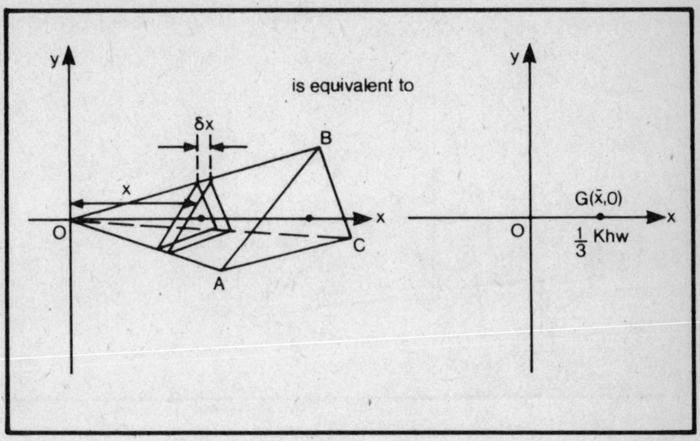

Figure 174.

Let the weight per unit volume be w. We take as an element of volume, a plane section of the tetrahedron of thickness δx and parallel to ABC, in order that its base is similar to ABC. The ratio of corresponding linear measurements in the two similar triangles is $x : h$, therefore, the ratio of their areas will be $x^2 : h^2$.

Hence, taking moments about the y-axis with the xy-plane horizontal, and comparing with the equivalent system,

$$w \int_0^h x \frac{x^2}{h^2} K \, dx = \frac{1}{3} Khw\bar{x},$$

where K is the area of the plane base ABC.

Thus, $\frac{3}{h^3}\left[\frac{x^4}{4}\right]_0^h = \bar{x} = \frac{3}{4}h.$

Therefore, **the centre of gravity of a uniform solid tetrahedron is $\frac{1}{4}$ of the way up the line joining its vertex to the centroid of its plane base, measured from the base.**

From this last result we can also deduce **the position of the centre of gravity of a uniform solid pyramid**:

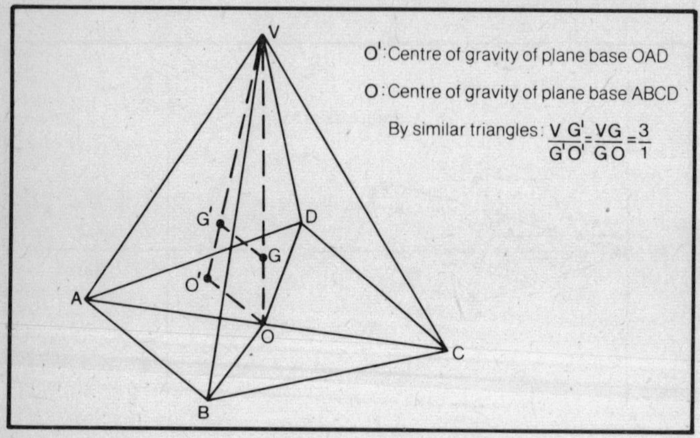

Figure 175.

As with the tetrahedron, we see that if we divide the pyramid into plane sections, parallel to the base, then we have similar figures, each with their centre of gravity at the centroid of the plane section. Hence, the centre of gravity, G, of the pyramid must lie on the line joining the vertex V to the centroid of the base, O.

We can also divide the pyramid into uniform solid tetrahedra, $OCDV$, $OBCV$, $OABV$, $OADV$ as shown in figure 175.

We have already seen that the centre of gravity, G, of a uniform tetrahedron lies $\frac{1}{4}$ of the way up the line joining the centre of gravity of its plane base to its vertex, measured from the base. Hence, by similar triangles, **the centre of gravity of the pyramid must lie $\frac{1}{4}$ of the way up *OV***, i.e., $OG : GV = 1 : 3$.

We may consider a **cone** to be a pyramid with an infinite number of sides. Hence, **the centre of gravity of a uniform cone lies on the line joining its vertex to the centre of gravity of its base, $\frac{1}{4}$ of the way up that line from the base**. (We have also shown this by integration.)

We shall now consider the **equilibrium positions of bodies freely suspended from a single point and of bodies resting on planes**.

Consider the body in figure 176. It is **freely suspended** from the point *P* and is in **equilibrium** under the action of the tension in the string, *T*, and its own weight, *W*. Clearly, $T = W$. There must be no turning effect so that *T* and *W* must act along the same line of action.

Hence, a freely suspended body will rest in equilibrium with its centre of gravity in a vertical line with the point of suspension.

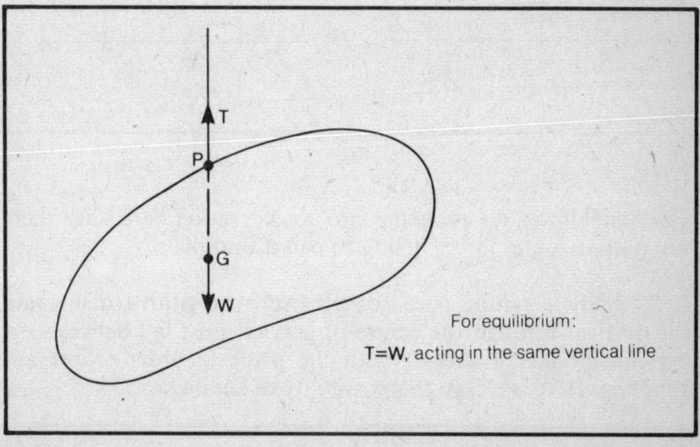

Figure 176.

Consider now a **body resting on a horizontal plane** as shown in figure 177. The forces acting on the body are its weight, acting through its centre of gravity, and the normal reactions of the plane on the body at its points of contact with the plane. The resultant normal reaction between the plane and the body lies between the two extreme points of contact, *A* and *B*.

If the body is resting on the plane in equilibrium, then the resultant normal reaction must be equal and opposite to the weight of the body, and must act in the same straight line.

Clearly, if the vertical through the centre of gravity falls outside the extremities of contact with the plane, then the normal reaction cannot act in the same line.

Hence, if this is the case, the body will topple about the extreme point of contact nearest to the line of action of the weight. Since toppling will occur about that point, then that is the point through which the resultant normal reaction will act (it will be the only point left in contact with the plane).

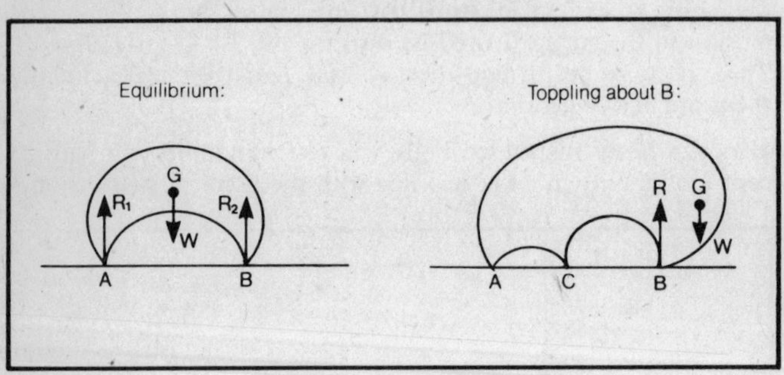

Figure 177.

Frictional forces do not come into consideration here since there are no horizontal forces acting to call it into play.

If the body is resting on a **rough inclined plane**, then again the vertical through the centre of gravity must fall between the extreme points of contact with the plane in order to prevent toppling. Here we have the possibility of sliding also.

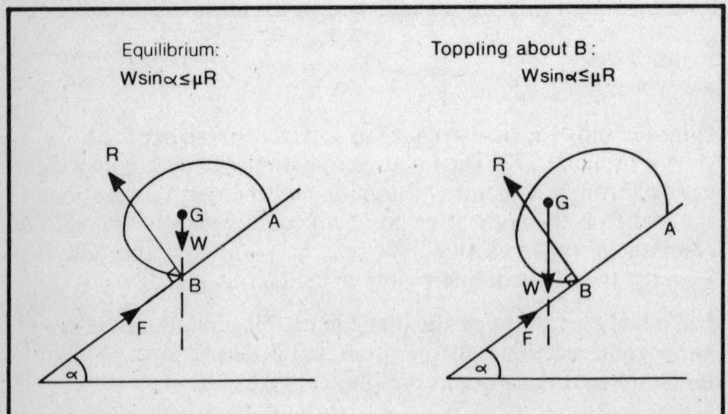

Figure 178.

Worked examples

Example 1 Show by integration that the centre of gravity of a uniform solid hemisphere of radius r is situated at a distance $3r/8$ from the centre of its plane face.

A solid composite uniform body consists of a right circular cylinder, of radius r and height $2r$, and a hemisphere of radius r, the plane face of the hemisphere coinciding with a circular end of the cylinder. Find the distance of the centre of gravity of the composite body from its plane face.

The plane face of this body is placed on a perfectly rough plane which is inclined at an angle θ to the horizontal. If the body is on the point of toppling, find the value of $\tan\theta$.

(A.E.B.)

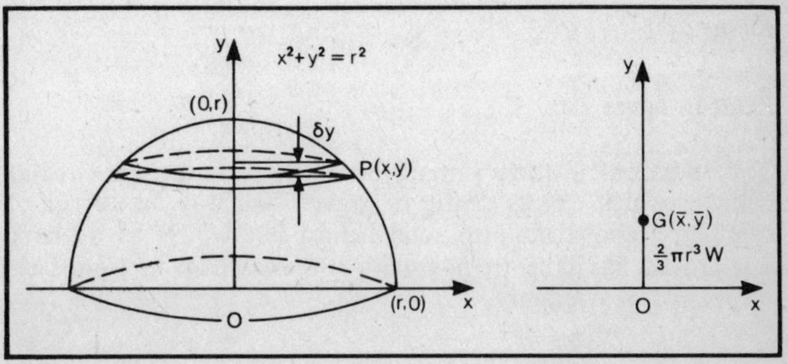

Figure 179.

Taking the axes as shown in figure 179, we see that, by symmetry, the centre of gravity will lie on Oy, i.e., $\bar{x} = \mathbf{0}$.

Let the constant weight per unit volume be W. Taking elements parallel to the plane face, of thickness δy and of volume approximating to that of a cylinder of height δy and radius x, then summing the moments and comparing with the equivalent system,

$$\int_0^r \pi x^2 W y \, dy = \frac{2}{3}\pi r^3 W \bar{y}.$$

Hence, $\frac{3}{2r^3}\int_0^r (r^2 - y^2)y\,dy = \bar{y} = \frac{3}{8r^3}\left[-(r^2 - y^2)^2\right]_0^r = \frac{\mathbf{3}}{\mathbf{8}}\boldsymbol{r}.$

The centre of gravity of a uniform solid hemisphere of radius *r* is situated at a distance 3*r*/8 from its plane face.

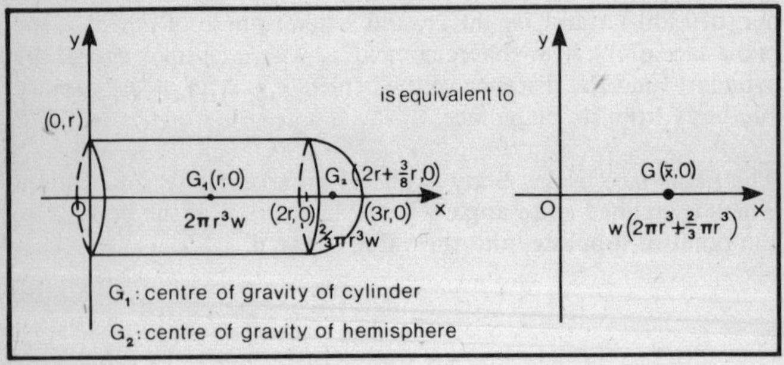

Figure 180.

Refer to figure 180.

The **composite body** is made up of the uniform right circular cylinder which has its centre of gravity half way up its axis of symmetry, and a uniform solid hemisphere which, as we have just proved, has its centre of gravity 3*r*/8 away from its plane base on its axis of symmetry.

Taking axes as shown in figure 180, the constant weight per unit volume to be *w*, taking moments about the *y*-axis and comparing with the equivalent system,

$$r2\pi r^3 w + \left(2r + \frac{3}{8}r\right)\frac{2}{3}\pi r^3 w = \bar{x}w\left(2\pi r^3 + \frac{2}{3}\pi r^3\right).$$

$$\text{Hence, } \bar{\boldsymbol{x}} = \frac{\mathbf{43}}{\mathbf{32}}\boldsymbol{r}.$$

Therefore, the distance of the centre of gravity of the composite body from its plane face is

$$\frac{\mathbf{43}}{\mathbf{32}}\boldsymbol{r}.$$

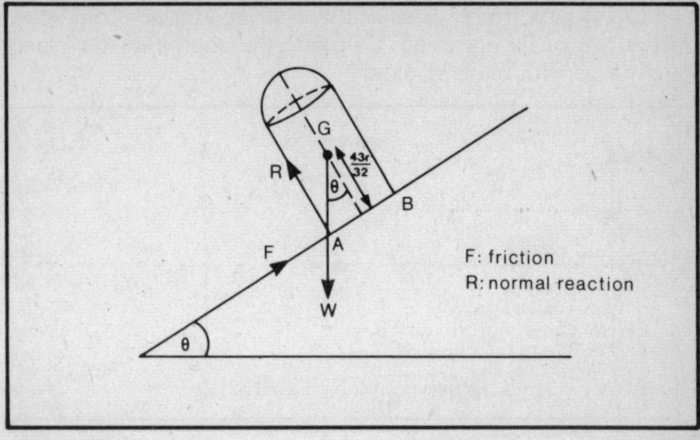

Figure 181.

Refer to figure 181.

If the surface is perfectly rough, then slipping can never occur, hence we need not take into account that the body may slip before it topples. Since the body is on the point of toppling, the vertical through the centre of gravity of the composite body passes through A, and the normal reaction, R, acts through A, as shown. Any increase in θ would cause the line of action of the weight to fall outside the surface of contact of the plane base with the inclined plane, and toppling would occur. From the diagram we see that

$$\tan\theta = \frac{32}{43}.$$

Example 2 A cylindrical can, made of thin material and open at the top, is of height $2r$ and the radius of the plane base is r. The mass per unit area of the uniform material in the plane base is twice that of the uniform material making up the curved surface of the can. Find the distance of the centre of gravity of this can from its plane base.

The can is placed on a rough inclined plane of angle of inclination θ to the horizontal, its plane base being in contact with the inclined plane. The surface is sufficiently rough to prevent sliding. Find the condition for θ that will mean that toppling will occur.

The can is now freely suspended by a string attached to a point on the rim of its open end. Calculate the angle that the plane base makes with the horizontal.

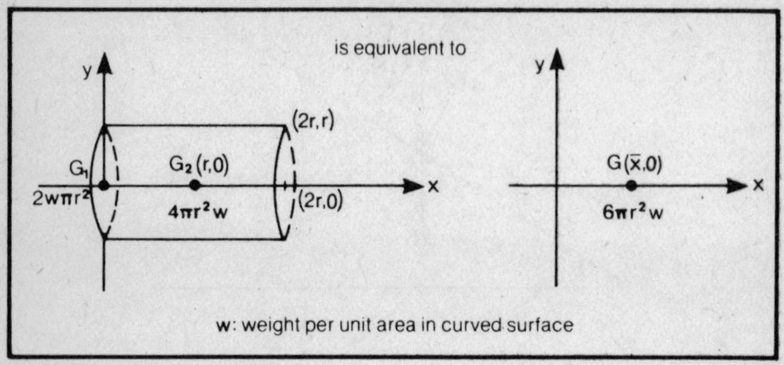

Figure 182.

Taking axes as shown in figure 182, by symmetry $\bar{y} = 0$. Taking moments about the y-axis for both equivalent systems,

$$r4\pi r^2 w = \bar{x}6\pi r^2 w,$$

therefore,
$$\frac{2}{3}r = \bar{x}.$$

Hence, **the centre of gravity of the can is $2r/3$ away from its plane base.**

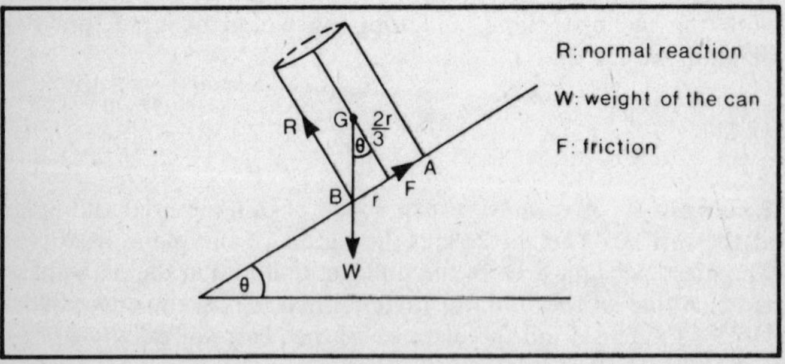

Figure 183.

The angle of inclination must be larger than that for which the can would be on the point of toppling to be sure that toppling will occur. If the can is on the point of toppling, then the line of action of its weight must pass through B, as shown in figure 183.

Hence, for toppling,

$$\tan\theta > \frac{3}{2}.$$

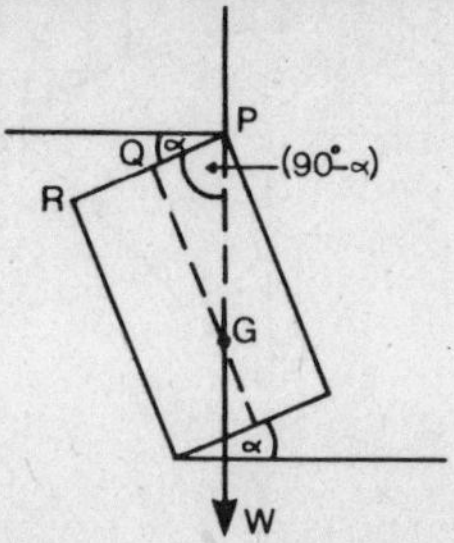

Figure 184.

If the can is freely suspended by a string attached to a point, say P, on its open end, then it will rest in equilibrium with its centre of gravity, G, vertically underneath the point of attachment.

Hence, from figure 184, if α is the angle of inclination of the plane base to the horizontal, $\tan(90° - \alpha) = \frac{4}{3}$ and $\alpha =$ **36·9°**.

Example 3 Prove that the centre of gravity of a uniform lamina in the shape of a sector of a circle, radius a and angle 2θ, is at a distance

$$\frac{2a\sin\theta}{3\theta}$$

from the centre of the circle. State this distance for the case where the sector is a semi-circle.

The lamina shown in figure 185, consists of a right-angled isosceles triangle ADB, together with the semi-circle AEB, radius r, centre C and diameter ACB. From this lamina is removed the sector $DAFB$ whose centre is D, leaving the shaded crescent $AEBF$. State the distances from D of the centres of gravity of

i) the triangle ADB,
ii) the semi-circle,
iii) the sector $DAFB$.

Hence, show that the centre of gravity of the crescent $AEBF$ is at a distance $\pi r/2$ from D.

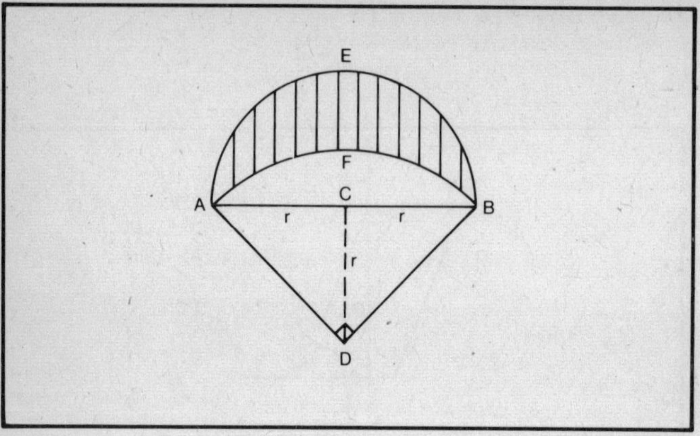

Figure 185.

(A.E.B.)

Earlier in the chapter we saw how to prove that the centre of gravity of a uniform lamina in the shape of a sector of a circle, radius a and angle 2θ, is at a distance $(2a \sin \theta)/3\theta$ from the centre of the circle. We shall not repeat it here.

For the case when this sector is a semi-circle, $2\theta = \pi$, $\theta = \pi/2$. Therefore, the centre of gravity is at a distance $\mathbf{4a/3\pi}$ from the centre.

(i) The distance from D of the centre of gravity of the isosceles triangular lamina is

$$\frac{2}{3}r \text{ (at its centroid).}$$

(ii) The distance from D of the centre of gravity of the semi-circle is

$$\left(r + \frac{4r}{3\pi}\right).$$

(iii) The centre of gravity of the sector $DAFB$ is at a distance

$$\frac{8r\sqrt{2}}{3\sqrt{2}\pi} \quad \text{since} \quad AD = \sqrt{2}r.$$

The crescent $AFBE$ may be considered to be made up of the isosceles triangle ABD + semi-circle ABE – sector $ADBF$.

Hence, taking moments about an axis parallel to AB through D, with the plane of the lamina horizontal, and letting the centre of gravity of the crescent be distance $\bar{y}$ from D,

$$\frac{2r}{3}r^2w + \left(r + \frac{4r}{3\pi}\right)\frac{\pi r^2}{2}w - \frac{8r}{3\pi}\frac{\pi}{4}(\sqrt{2}\,r)^2w = \bar{y}\left(r^2 + \frac{\pi r^2}{2} - \frac{\pi r^2}{2}\right)w,$$

where w is the constant weight per unit area.

$$\text{Hence, } \bar{y} = \frac{\pi r}{2}.$$

Key terms

The **centre of gravity** of a body is the fixed point through which the line of action of its weight passes whatever the orientation of the body.

The **centre of mass** of a body is the point at which the mass of a body may be considered to be concentrated. The centre of mass of a system of particles, $P_1, P_2, P_3, \ldots$ with position vectors $r_1, r_2, r_3, \ldots$ and masses $m_1, m_2, m_3, \ldots$ respectively is defined as

$$\boldsymbol{r}_m = \frac{\sum \boldsymbol{r}_p m_p}{\sum m_p}.$$

The **centre of gravity** of a body may be found by using the fact that the sum of the moments of the weights of its parts about any axis is equal to the moment of the weight of the body about the same axis.

The **centre of mass** of a uniform body lies on every line or axis of symmetry for the body.

The centre of gravity and the centre of mass coincide in a uniform gravitational field.

The centroid of a body is that point about which the area, in the case of a lamina, and the volume, in the case of a solid, is uniformly distributed. It is the geometric centre. The centre of mass and the centroid of a body are coincident when the mass is uniformly distributed for the body.

When a body is freely suspended from a point it will rest in equilibrium with its centre of gravity directly below the point of suspension.

When a body rests on a plane, it will not topple provided the vertical through its centre of gravity falls inside the extremities of the contact of the body with the plane.

We summarise the positions of the centres of gravity and mass of some standard uniform bodies, below.

Body	**Position of centre of gravity or mass along axis or line of symmetry**
Solid hemisphere	$\frac{3a}{8}$ from the base
Hollow hemisphere	$\frac{a}{2}$ from the plane face
Semi-circular lamina	$\frac{4a}{3\pi}$ from the centre of the circle.
Sector subtending angle 2α at the centre	$\frac{2a \sin \alpha}{3\alpha}$ from the centre of the circle.
Arc subtending angle 2α at the centre	$\frac{a \sin \alpha}{\alpha}$ from the centre of the circle.
Solid cone/ tetrahedron/ pyramid	$\frac{h}{4}$ from the base.
Hollow cone/ tetrahedron/ pyramid	$\frac{h}{3}$ from the base.

Chapter 16
Vector Moment. Resultant of Forces in 3-Dimensions

Vector moment

We often refer to the moment of a force about a **point**, where what is actually meant, is the moment of the force about an axis passing through that point perpendicular to the plane containing the point and the force.

Consider the moment of a force **F** about the point P, as shown in figure 186.

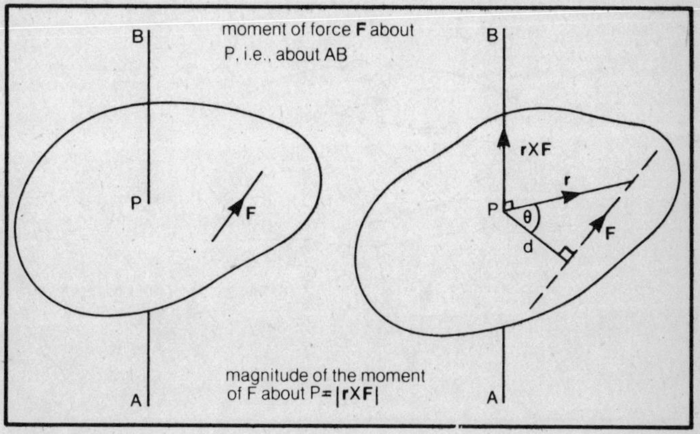

Figure 186.

The **magnitude** of the moment of the force about AB is given by $|\mathbf{r} \times \mathbf{F}| = rF \sin\theta = dF$ where **r** is the position vector relative to P of **any** point on the line of action of **F**. The vector $\mathbf{r} \times \mathbf{F}$ is parallel to AB.

Thus the vector $\mathbf{r} \times \mathbf{F}$ has magnitude equal to the magnitude of the moment of **F** about P and direction equal to that of the axis of rotation in the sense of the motion of a right-handed screw from **r** to **F**.

$\mathbf{r} \times \mathbf{F}$ is called the **vector moment** of **F** about P.

If the **vector moment** of **F** about P is **a**, then $\mathbf{r} \times \mathbf{F} = \mathbf{a}$ is a **vector equation of the line of action of F**.

Let us consider the **vector moment of a couple**. We shall first look at the **vector moment about a point *P* in the plane of the couple**. Let $\mathbf{p}_1$ and $\mathbf{p}_2$ be the position vectors of two points, A and B, on the lines of action of **F** and $-\mathbf{F}$ respectively. Refer to figure 187.

The vector moment of the couple about *P* is,

$$\mathbf{p}_1 \times \mathbf{F} + \mathbf{p}_2 \times (-\mathbf{F}) = (\mathbf{p}_1 - \mathbf{p}_2) \times \mathbf{F} = \mathbf{BA} \times \mathbf{F}$$

i.e., it has **magnitude** equal to the **magnitude** of the moment of the couple, and direction perpendicular to the plane of the couple, in the sense of the motion of a right-handed screw from **BA** to **F**.

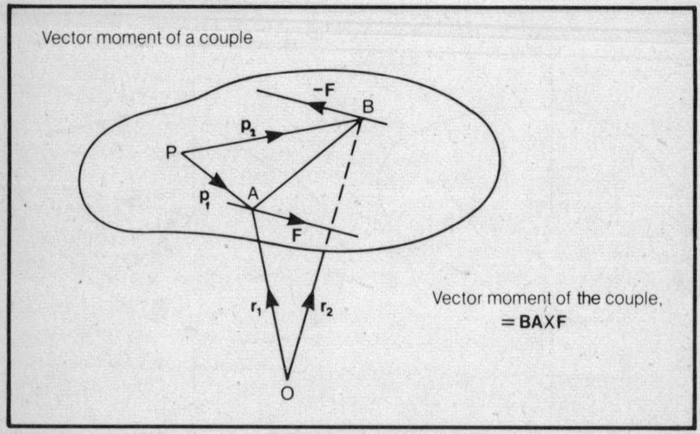

Figure 187.

Let us now consider the **vector moment of the couple about a point *O* not in the plane of the couple**. Let $\mathbf{r}_1$ and $\mathbf{r}_2$ respectively be the position vectors of A and B relative to O.

The **vector moment of the couple about *O*** is

$$\mathbf{r}_1 \times \mathbf{F} + \mathbf{r}_2 \times (-\mathbf{F}) = (\mathbf{r}_1 - \mathbf{r}_2) \times \mathbf{F} = \mathbf{BA} \times \mathbf{F}$$

Hence the **vector moment of a couple is constant about any point**, and is **independent of the point about**

which moments are taken. The **vector moment** of a couple consisting of forces $\mathbf{F}$ and $-\mathbf{F}$ is $\mathbf{BA} \times \mathbf{F}$ where B is **any** point on the line of action of $-\mathbf{F}$, and A is **any** point on the line of action of $\mathbf{F}$.

We say that two couples are **equal** if their **vector moments** are the same. This is only possible if they act in **parallel planes** because the vector moment of a couple is perpendicular to the plane containing the couple.

The **resultant vector moment** of a system of forces is the sum of the vector moments of the individual forces. Consider the resultant vector moment of a set of forces $\mathbf{F}_1$, $\mathbf{F}_2$, $\mathbf{F}_3$, ... about a point P, where $\mathbf{r}_1$, $\mathbf{r}_2$, $\mathbf{r}_3$, ... are the position vectors of points on the lines of action of the forces respectively, relative to P. The **resultant vector moment** about P is $(\mathbf{r}_1 \times \mathbf{F}_1) + (\mathbf{r}_2 \times \mathbf{F}_2) + (\mathbf{r}_3 \times \mathbf{F}_3) + \dots$

Resultants of force in three dimensions

As we have already stated in Chapter 14, two systems of forces are equivalent if their effect on a body is the same in all respects. The resultant of a set of forces is the simplest equivalent system that it may be reduced to.

We have seen that a set of coplanar forces is either in equilibrium, or it reduces to a single force, or it reduces to a couple. If we now consider forces in **three dimensions**, we see that a fourth possibility exists. If we have two forces whose lines of action are skew, then the resultant of this system will involve a combination of linear motion and rotation, not in the same plane. Hence, we may reduce this system to a **single force and a non-coplanar couple**.

Let us examine the conditions under which each of the four possible resultants of systems of non-coplanar forces occur.

i) The system reduces to a **single resultant force R** acting through a point with position vector $\mathbf{a}$ relative to the origin.

Let us consider two non-parallel forces $\mathbf{F}_1$ and $\mathbf{F}_2$ as shown in figure 188(i). If they reduce to a single force $\mathbf{R}$, then their lines of action must intersect at some point. Let the position vector of this point be $\mathbf{a}$ relative to O. The resultant vector moment of $\mathbf{F}_1$ and $\mathbf{F}_2$ about O is

$$\mathbf{a} \times \mathbf{F}_1 + \mathbf{a} \times \mathbf{F}_2 = \mathbf{a} \times (\mathbf{F}_1 + \mathbf{F}_2) = \mathbf{a} \times \mathbf{R}.$$

ence the resultant vector moment of the system is equal to the vector moment of the resultant force about the same point. This argument may be applied successively to systems of any number of non-parallel forces.

Before we generalise this result to any system which reduces to a single force, we must examine the case of **parallel forces.**

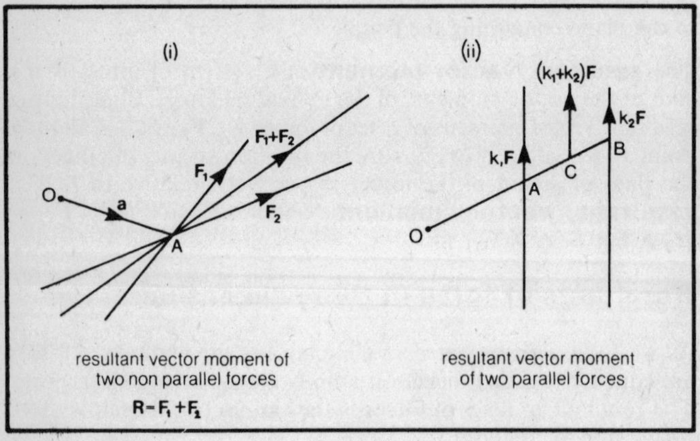

Figure 188.

Referring to figure 188(ii), where OB is a transversal of the two lines of action of k_1 **F** and k_2 **F**. The resultant vector moment about O of k_1 **F** and k_2 **F** is

$$\mathbf{OA} \times k_1 \mathbf{F} + \mathbf{OB} \times k_2 \mathbf{F} = (k_1 \mathbf{OA} + k_2 \mathbf{OB}) \times \mathbf{F}$$

$$= \left(\frac{k_1 \mathbf{OA} + k_2 \mathbf{OB}}{k_2 + k_1}\right) \times (k_2 + k_1)\mathbf{F}.$$

However, $\dfrac{k_1 \mathbf{OA} + k_2 \mathbf{OB}}{k_2 + k_1}$

is the position vector of a point dividing AB in the ratio $k_2 : k_1$ and $(k_2 + k_1)\mathbf{F} = k_2 \mathbf{F} + k_1 \mathbf{F}$. Hence **the resultant vector moment of k_1 F and k_2 F is the vector moment of the resultant force about the same point.** Again this argument may be applied successively to further parallel forces.

Hence the general result that **if a set of forces reduces to a single force, the vector sum of the moments of all the forces about a point, is equal to the vector moment of the resultant force about the same point.**

If a system of forces reduces to a **single resultant force** then $\sum \mathbf{F} = \mathbf{R}$ and $\sum \mathbf{r} \times \mathbf{F} = \mathbf{a} \times \mathbf{R}$ where **a** is the position vector of a point on the line of action of **R**.

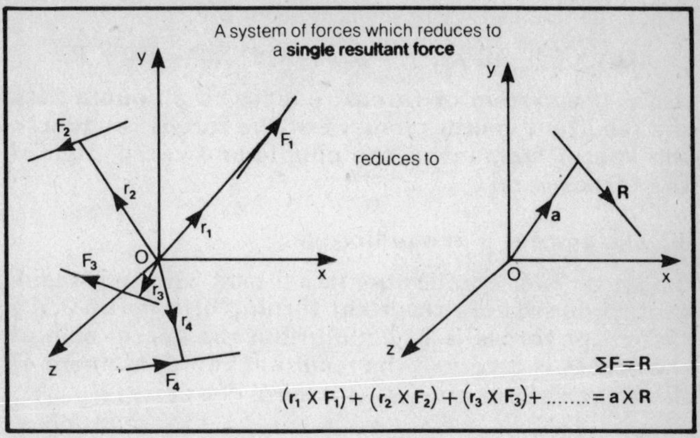

Figure 189.

ii) **The system of forces reduces to a couple of vector moment G.**

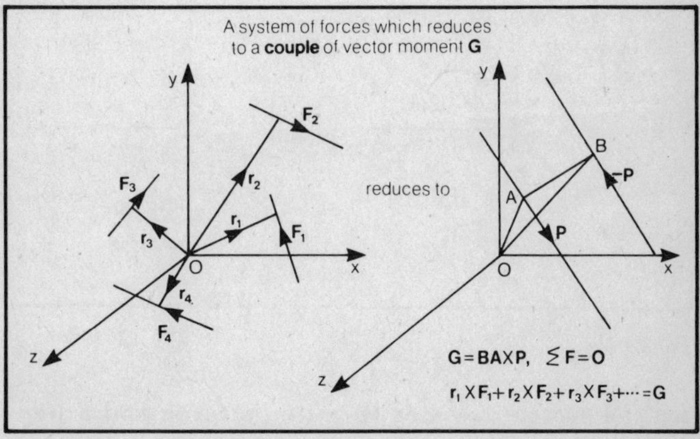

Figure 190.

If a system of forces reduces to a couple of vector moment $\mathbf{G} = \mathbf{BA} \times \mathbf{P}$ as shown in figure 190, then we can divide the original system into two subsets, one which contains all the forces which reduce to the single force **P** passing through A, and the other which contains the other forces which reduce to the single force $-\mathbf{P}$, passing through B.

Hence, the resultant vector moment about O is,

$$\mathbf{OA} \times \mathbf{P} + \mathbf{OB} \times -\mathbf{P} = (\mathbf{OA} - \mathbf{OB}) \times \mathbf{P} = \mathbf{BA} \times \mathbf{P}.$$

Hence, **if a system of forces reduces to a couple then the resultant vector moment of the forces is equal to the vector moment of the couple and vector sum of the forces is zero.**

iii) **The system is in equilibrium.**

If a system is in **equilibrium** then it must have **zero resultant force** and **zero resultant turning effect**. Hence, **if a system of forces is in equilibrium the vector sum of the forces is zero and the resultant vector moment of the forces about any point is zero.** (See figure 191.)

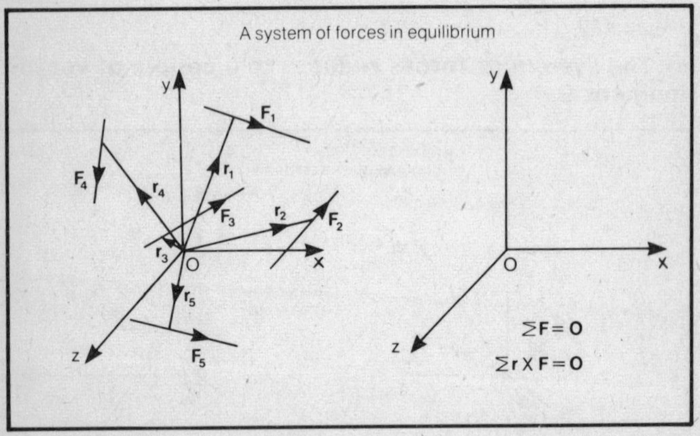

Figure 191.

iv) **The system reduces to a single force and a non-coplanar couple.**

As we have said earlier, this situation arises when the system reduces to two forces whose lines of action are **skew**. To establish for a general system of forces that this is the case, requires vector techniques more advanced than A-level standard. If we are asked to solve problems involving this type of situation we shall be given extra information enabling us to do so.

We have examined above the results that follow when we are **given** the resultant of a system of forces. Let us now examine how we could **prove** that the resultant of a system is one of the above cases.

To prove that a system of forces is in **equilibrium it** is necessary to show **both that** $\sum \mathbf{F} = \mathbf{O}$ **and that** $\sum \mathbf{r} \times \mathbf{F} = \mathbf{O}$.

To prove that a system of forces reduces to a **couple** it is necessary to show **both that** $\sum \mathbf{F} = \mathbf{O}$ **and that** $\sum \mathbf{r} \times \mathbf{F} \neq \mathbf{O}$.

To prove that a system of forces reduces to a **single force**, presents difficulties. If we find that the **vector sum of the forces is not zero**, then the system may reduce to **either a single resultant force or a single force and a non-coplanar couple**. Again at this level we would need some extra information to distinguish between these two cases.

Worked examples

Example 1 Three forces $\mathbf{F}_1 = 4\mathbf{i} + \mathbf{j}$, $\mathbf{F}_2 = 2\mathbf{i} + \mathbf{j} + 2\mathbf{k}$, and $\mathbf{F}_3 = 5\mathbf{i} + \mathbf{j} - \mathbf{k}$ act at the point whose position vector is $\mathbf{r} = \mathbf{i} + \mathbf{k}$. Show that the forces are coplanar. Find:
i) the magnitude of the resultant of the forces and vector and cartesian equations of the line of action of the resultant,
ii) the cosine of the angle between the lines of action of the forces $\mathbf{F}_1$ and $\mathbf{F}_2$.
Another force $\mathbf{F}_4$ has magnitude $3\sqrt{2}$ units and is in a direction perpendicular to $\mathbf{F}_1$, $\mathbf{F}_2$ and $\mathbf{F}_3$. Find $\mathbf{F}_4$ if it makes an acute angle with the direction $\mathbf{j}$. (A.E.B.)

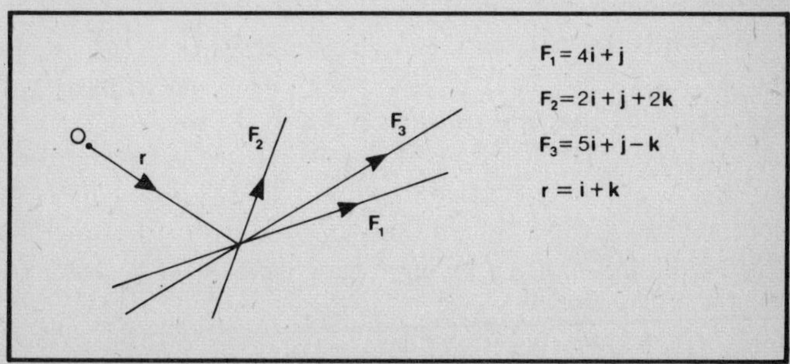

Figure 192.

We are asked to show that the forces are coplanar. We are told that the lines of action of the forces intersect at $\mathbf{r} = \mathbf{i} + \mathbf{k}$. This being the case, if we can show that the vector product of direction vectors for $\mathbf{F}_1$ and $\mathbf{F}_2$ has the same direction as the vector product of direction vectors for $\mathbf{F}_1$ and $\mathbf{F}_3$ (or $\mathbf{F}_2$ and $\mathbf{F}_3$), then the forces are coplanar. Direction vectors for $\mathbf{F}_1$, $\mathbf{F}_2$, $\mathbf{F}_3$, are $4\mathbf{i} + \mathbf{j}$, $2\mathbf{i} + \mathbf{j} + 2\mathbf{k}$, $5\mathbf{i} + \mathbf{j} - \mathbf{k}$, respectively.

$$(4\mathbf{i} + \mathbf{j}) \times (2\mathbf{i} + \mathbf{j} + 2\mathbf{k}) = \begin{vmatrix} \mathbf{i} & \mathbf{j} & \mathbf{k} \\ 4 & 1 & 0 \\ 2 & 1 & 2 \end{vmatrix} = 2\mathbf{i} - 8\mathbf{j} + 2\mathbf{k} \text{ and}$$

$$(4\mathbf{i} + \mathbf{j}) \times (5\mathbf{i} + \mathbf{j} - \mathbf{k}) = \begin{vmatrix} \mathbf{i} & \mathbf{j} & \mathbf{k} \\ 4 & 1 & 0 \\ 5 & 1 & -1 \end{vmatrix} = -\mathbf{i} + 4\mathbf{j} - \mathbf{k}$$

These two vector products have equal direction ratios, $-1:4:-1$. Therefore they are parallel vectors and **the forces are coplanar**.

(i) The magnitude and direction of the resultant force, $\mathbf{F}$, is given by,

$$\begin{aligned} \mathbf{F} &= \mathbf{F}_1 + \mathbf{F}_2 + \mathbf{F}_3 \\ &= (4\mathbf{i} + \mathbf{j}) + (2\mathbf{i} + \mathbf{j} + 2\mathbf{k}) + (5\mathbf{i} + \mathbf{j} - \mathbf{k}) \\ &= 11\mathbf{i} + 3\mathbf{j} + \mathbf{k}. \end{aligned}$$

Since this force is the resultant of three concurrent forces, its line of action also passes through their point of intersection, $\mathbf{r}$.

The magnitude of $\mathbf{F}$ is $\sqrt{11^2 + 3^2 + 1^2} = \sqrt{131}$.

The equation of its line of action, in vector form, is $\mathbf{r} = \mathbf{i} + \mathbf{k} + \lambda(11\mathbf{i} + 3\mathbf{j} + \mathbf{k})$. ($(\mathbf{i} + \mathbf{k})$ is the position vector of a point on the line of action, and $11\mathbf{i} + 3\mathbf{j} + \mathbf{k}$ is a direction vector for the line.)

Since $x = 1 + 11\lambda$, $y = 3\lambda$, $z = 1 + \lambda$, then the **Cartesian equation of the line of action** of $\mathbf{F}$ is

$$\frac{x-1}{11} = \frac{y}{3} = z - 1$$

(ii) The cosine of the angle, say θ, between the lines of action of $\mathbf{F}_1$ and $\mathbf{F}_2$, is given by

$$\frac{(4\mathbf{i} + \mathbf{j}) \cdot (2\mathbf{i} + \mathbf{j} + 2\mathbf{k})}{|4\mathbf{i} + \mathbf{j}|\,|2\mathbf{i} + \mathbf{j} + 2\mathbf{k}|} = \cos\theta \quad \text{i.e., } \cos\theta = \frac{3}{\sqrt{17}}$$

If the force $\mathbf{F}_4$ is in a direction perpendicular to $\mathbf{F}_1$, $\mathbf{F}_2$, and $\mathbf{F}_3$, then its direction ratios must be that of $\mathbf{F}_1 \times \mathbf{F}_2$, i.e., $-1:4:-1$. Hence $\mathbf{F}_4$ may be written as

$$\mathbf{F}_4 = \pm\frac{3\sqrt{2}}{\sqrt{18}}(-\mathbf{i} + 4\mathbf{j} - \mathbf{k}) = \pm(-\mathbf{i} + 4\mathbf{j} - \mathbf{k})$$

We decide between the plus and minus signs by using the information that $\mathbf{F}_4$ makes an acute angle with the direction $\mathbf{j}$.

$$\text{Either} \quad \frac{(-\mathbf{i} + 4\mathbf{j} - \mathbf{k}) \cdot \mathbf{j}}{\sqrt{18}} \quad \text{or} \quad \frac{-(-\mathbf{i} + 4\mathbf{j} - \mathbf{k}) \cdot \mathbf{j}}{\sqrt{18}}$$

must be less than 1 but greater than 0, since these expressions give the cosines of the angles between the vectors.

$$\text{Hence} \qquad \mathbf{F}_4 = -\mathbf{i} + 4\mathbf{j} - \mathbf{k}.$$

Example 2 Two forces $\mathbf{F}_1 = -6\mathbf{i} - \mathbf{j} + 2\mathbf{k}$ and $\mathbf{F}_2 = 6\mathbf{i} + 3\mathbf{j} - 5\mathbf{k}$ act through the points A and B which have position vectors $\mathbf{r}_1 = 5\mathbf{j} + \mathbf{k}$ and $\mathbf{r}_2 = -2\mathbf{i} + 3\mathbf{j} + 2\mathbf{k}$ respectively. Find the sum of the vector moments of $\mathbf{F}_1$ and $\mathbf{F}_2$ about the origin and show that there is no point about which this sum is zero.

A third force $\mathbf{F}_3$ which acts parallel to AB is added to the system; the three forces are equivalent to a force which acts in the $\mathbf{i}$-$\mathbf{j}$ plane, together with a couple. Find $\mathbf{F}_3$.

(A.E.B.)

The sum of the vector moments about the origin is,

$$(\mathbf{r}_1 \times \mathbf{F}_1) + (\mathbf{r}_2 \times \mathbf{F}_2) = \begin{vmatrix} \mathbf{i} & \mathbf{j} & \mathbf{k} \\ 0 & 5 & 1 \\ -6 & -1 & 2 \end{vmatrix} + \begin{vmatrix} \mathbf{i} & \mathbf{j} & \mathbf{k} \\ -2 & 3 & 2 \\ 6 & 3 & -5 \end{vmatrix}$$

$$= (11\mathbf{i} - 6\mathbf{j} + 30\mathbf{k}) + (-21\mathbf{i} + 2\mathbf{j} - 24\mathbf{k})$$

$$= -10\mathbf{i} - 4\mathbf{j} + 6\mathbf{k}.$$

This resultant vector moment is not a zero vector. If there is no point about which the sum of the vector moments is zero, then either the system must reduce to a couple, or to a single force together with a non-coplanar couple. The latter case occurs when the system reduces to two forces with skew lines of action. We must check for both of these cases.

If the system reduces to a couple, then the vector sum of $\mathbf{F}_1$ and $\mathbf{F}_2$ would be zero. This is not the case.

The line of action of $\mathbf{F}_1$ is given by,

$$\mathbf{r} = (5\mathbf{j} + \mathbf{k}) + \lambda(-6\mathbf{i} - \mathbf{j} + 2\mathbf{k})$$

and of $\mathbf{F}_2$ by, $\mathbf{r} = (-2\mathbf{i} + 3\mathbf{j} + 2\mathbf{k}) + \mu(6\mathbf{i} + 3\mathbf{j} - 5\mathbf{k})$.

These two lines of action are not parallel since their direction vectors have different direction ratios. Hence, unless the lines are skew, we should be able to find a λ and μ, such that

$$(5\mathbf{j} + \mathbf{k}) + \lambda(-6\mathbf{i} - \mathbf{j} + 2\mathbf{k}) = (-2\mathbf{i} + 3\mathbf{j} + 2\mathbf{k}) + \mu(6\mathbf{i} + 3\mathbf{j} - 5\mathbf{k}).$$

Equating coefficients of $\mathbf{i}$, $-6\lambda = -2 + 6\mu$.

Equating coefficients of $\mathbf{j}$, $5 - \lambda = 3 + 3\mu$.

Equating coefficients of $\mathbf{k}$, $1 + 2\lambda = 2 - 5\mu$.

From the first two equations, $\mu = \frac{5}{6}$ and $\lambda = -\frac{1}{2}$.

Substituting these values into the third equation,

the L.H.S $= 1 + 2\lambda = 0$, and the R.H.S $= 2 - 5\mu \neq 0$.

Hence, we must conclude that these values of μ and λ are not consistent with the lines intersecting. The lines of action of forces $\mathbf{F}_1$ and $\mathbf{F}_2$ are skew. Hence, forces $\mathbf{F}_1$ and $\mathbf{F}_2$ will not reduce to a single force. Therefore, since the forces are not in equilibrium either, **there is no point about which the sum of the vector moments is zero**. (This can only occur when moments are taken about a point on the line of action of a single resultant force, or when the system is in equilibrium.

We are told that a third force, $\mathbf{F}_3$, which acts parallel to AB, is added to the system and that the new system is equivalent to a system consisting of a force which acts in the $\mathbf{i}$-$\mathbf{j}$ plane together with a couple.

If the two systems are equivalent, the vector sum of the forces is the same in both systems.

$$\mathbf{AB} = \mathbf{r}_2 - \mathbf{r}_1 = -2\mathbf{i} - 2\mathbf{j} + \mathbf{k}.$$

Since $\mathbf{F}_3$ is parallel to $\mathbf{AB}$, then $\mathbf{F}_3 = \lambda(-2\mathbf{i} - 2\mathbf{j} + \mathbf{k})$, where λ is a constant.

We shall let the resultant force in the **i**-**j** plane be $a\mathbf{i} + b\mathbf{j}$.

Hence,

$$(-6\mathbf{i} - \mathbf{j} + 2\mathbf{k}) + (6\mathbf{i} + 3\mathbf{j} - 5\mathbf{k}) + \lambda(-2\mathbf{i} - 2\mathbf{j} + \mathbf{k}) = a\mathbf{i} + b\mathbf{j}$$

(the vector sum of the couple is zero).

Equating coefficients of k, $\lambda = +3$.

$$\text{Hence, } \mathbf{F}_3 = -6\mathbf{i} - 6\mathbf{j} + 3\mathbf{k}.$$

Example 3 Forces $\mathbf{F}_1 = 2\mathbf{i} - \mathbf{j} + 3\mathbf{k}$, $\mathbf{F}_2 = 4\mathbf{i} + \mathbf{j} + 5\mathbf{k}$, $\mathbf{F}_3 = -6\mathbf{i} - 8\mathbf{k}$, act respectively at points with position vectors $\mathbf{i} - 2\mathbf{j} - \mathbf{k}$, $7\mathbf{i} - 2\mathbf{j} + 7\mathbf{k}$, $3\mathbf{i} - 3\mathbf{j} + \mathbf{k}$. Show that the lines of action of $\mathbf{F}_1$ and $\mathbf{F}_2$ intersect and give the position vector of their point of intersection.

Show that the system of forces is equivalent to a couple and find its magnitude.

If the force $\mathbf{F}_3$ is now replaced by a force $\mathbf{F}_4$ such that the system is now in equilibrium, find $\mathbf{F}_4$ and give an equation of its line of action in vector form.

An equation of the line of action of $\mathbf{F}_1$ is

$$\mathbf{r} = (\mathbf{i} + 2\mathbf{j} - \mathbf{k}) + \lambda(2\mathbf{i} - \mathbf{j} + 3\mathbf{k})$$

and an equation of the line of action of $\mathbf{F}_2$ is

$$\mathbf{r}_2 = (7\mathbf{i} - 2\mathbf{j} + 7\mathbf{k}) + \mu(4\mathbf{i} + \mathbf{j} + 5\mathbf{k}).$$

If these two lines intersect, there will be values of λ and μ such that

$$(\mathbf{i} - 2\mathbf{j} - \mathbf{k}) + \lambda(2\mathbf{i} - \mathbf{j} + 3\mathbf{k}) = (7\mathbf{i} - 2\mathbf{j} + 7\mathbf{k}) + \mu(4\mathbf{i} + \mathbf{j} + 5\mathbf{k}).$$

Equating coefficients of **i**, $1 + 2\lambda = 7 + 4\mu$.

Equating coefficients of **j**, $-2 - \lambda = -2 + \mu$.

Equating coefficients of **k**, $-1 + 3\lambda = 7 + 5\mu$.

From the first two equations, $\lambda = 1$ and $\mu = -1$.

Substituting these values into the third equation, L.H.S = 2 and R.H.S = 2. Hence these values of λ and μ are consistent with intersection.

The position vector of the point of intersection is given by substituting $\lambda = 1$ or $\mu = -1$ into the appropriate equation of the line of action, i.e., they intersect at $\mathbf{r} = 3\mathbf{i} - 3\mathbf{j} + 2\mathbf{k}$.

The vector sum of the forces $(\mathbf{F}_1 + \mathbf{F}_2 + \mathbf{F}_3)$

$$= (2\mathbf{i} - \mathbf{j} + 3\mathbf{k}) + (4\mathbf{i} + \mathbf{j} + 5\mathbf{k}) + (-6\mathbf{i} - 8\mathbf{k}) = 0$$

The resultant vector moment,

$$(\mathbf{i} - 2\mathbf{j} - \mathbf{k}) \times (2\mathbf{i} - \mathbf{j} + 3\mathbf{k}) + (7\mathbf{i} - 2\mathbf{j} + 7\mathbf{k}) \times (4\mathbf{i} + \mathbf{j} + 5\mathbf{k}) + (3\mathbf{i} - 3\mathbf{j} + \mathbf{k}) \times (-6\mathbf{i} - 8\mathbf{k})$$

$$= \begin{vmatrix} \mathbf{i} & \mathbf{j} & \mathbf{k} \\ 1 & -2 & -1 \\ 2 & -1 & 3 \end{vmatrix} + \begin{vmatrix} \mathbf{i} & \mathbf{j} & \mathbf{k} \\ 7 & -2 & 7 \\ 4 & 1 & 5 \end{vmatrix} + \begin{vmatrix} \mathbf{i} & \mathbf{j} & \mathbf{k} \\ 3 & -3 & 1 \\ -6 & 0 & -8 \end{vmatrix} = 6\mathbf{j}$$

Since the vector sum of the forces is zero and the resultant vector moment is not zero, **the system of forces is equivalent to a couple of magnitude $|6\mathbf{j}| = 6$ units.**

If $\mathbf{F}_3$ is replaced by $\mathbf{F}_4$, so that the system is now in equilibrium, $\mathbf{F}_1 + \mathbf{F}_2 + \mathbf{F}_4 = 0$.

Letting $\mathbf{F}_4 = a\mathbf{i} + b\mathbf{j} + c\mathbf{k}$, then

$$(2\mathbf{i} - \mathbf{j} + 3\mathbf{k}) + (4\mathbf{i} + \mathbf{j} + 5\mathbf{k}) + (a\mathbf{i} + b\mathbf{j} + c\mathbf{k}) = 0.$$

Equating coefficients of $\mathbf{i}$, $\mathbf{j}$ and $\mathbf{k}$, $a = -6$, $b = 0$ and $c = -8$. Hence, $\mathbf{F}_4 = -6\mathbf{i} - 8\mathbf{k}$.

We are in a three force system in equilibrium. $\mathbf{F}_1$ and $\mathbf{F}_2$ intersect at $\mathbf{r} = 3\mathbf{i} - 3\mathbf{j} + 2\mathbf{k}$, therefore their resultant will also pass through that point.

Hence, $\mathbf{F}_4 = -(\mathbf{F}_1 + \mathbf{F}_2) = -6\mathbf{i} - 8\mathbf{k}$ and must pass through $3\mathbf{i} - 3\mathbf{j} + 2\mathbf{k}$.

A vector equation of the line of action of $\mathbf{F}_4$ is

$$\mathbf{r} = (3\mathbf{i} - 3\mathbf{j} + 2\mathbf{k}) + \lambda(-6\mathbf{i} - 8\mathbf{k}).$$

Example 4 A uniform sphere, of radius a and weight W, is supported by three equal rough rods joined at their ends to form a horizontal equilateral triangle of side b. Show that the centre of the sphere is at a height

$$\sqrt{\left(a^2 - \frac{b^2}{12}\right)}$$

above the level of the rods.

A couple is applied to the sphere about a vertical axis through the centre of the sphere. If the sphere is on the point of slipping, show that the moment of the couple about the vertical axis is

$$\frac{\mu W a b}{\sqrt{12a^2 - b^2}}$$

where μ is the coefficient of friction between each rod and the sphere.

(This problem is best solved by not using vector methods, merely the conditions for equilibrium in three dimensions).

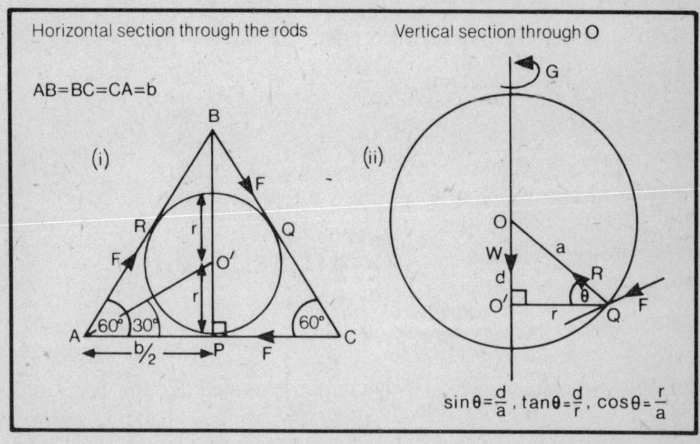

Figure 193.

The horizontal section of the sphere at the level of the rods must be the inscribed circle of the equilateral triangle of side b, formed by the rods. The centre of the inscribed circle lies at the intersection of the bisectors of the angles of the triangle.

Referring to figure 193(i), r being the radius of the inscribed circle,

$$r = \tan 30^\circ \frac{b}{2} = \frac{b}{2\sqrt{3}}.$$

Referring to diagram (ii), $\cos\theta = \dfrac{r}{a} = \dfrac{b}{2\sqrt{3}\,a}$.

If $\cos\theta = \dfrac{b}{2\sqrt{3}\,a}$, then $\tan\theta = \dfrac{\sqrt{12a^2 - b^2}}{b}$, and since $d = r \tan\theta$,

$$\boldsymbol{d} = \frac{b}{2\sqrt{3}} \frac{\sqrt{12a^2 - b^2}}{b} = \sqrt{\boldsymbol{a}^2 - \frac{\boldsymbol{b}^2}{\mathbf{12}}}, \quad \text{as required.}$$

The sphere is supported entirely by its contact with the rods in a symmetrical fashion. The normal reactions of the rods on the sphere will pass through O, the centre of the sphere. When the couple is applied to the sphere, it is on the point of motion so that the frictional forces, F, at each point of contact, acting along the rods in the directions opposing motion, are limiting.

Resolving **vertically** for the sphere, $3R \sin\theta = W$.

At each point of contact,

$$F = \mu R = \frac{\mu W}{3 \sin\theta} = \frac{\mu W\, 2\sqrt{3}\, a}{3\sqrt{12a^2 - b^2}}$$

The resultant moment about the vertical axis must be zero.

Therefore, $$\boldsymbol{G} = 3rF = \frac{3b}{2\sqrt{3}} \frac{\mu W 2\sqrt{3}\, a}{3\sqrt{12a^2 - b^2}}$$

$$= \frac{\boldsymbol{\mu W a b}}{\sqrt{\mathbf{12}\boldsymbol{a}^2 - \boldsymbol{b}^2}} \textbf{ units,}$$

in the opposite sense to the couple provided by the frictional forces.

Key terms

The **moment** of a force about a **point** is the moment of that force about an axis perpendicular to the plane containing the force and the point, passing through that point.

The **vector moment** of a force **F** about a point P is **r** $\times$ **F** where **r** is the position vector, relative to P, of a point on the line of action of the force. Its magnitude is that of the magnitude of its moment about P and its direction is perpendicular to the plane containing **r** and **F** in the sense of the motion of a right handed screw from **r** to **F**, i.e., its direction is parallel to the axis of rotation.

The **vector moment of a couple** consisting of equal unlike parallel forces **F** and −**F** is given by **BA** $\times$ **F** where B is any point on the line of action of −**F** and A is any point on the line of action of **F**. Thus, it is a vector, perpendicular to the plane of the couple. The **vector moment of a couple** is **independent** of the point about which moments are taken and is **constant**.

Two couples are **equal** if their **vector moment is equal**, hence equal couples act in **parallel planes**.

The **resultant vector moment** of a set of forces is the vector sum of the vector moments of the individual forces.

Two sets of forces are **equivalent** if their effect on a body is the same in all respects.

The **resultant** of a system of forces is the **simplest possible equivalent system** that it may be reduced to.

A system of forces in three dimensions reduces to **equilibrium**, or to a **single resultant force**, or to a **couple**, or to a **single force and a non-coplanar couple**. The latter case arises when the system reduces to two forces whose lines of action are skew.

If a system is in **equilibrium** then $\sum \mathbf{F} = \mathbf{O}$ and $\sum (\mathbf{r} \times \mathbf{F}) = \mathbf{O}$.

If the system reduces to a **single resultant force**, **R**, then $\sum \mathbf{F} = \mathbf{R}$ and the line of action may be found by comparing its turning effect with that of the resultant vector moment of the original system. We can also find its line of action if we know the position vector of a point on its line of action since its direction vector will be that of **R**.

If the system reduces to a **couple** then $\sum \mathbf{F} = \mathbf{O}$ and $\sum (\mathbf{r} \times \mathbf{F}) \neq \mathbf{O}$.

If the system reduces to a **single force, R, and a non-coplanar couple**, then $\sum \mathbf{F} = \mathbf{R}$ and the moments about any point will not be zero, but will not, in general, be constant.

Chapter 17
Motion of the Centre of Mass

We shall examine the motion of the centre of mass of a system of particles. Let a typical particle P be of mass m_p, with, at time t, position vector $\mathbf{r}_p$, velocity vector $\mathbf{v}_p$, acceleration vector $\mathbf{a}_p$, and have single resultant force $\mathbf{F}_p$ acting on it. Let the **initial** position, velocity and acceleration vectors be $\mathbf{R}_p$, $\mathbf{V}_p$, $\mathbf{A}_p$ respectively.

Let the position vector of the centre of mass be, at time t, $\mathbf{r}_m$ and the velocity and acceleration vectors be $\mathbf{v}_m$, $\mathbf{a}_m$, respectively. Initially, the position, velocity and acceleration vectors of the centre of mass of the system are $\mathbf{R}_m$, $\mathbf{V}_m$, $\mathbf{A}_m$, respectively.

By definition $\mathbf{r}_m \sum m_p = \sum m_p \mathbf{r}_p$. Hence, differentiating with respect to time,

$$\frac{d\mathbf{r}_m}{dt} \sum m_p = \sum m_p \frac{d\mathbf{r}_p}{dt}, \quad \text{i.e., } \mathbf{v}_m \sum m_p = \sum m_p \mathbf{v}_p \quad \text{and}$$

$$\frac{d^2\mathbf{r}_m}{dt^2} \sum m_p = \sum m_p \frac{d^2\mathbf{r}_m}{dt^2}, \quad \text{i.e., } \mathbf{a}_m \sum m_p = \sum m_p \mathbf{a}_p.$$

Applying Newton's second law to the particle P, $\mathbf{F}_p = m_p \mathbf{a}_p$ and summing for the set, $\sum \mathbf{F}_p = \sum m_p \mathbf{a}_p$.

Hence, $\mathbf{a}_m \sum m_p = \sum \mathbf{F}_p$, where $\sum \mathbf{F}_p$ is the vector sum of the forces acting on the system. The summation of any internal forces over the whole system will be zero. Hence, $\sum \mathbf{F}_p$ is the vector sum of the external forces.

Therefore, **the acceleration of the centre of mass of a system of particles is that which would be produced by the resultant force acting on a particle of mass equal to the total mass of the system.**

Consider now the impulse, I_p, acting on each particle. We know that 'impulse = change in momentum,' hence, for P,

$$I_p = m_p \mathbf{v}_p - m_p \mathbf{V}_p.$$

Summing for the system, $\sum I_p = \sum m_p \mathbf{V}_p - \sum m_p \mathbf{V}_p$.

$\sum \mathrm{I}_p$ is the resultant external impulse acting on the system, since equal, opposite internal impulses will cancel out in the summation.

We know that $\mathbf{v}_m \sum m_p = \sum m_p \mathbf{v}_p$ and $\mathbf{V}_m \sum m_p = \sum m_p \mathbf{V}_p$.

Therefore, $\sum \mathrm{I}_p = \mathbf{v}_m \sum m_p - \mathbf{V}_m \sum m_p$.

This result demonstrates that **the change in the resultant momentum of a system, is the same as that which would be produced by the resultant impulse acting on a particle of the same mass as that of the total system**.

These two results for the **acceleration** of the centre of mass and for the **change in momentum** of the centre of mass are general results for the centre of mass of a system of particles and forces. In particular they apply to a **rigid body** which is considered to be a large number of particles held together by internal forces which maintain the particles in a constant relative position to one another.

We shall now consider some particular results for the motion of the centre of mass in a **uniform gravitational field** (or in any system where all the forces act in the same direction).

If the only forces acting on the particles are their weights, then, with the usual notation, $\mathbf{F}_p = m_p \mathbf{g}$. The work done in time t on particle P is $\mathbf{F}_p . (\mathbf{r}_p - \mathbf{R}_p)$.

Summing for the system, the total work done is

$$\sum \mathbf{F}_p . (\mathbf{r}_p - \mathbf{R}_p) = \sum m_p \mathbf{g} . (\mathbf{r}_p - \mathbf{R}_p)$$

$$= \sum \mathbf{g} . (m_p \mathbf{r}_p - m_p \mathbf{R}_p) = \mathbf{g} . \sum (m_p \mathbf{r}_p - m_p \mathbf{R}_p).$$

Let us now consider the work that would be done on a particle of mass $\sum m_p$ in moving it from $\mathbf{R}_m$ to $\mathbf{r}_m$ by the force $\sum \mathbf{F}_p$.

We know that the work done is

$$(\sum \mathbf{F}_p) . (\mathbf{r}_m - \mathbf{R}_m) = (\sum m_p \mathbf{g}) . \left(\frac{\sum m_p \mathbf{r}_p}{\sum m_p} - \frac{\sum m_p \mathbf{R}_p}{\sum m_p} \right)$$

$$= \mathbf{g} . \sum (m_p \mathbf{r}_p - m_p \mathbf{R}_p)$$

Hence, **the total work done by the individual gravitational forces in displacing the centre of mass from $\mathbf{R}_m$**

to $\mathbf{r}_m$ is equal to the work that would be done by the resultant gravitational force acting on a particle of mass $\sum m_p$ in giving it the same displacement.

Since the change in kinetic energy of P is the work done by the resultant force acting on it, **the total change in kinetic energy of this system is equal to the change in kinetic energy of a particle of mass $\sum m_p$ placed at the centre of mass and acted on by the resultant weight over the corresponding time interval.**

We conclude that for a system of particles moving under gravity in a **uniform gravitational field**, the properties of their linear motion corresponds to those which would be produced by the resultant weight acting on a particle whose mass is that of the total mass of the system placed at the centre of mass.

In a uniform gravitational field, the centre of mass coincides with the centre of gravity.

(Generally, the total work done by all the forces, $\mathbf{F}_p$, acting on a system of particles is

$$\sum \mathbf{F}_p \cdot (\mathbf{r}_p - \mathbf{R}_p) = \sum \mathbf{F}_p \cdot \mathbf{r}_p - \sum \mathbf{F}_p \cdot \mathbf{R}_p,$$

and the work done by a force of $\sum \mathbf{F}_p$ acting on a particle of mass $\sum m_p$, placed at the centre of mass, in moving it from $\mathbf{R}_m$ to $\mathbf{r}_m$ is

$$\left(\sum \mathbf{F}_p\right) \cdot (\mathbf{r}_m - \mathbf{R}_m) = \left(\sum \mathbf{F}_p\right) \cdot \left(\frac{\sum m_p \mathbf{r}_p}{\sum m_p} - \frac{\sum m_p \mathbf{R}_p}{\sum m_p}\right).$$

We see that, in general, these two quantities of work will not be the same.)

Worked examples

Example 1 Define the centre of mass of a system of particles $P_1, P_2, \ldots$ of masses $m_1, m_2, \ldots$ and with position vectors $\mathbf{r}_1, \mathbf{r}_2, \ldots$ respectively. Show that the position of the centre of mass is independent of the choice of origin.

Find the position vector of the centre of mass of particles of masses 2, 4, 6, 3 units and respective position vectors $\mathbf{i} + \mathbf{j}$, $2\mathbf{i} - \mathbf{j}$, $\mathbf{i} - 2\mathbf{j}$, $3\mathbf{i} + \mathbf{j}$.

If each mass is acted on by a force proportional to its distance from the origin and directed towards the origin, find the direction vector of the original acceleration of the centre of mass.

The **centre of mass** of the set of particles is defined as that point with position vector $\mathbf{r}_m$,

where
$$\mathbf{r}_m = \frac{\sum_p m_p \mathbf{r}_p}{\sum_p m_p}.$$

Suppose that, relative to origin O, the centre of mass of the system is at point C with position vector $\mathbf{r}_m$, and that relative to origin O' it is at point C' with position vector $\mathbf{s}_m$. We shall show that C' and C are coincident. Let the position vector of O' relative to O be $\mathbf{d}$ and the position vectors of $P_1, P_2, \ldots$ be $\mathbf{s}_1, \mathbf{s}_2, \ldots$ relative to O'.

For each particle $\mathbf{r}_p = \mathbf{d} + \mathbf{s}_p$.

Therefore,
$$\mathbf{r}_m = \frac{\sum_p m_p \mathbf{r}_p}{\sum_p m_p} = \frac{\sum_p m_p(\mathbf{d} + \mathbf{s}_p)}{\sum_p m_p}$$
$$= \frac{\mathbf{d}\sum_p m_p + \sum_p m_p \mathbf{s}_p}{\sum_p m_p} = \mathbf{d} + \frac{\sum_p m_p \mathbf{s}_p}{\sum_p m_p},$$

but $\mathbf{s}_m = \dfrac{\sum_p m_p \mathbf{s}_p}{\sum_p m_p}$. Thus, $\mathbf{r}_m = \mathbf{d} + \mathbf{s}_m$.

Hence, C and C' coincide, and we see that **the position of the centre of mass is independent of the choice of origin.**

The position vector, $\mathbf{r}_m$, of the centre of mass of particles of mass 2, 4, 6, 3 units and position vectors $(\mathbf{i} + \mathbf{j})$, $(2\mathbf{i} - \mathbf{j})$, $(\mathbf{i} - 2\mathbf{j})$, $(3\mathbf{i} + \mathbf{j})$ respectively, is, by definition,

$$\mathbf{r}_m = \frac{2(\mathbf{i} + \mathbf{j}) + 4(2\mathbf{i} - \mathbf{j}) + 6(\mathbf{i} - 2\mathbf{j}) + 3(3\mathbf{i} + \mathbf{j})}{15}$$
$$= \frac{5\mathbf{i}}{3} - \frac{11\mathbf{j}}{15}.$$

If each particle is acted on by a force, $\mathbf{F}_p$, directed towards the origin and proportional to the distance of the particle from the origin, then, in general, with the usual notation,

$$\mathbf{F}_p = -k|\mathbf{r}_p|\hat{\mathbf{r}}_p = -k\mathbf{r}_p$$

where k is the positive constant of proportionality.

The initial acceleration of the centre of mass, $\mathbf{a}_m$

$$= \frac{\sum \mathbf{F}_p}{\sum m_p}$$

$$= -\frac{k}{15}(\mathbf{i}+\mathbf{j}+2\mathbf{i}-\mathbf{j}+\mathbf{i}-2\mathbf{j}+3\mathbf{i}+\mathbf{j})$$

$$= -\frac{k}{15}(7\mathbf{i}-\mathbf{j}).$$

Hence, the initial acceleration of the centre of mass has direction vector $(-7\mathbf{i}+\mathbf{j})$.

Example 2 Three particles are at rest on a smooth horizontal plane, their masses being 2, 6, 2 units and their position vectors being $3\mathbf{i}+4\mathbf{j}$, $\mathbf{i}+\mathbf{j}$, $5\mathbf{i}-12\mathbf{j}$, respectively. Find the position vector of their centre of mass.

Each mass is acted on by a constant force in the horizontal plane directed away from the origin and of magnitudes 10, $3\sqrt{2}$, 26 units respectively. Find the acceleration of the centre of mass. Find also the resultant linear momentum of the particles when the total work done by the forces is 30 units.

The position vector, $\mathbf{r}_m$, of the centre of mass is, by definition,

$$\mathbf{r}_m = \frac{\sum m_p \mathbf{r}_p}{\sum m_p} = \frac{2(3\mathbf{i}+4\mathbf{j})+6(\mathbf{i}+\mathbf{j})+2(5\mathbf{i}-12\mathbf{j})}{10}$$

$$= \frac{11}{5}\mathbf{i}-\mathbf{j}.$$

Each mass is acted on by a **constant** force. This force will be the **resultant force** acting on each particle. The particles are on a **smooth** horizontal plane, hence there will be no frictional forces and the normal reaction of the plane will be equal and opposite to the weights. The forces will have the direction of the respective position vectors and the magnitudes given, i.e., they are,

$$\frac{10}{5}(3\mathbf{i}+4\mathbf{j}),\ 3\frac{\sqrt{2}}{\sqrt{2}}(\mathbf{i}+\mathbf{j}),\ \frac{26}{13}(5\mathbf{i}-12\mathbf{j}),\ \text{respectively.}$$

The acceleration of the centre of mass, $\mathbf{a}_m$, is that which would be produced by the resultant force acting on a particle of mass equal to the total mass of the system.

Hence,
$$\mathbf{a}_m = \frac{1}{10}(6\mathbf{i}+8\mathbf{j}+3\mathbf{i}+3\mathbf{j}+10\mathbf{i}-24\mathbf{j})$$

$$= \frac{19}{10}\mathbf{i} - \frac{13}{10}\mathbf{j}.$$

The **resultant linear momentum** of the particles at the instant when the total work done is 30 units, is equal to the impulse of the resultant force over the time, t, for which it was acting. The initial momentum was zero and 'impulse = change in momentum'). Therefore the resultant linear momentum at time $t = (19\mathbf{i} - 13\mathbf{j})t$.

From Newton's second law, the acceleration of each mass is $(3\mathbf{i}+4\mathbf{j})$, $\frac{1}{2}(\mathbf{i}+\mathbf{j})$, $(5\mathbf{i}-12\mathbf{j})$, respectively. Since their accelerations are constant, from '$\mathbf{v} = \mathbf{u} + \mathbf{a}t$,' their velocity vectors are $(3\mathbf{i}+4\mathbf{j})t$, $\frac{1}{2}(\mathbf{i}+\mathbf{j})t$, $(5\mathbf{i}-12\mathbf{j})t$ respectively.

The **total work done** by the forces may be measured by the **change in kinetic energy** they produce, i.e., **total work done**

$$= \frac{1}{2}\cdot 2\cdot |(3\mathbf{i}+4\mathbf{j})|^2\, t^2$$

$$+ \frac{1}{2}\cdot 6\cdot |\tfrac{1}{2}(\mathbf{i}+\mathbf{j})|^2\, t^2 + \frac{1}{2}\cdot 2\cdot |(5\mathbf{i}-12\mathbf{j})|^2\, t^2$$

$$= t^2\left(25 + \frac{3}{2} + 169\right) = \mathbf{\frac{391}{2}}\, \boldsymbol{t^2}.$$

We are told that the total work done is 30 units, therefore,

$$t = \sqrt{\frac{60}{391}}$$

and the **resultant linear momentum at this time**

$$\text{is} \quad (19\mathbf{i} - 13\mathbf{j})\sqrt{\frac{60}{391}}.$$

Key terms

The **acceleration of the centre of mass** of a system is that which would be produced by the resultant force acting on a particle of the same mass as the total mass of the system.

The **change in resultant momentum** of a system is the same as that which would be produced by the resultant impulse acting on a particle of the same mass as the total mass of the system.

The properties of the linear motion of a system of particles moving under gravity in a **uniform gravitational field** correspond with those which would be produced by the resultant weight acting on the total mass concentrated at the centre of mass.

Chapter 18
Stability of Equilibrium

A system is said to be in **stable equilibrium if**, when it is given an arbitrary small disturbance, it returns to its original position of equilibrium. If this is not the case and it continues to move away from its initial position, then it is said to be in **unstable equilibrium**. If it remains at the position to which it was disturbed then it is in **neutral equilibrium**.

If a body is disturbed from a position of **unstable equilibrium** it must gain kinetic energy. In a **conservative system of forces**, this must be at the expense of the potential energy of the system. When a body is displaced from a position of **stable equilibrium** in a **system of conservative forces**, then in order to have the energy to return to its position of stable equilibrium, it must have gained potential energy.

Hence we can say that, **in a position of unstable equilibrium, the potential energy of the system, in a conservative system of forces, is a maximum. In a position of stable equilibrium, in a conservative system of forces, the potential energy is a minimum.**

Let us illustrate these ideas with reference to a bead threaded onto a smooth wire as shown in figure 194.

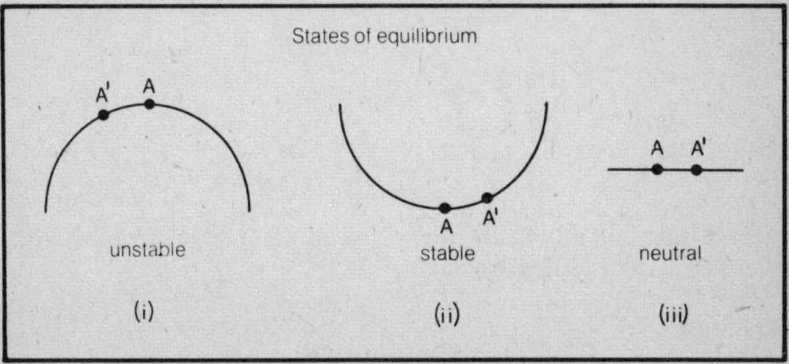

Figure 194.

In **diagram (i)**, the bead is in equilibrium at A, and when slightly displaced to A', then we see that it will continue to move away from A. Hence it is in **unstable** equilibrium. We see that its **potential energy is a maximum** at A.

In **diagram (ii)**, the bead is in equilibrium at A, and when slightly disturbed to A', we see that it will return to A. Hence it is in **stable** equilibrium and we see that **its potential energy is a minimum.**

In **diagram (iii)**, the bead is in equilibrium at A, and when it is slightly disturbed to A', it is also in equilibrium there. This state is described as **neutral equilibrium**, and the **potential energy is constant.**

We shall now formalise these ideas. Consider a conservative system of forces where the total mechanical energy of the system is constant. Let P.E. + K.E. = λ, where λ is a constant. Since K.E. = $\frac{1}{2}mv^2$, then P.E. = $\lambda - \frac{1}{2}mv^2$. Differentiating with respect to time,

$$\frac{\mathrm{d(P.E.)}}{\mathrm{d}t} = -mv\frac{\mathrm{d}v}{\mathrm{d}t}.$$

In a position of equilibrium, $\frac{\mathrm{d}v}{\mathrm{d}t} = 0$,

since the resultant force and therefore the resultant acceleration is zero.

Hence, in equilibrium, $\frac{\mathrm{d(P.E.)}}{\mathrm{d}t} = 0.$

It is often convenient to express P.E. in terms of another variable, say θ.

Since $\frac{\mathrm{d(P.E.)}}{\mathrm{d}t} = \frac{\mathrm{d(P.E.)}}{\mathrm{d}\theta}\cdot\frac{\mathrm{d}\theta}{\mathrm{d}t}$,

then if $\frac{\mathrm{d(P.E.)}}{\mathrm{d}\theta} = 0$, $\frac{\mathrm{d(P.E.)}}{\mathrm{d}t} = 0.$

As we have already said, in a position of **stable equilibrium** the **P.E. is a minimum.**

Therefore $\frac{\mathrm{d(P.E.)}}{\mathrm{d}\theta} = 0$ and $\frac{\mathrm{d^2(P.E.)}}{\mathrm{d}\theta^2} > 0.$

In a position of **unstable equilibrium**, the **P.E. is a maximum.**

Therefore $\dfrac{\mathrm{d(P.E.)}}{\mathrm{d}\theta} = 0$ and $\dfrac{\mathrm{d^2(P.E.)}}{\mathrm{d}\theta^2} < 0.$

Worked examples

Example 1 A hemisphere of radius a is constructed so that its centre of mass G lies on the radius of symmetry, at a distance c from the centre P of its plane face. It is placed with its plane face horizontal and its curved surface touching the highest point of a fixed sphere of radius b and centre O. It is assumed that no slipping can take place between the surfaces. The hemisphere is now rolled so that O and P always lie in a fixed vertical plane. Show that when the angle between OP and the vertical is θ,

$$\left(0 < \theta \leq \frac{\pi a}{2b}\right), \quad O\hat{P}G = \frac{b\theta}{a}.$$

Hence show that the hemisphere will remain in equilibrium in this position provided

$$a \sin \theta = c \sin\left[\left(1 + \frac{b}{a}\right)\theta\right].$$

If $b = a$, show that such a position exists only if $2c > a$. If this condition is satisfied, investigate the stability of this position of equilibrium.

(W.J.E.C.)

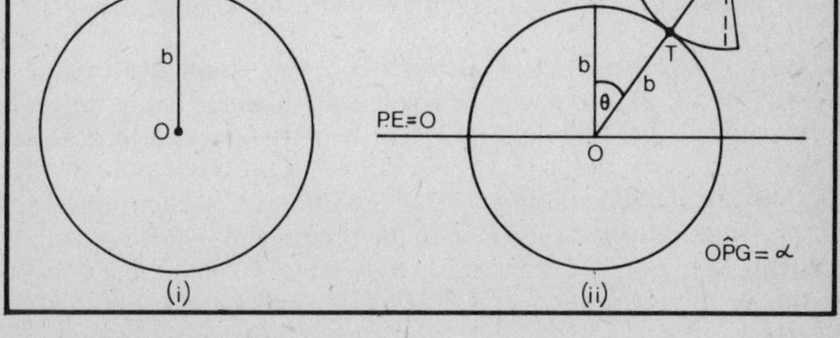

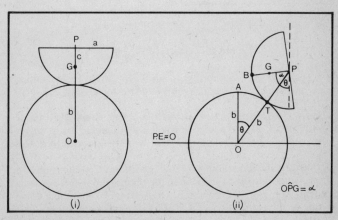

Figure 195.

Referring to figure 195, we see that since the hemisphere rolled from the position in diagram (i) to the position in diagram (ii) with no slipping, then the arc BT of the hemisphere must have the same length as the arc AT of the sphere.

Hence $a\alpha = b\theta$ and so $\boldsymbol{O\hat{P}G = \frac{b}{a}\theta}$.

(The condition $0 < \theta \leq \frac{\pi}{2b}$ is to make sure that $\alpha \leq \frac{\pi}{2}$, so that the hemisphere is still moving on its curved surface). This will be a position of equilibrium if the vertical through G passes through T, the point of contact, otherwise toppling will occur. If this is the case, then triangle GPT has GT vertical, and angle $GTP = \theta$ (see figure 196).

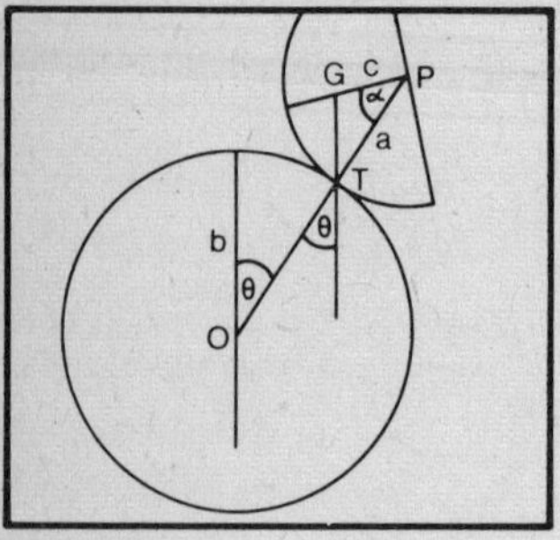

Figure 196.

Applying the sine rule to triangle GTP,

$$\frac{c}{\sin(G\hat{T}P)} = \frac{a}{\sin(180° - (\alpha + G\hat{T}P))}$$

Hence, if GT is vertical,
$a \sin\theta = c \sin(\alpha + \theta)$

and since $\alpha = \frac{b}{a}\theta$, then

$$\boldsymbol{a \sin\theta = c \sin\left[\left(1 + \frac{b}{a}\right)\theta\right]}.$$

If $b = a$ then, for equilibrium, $a \sin\theta = c \sin 2\theta$. Hence, either $\sin\theta = 0$, which is the initial position, or $\cos\theta = a/2c$.

Thus, for **an equilibrium position other than the initial position to exist**, since $0 < \cos\theta < 1$, (a and c are positive, therefore $\cos\theta \leq 1$, and $\cos\theta = 1$ is the initial position) $\boldsymbol{2c > a}$.

To investigate the stability of this position of equilibrium, we need to find a general expression for the potential energy of the system. We are in a conservative system of forces: the force acting at the point of contact of the hemisphere and sphere does no work since the point of contact is instantaneously stationary.

Referring to figure 195(ii) and taking the zero potential energy level as the horizontal through O, (we must always take the level of P.E. = 0 as one which does not vary) then,

$$\text{P.E.} = ((a + b)\cos\theta - c\sin(90^\circ - (\theta + \alpha)))mg$$

where m is the mass of the hemisphere.

Since $b = a$ and thus $\theta = \alpha$, then,

$$\text{P.E.} = (2a\cos\theta - c\cos 2\theta)mg$$

$$\text{and}\quad \frac{\text{d(P.E.)}}{\text{d}\theta} = (-2a\sin\theta + 2c\sin 2\theta)mg.$$

At equilibrium $\dfrac{\text{d(P.E.)}}{\text{d}\theta} = 0$, $\sin\theta = 0$ or $\cos\theta = \dfrac{a}{2c}$,

as we already know.

We shall now investigate whether we have stable or unstable equilibrium at

$$\cos\theta = \frac{a}{2c}\quad\text{by inspecting}\quad \frac{\text{d}^2\text{(P.E.)}}{\text{d}\theta^2}.$$

Since $$\frac{\text{d}^2\text{(P.E.)}}{\text{d}\theta^2} = (-2a\cos\theta + 4c\cos 2\theta)mg,$$

then when

$$\cos\theta = \frac{a}{2c},\quad \frac{\text{d}^2\text{(P.E.)}}{\text{d}\theta^2} = -\frac{a^2}{c} + \frac{2a^2}{c} - 4c = \frac{a^2 - 4c^2}{c}.$$

We know that $2c > a$; since both a and c are positive, $a^2 - 4c^2 < 0$. **Hence the equilibrium position is unstable since the potential energy is a maximum.**

Example 2 A uniform rod AB, of mass m and length $2a$, can turn about a fixed smooth pivot at A. A light inextensible string of length l attached to the other end of the rod passes through a smooth ring fixed at a point C vertically above A so that $AC = 2a$. A particle P of mass $m/2$ hangs from the other end of the string. $0 \leq C\hat{A}B \leq \pi$. Find the potential energy of the system. Find the two positions of equilibrium and investigate their stability.

The system is conservative since the external force R at the hinge does no work when θ varies and the tensions are internal forces.

Referring to figure 197,

$$BC = 4a \sin\frac{\theta}{2} \quad \text{and} \quad CP = l - 4a \sin\frac{\theta}{2}.$$

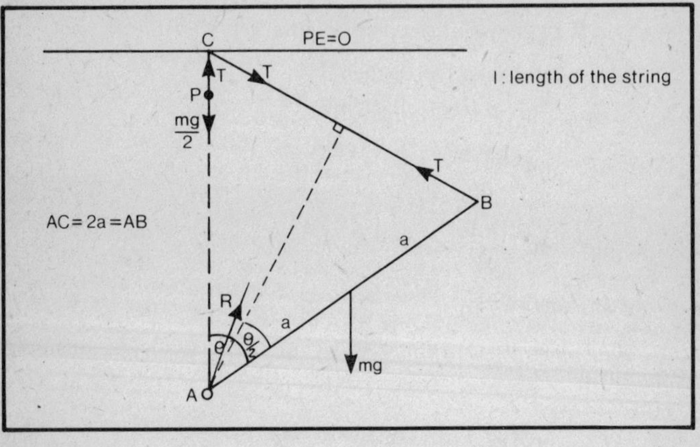

Figure 197.

Taking the zero potential energy level as the horizontal through C, (again a level which does not vary),

$$\text{P.E.} = -\left(l - 4a \sin\frac{\theta}{2}\right)\frac{mg}{2} - (2a - a\cos\theta)mg$$

$$= \mathbf{2amg \sin\frac{\theta}{2} + mga \cos\theta - \frac{mgl}{2} - 2amg.}$$

To find the positions of equilibrium, we find the values of θ for which $\mathrm{d}(\text{P.E.})/\mathrm{d}\theta = 0$.

Since $$\frac{\mathrm{d}(\text{P.E.})}{\mathrm{d}\theta} = amg \cos\frac{\theta}{2} - mga \sin\theta,$$

then, $$\frac{\mathrm{d}(\text{P.E.})}{\mathrm{d}\theta} = 0 \quad \text{when} \quad \cos\frac{\theta}{2} = \sin\theta,$$

i.e., $$\cos\frac{\theta}{2} = 0 \quad \text{or} \quad \sin\frac{\theta}{2} = \frac{1}{2}.$$

Hence the system is in equilibrium when $\theta = \pi$ and when $\theta = \pi/3$.

To investigate the stability of these positions, we examine

$$\frac{d^2(\text{P.E.})}{d\theta^2}.$$

$$\frac{d^2(\text{P.E.})}{d\theta^2} = -\frac{amg}{2}\sin\frac{\theta}{2} - mga\cos\theta.$$

When $\theta = \pi$, $\dfrac{d^2(\text{P.E.})}{d\theta^2} = -\dfrac{amg}{2} + mga > 0$,

therefore the equilibrium is **stable**.

When $\theta = \dfrac{\pi}{3}$, $\dfrac{d^2(\text{P.E.})}{d\theta^2} = -\dfrac{amg}{4} - \dfrac{mga}{2} < 0$,

therefore the equilibrium is **unstable**.

Example 3 A smooth circular wire of radius a is fixed in a vertical plane. A small ring of mass m is threaded onto the wire. An elastic string, obeying Hooke's Law, of modulus $3mg$ and natural length a, has one end attached to the ring and the other end to the highest point of the wire. Find the potential energy of the system when the string is taut and makes an angle θ with the vertical through the centre.

Find the positions of equilibrium with the string taut and investigate their stability.

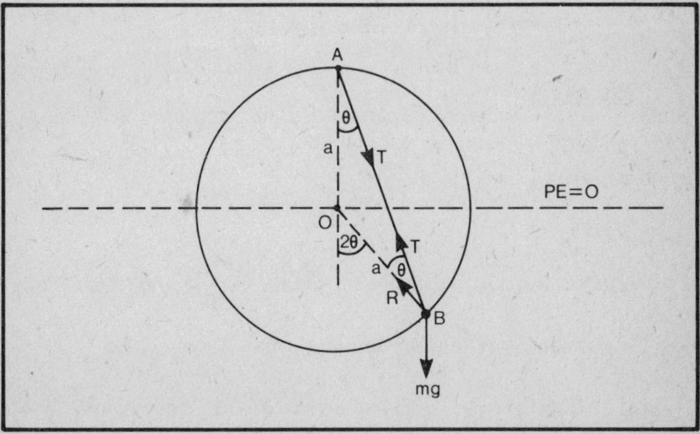

Figure 198.

The system of forces is conservative since the external reaction of the wire on the ring does no work: the displacement of the point of application of the force is always perpendicular to the force.

When evaluating the potential energy of the system, we have to take into account both the gravitational potential energy and the elastic potential energy.

Refer to figure 198.

For the string to be taut, $AB \geq a$, therefore $\theta \leq \frac{\pi}{3}$.

For the general position, taking the zero potential energy level as the horizontal through O, the gravitational potential energy = $-mga \cos 2\theta$ and the elastic potential energy

$$= \frac{3mg}{2a}(2a \cos \theta - a)^2.$$

Hence, the **total P.E.**

$$= \frac{3mga}{2}(4 \cos^2 \theta + 1 - 4 \cos \theta) - mga \cos 2\theta$$

$$= mga\left(3 \cos 2\theta + \frac{9}{2} - 6 \cos \theta - \cos 2\theta\right)$$

$$= \mathbf{2mga\left(\cos 2\theta - 3 \cos \theta + \frac{9}{4}\right)}$$

Since $\dfrac{d(\text{P.E.})}{d\theta} = 2mga(-2 \sin 2\theta + 3 \sin \theta)$, then

$$\frac{d^2(\text{P.E.})}{d\theta^2} = 2mga(-4 \cos 2\theta + 3 \cos \theta).$$

Positions of equilibrium occur when $\dfrac{d(\text{P.E.})}{d\theta} = 0$, i.e., when $3 \sin \theta = 2 \sin 2\theta$. Therefore either $\sin \theta = 0$ or $\cos \theta = \frac{3}{4}$.

Hence, **equilibrium** occurs when $\boldsymbol{\theta = 0}$ and when $\boldsymbol{\theta = \arccos\left(\frac{3}{4}\right)}$.

When $\theta = 0$, $\dfrac{d^2(\text{P.E.})}{d\theta^2} < 0$

and it is a position of **unstable** equilibrium.

When $\theta = \arccos\left(\dfrac{3}{4}\right)$,

$$\frac{d^2(\text{P.E.})}{d\theta^2} = 2mga\left(-4\left(2 \cdot \frac{9}{16} - 1\right) + \frac{9}{4}\right) > 0$$

and it is a position of **stable** equilibrium.

Key terms

A system is in **stable** equilibrium if, when given an arbitrary small displacement from its equilibrium position, it returns to that equilibrium position. If this is not the case, and motion continues, the system is in **unstable** equilibrium. If the system remains in the position to which it was displaced, it is in **neutral** equilibrium.

In a **conservative system of forces**, when the P.E. of the system is expressed in terms of a single variable, θ, then **equilibrium** occurs when

$\dfrac{d(\text{P.E.})}{d\theta} = 0$. If the P.E. is a **minimum**, i.e., $\dfrac{d^2(\text{P.E.})}{d\theta^2} > 0$

then the system is in **stable** equilibrium. If the P.E. is a **maximum**, i.e.,

$$\frac{d^2(\text{P.E.})}{d\theta^2} < 0$$

then the system is in **unstable** equilibrium.

Chapter 19
Rotation of a Rigid Body about a Fixed Axis. Calculation of Moments of Inertia

In this chapter we shall investigate the properties of a **rigid body rotating about a fixed axis**. When a **torque** is applied to a rigid body its **angular** motion changes. Experimental evidence suggests that when a torque is applied to a body, free to rotate about a particular axis, the **angular acceleration produced is proportional to the torque**. The **angular acceleration** produced by a particular torque is found to be affected by a change in **size, shape, mass or axis of rotation** of the body.

Consider a rigid body freely rotating about a fixed axis with angular acceleration $\ddot{\theta}$ and angular velocity $\dot{\theta}$. A rigid body is made up of a large number of particles whose relative positions are fixed. Let P be a typical constituent particle of mass m_p and distance r_p from the axis of rotation. Figure 199 below shows a cross section of the body perpendicular to the axis of rotation.

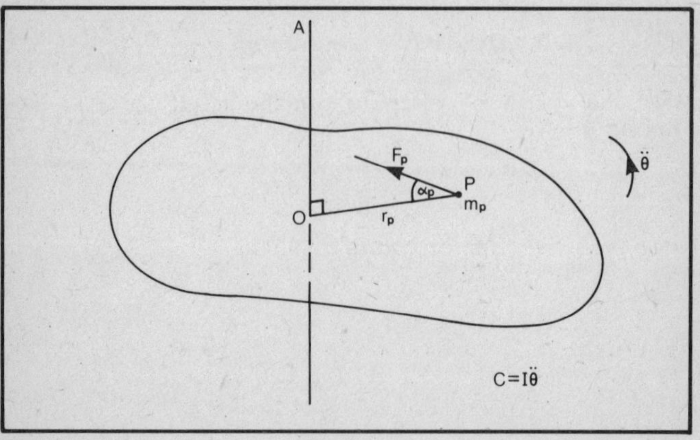

Figure 199.

Each particle, P, will be rotating about O in a **circular path**, radius r_p. The linear velocity of P is $r_p\dot{\theta}$. Since P is moving in the circle, centre O, the resultant force, F_p, acting on it must be in the plane of the above cross-section. Applying Newton's second law perpendicular to OP, in the plane of the cross-section,

$$F_p \sin \alpha_p = m_p r_p \ddot{\theta}, \text{ giving } r_p \sin \alpha_p F_p = m_p r_p^2 \ddot{\theta}.$$

$r_p \sin \alpha_p F_p$ is the moment of F_p about O.

Summing for the whole body: $\sum \boldsymbol{r_p \sin \alpha_p F_p} = \sum \boldsymbol{m_p r_p^2 \ddot{\theta}}$.

Since $\ddot{\theta}$ is the same for each particle in the body, this may be written as $\boldsymbol{C} = (\sum \boldsymbol{m_p r_p}^2)\ddot{\theta}$ where C is the sum of moments of the **external** forces acting on the body about the axis of rotation. (Any **internal** forces included in F_p will not affect the sum since they occur in equal opposite pairs.)

The quantity $\sum \boldsymbol{m_p r_p}^2$ is called the **moment of inertia** of the body about the given axis and gives a measure of both the mass of the body and the distribution of mass about the axis.

Hence we arrive at the equation of motion $\boldsymbol{C} = \boldsymbol{I}\ddot{\theta}$, where $I = \sum m_p r_p^2$. The **unit** of I is 1 kg m^2.

Let us consider this equation of rotational motion, $C = I\ddot{\theta}$, more closely and compare it with the equation of linear motion, $F = ma$.

F is the **resultant force** acting on a body of **mass *m*** which is necessary to give it a **linear acceleration of *a***. ***C*** is the **resultant torque** acting on a body of **moment of inertia *I*** which is necessary to give it an **angular acceleration** $\ddot{\theta}$ about a particular axis.

We may consider a body's **moment of inertia**, I, as a measure of the body's resistance to change in its **rotational** motion, in the same way as its **mass**, m, is a measure of its resistance to change in its **linear** motion.

Before we discuss further properties of a rigid body rotating about a fixed axis, we shall look more closely at moments of inertia and how to calculate them for various standard bodies.

Calculation of moments of inertia

The evaluation of $\sum mr^2$ for a particular body about a particular axis is done in one of two ways. If the body consists of a finite number of particles, then the summation is straight addition. If, as in a solid body, we assume a continuous distribution of matter, or an infinite number of particles, then we do the summation by integration. We shall now evaluate the moments of inertia of some standard bodies about the stated axes.

Thin uniform rod of mass *M* and length 2*a*

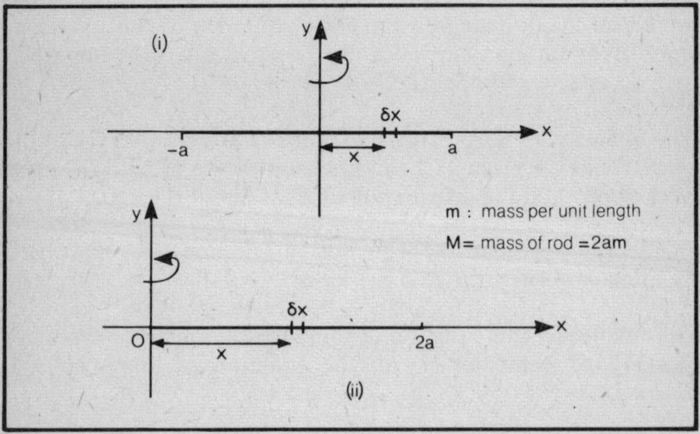

Figure 200.

We shall first consider the **moment of inertia of the rod about an axis perpendicular to the rod through the midpoint**.

Referring to figure 200(i), and taking m as the constant mass per unit length, the moment of inertia of an element of the rod of length δx and distance x from the axis of rotation Oy is $m\,\delta x\,x^2$.

Hence the total moment of inertia, I, is the limit of

$$\sum m\,\delta x\,x^2 \quad \text{as} \quad \delta x \to 0, \quad \text{i.e.,} \quad \boldsymbol{I} = \int_{-a}^{a} mx^2\,dx = \frac{1}{3}m\left[x^3\right]_{-a}^{a}$$

$$= \frac{2}{3}ma^2 = \frac{\mathbf{1}}{\mathbf{3}}\boldsymbol{Ma^2}, \quad \text{since } M = 2am.$$

In a similar way, we could show that **the moment of inertia of the rod about a parallel axis through one end is**

$$\frac{4}{3}Ma^2, \text{ see figure 200(ii).}$$

The moment of inertia of the rod about an axis parallel to its length, distance p away, is Mp^2 since every element of the rod is at a distance p from the axis and $\sum m\,\delta x\,p^2 = Mp^2$, see figure 201(i).

The moment of inertia of the rod about an axis, through its centre at an angle θ with its length is given by

$$I = \int_{-a}^{+a} m(x \sin\theta)^2\,dx,$$

since $x \sin\theta$ is the distance of each element of the rod away from the axis.

Therefore $I = \dfrac{Ma^2}{3}\sin^2\theta$, see figure 201(ii).

We are now in a position to calculate the moments of inertia of bodies which may be divided into **rod-like elements**.

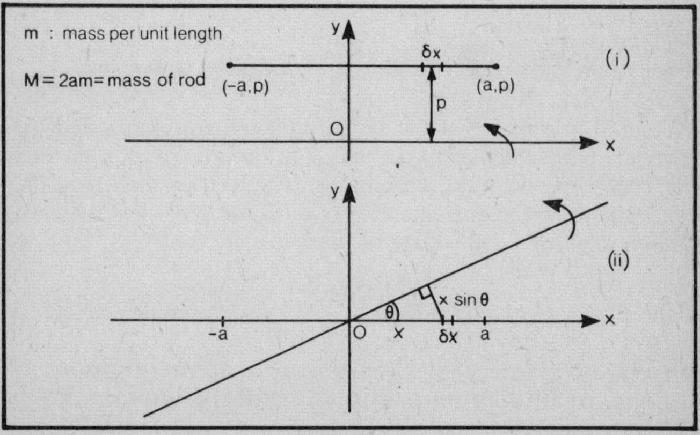

Figure 201.

Uniform rectangular lamina of sides 2*a* and 2*b* and mass *M*.

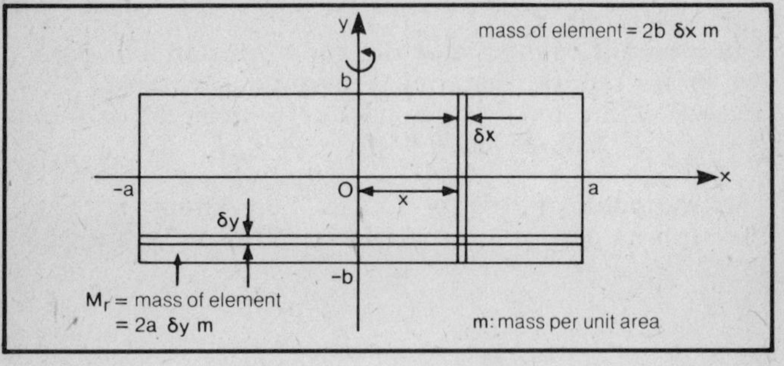

Figure 202.

Refer to figure 202.

Let us consider the moment of inertia about **an axis passing through the midpoint of the lamina, parallel to the sides of length 2*b***. We divide the lamina into rod-like elements of mass $2bm\ \delta x$ parallel to the axis, distance x away, with moment of inertia $2bm\ \delta x\ x^2$ about the axis.

Thus,

$$I = \int_{-a}^{a} 2bmx^2\, \mathrm{d}x = 2b\frac{2a^3}{3}m = \frac{Ma^2}{3}, \quad \text{since} \quad M = 4abm.$$

We could also arrive at the same result by considering rod-like elements perpendicular to the axis, and the axis passing through the centre of the rods. The moment of inertia of the rod-like element about Oy (see figure 202) is, as we have just deduced, $\frac{1}{3}M_r a^2$ where M_r is the mass of the rod.

Hence, $$I = \int_{-b}^{b} \frac{1}{3}a^2\, 2am\, \mathrm{d}y = \frac{Ma^2}{3}, \quad \text{as we should expect.}$$

We may also deduce, in a similar way, that **the moment of inertia of the lamina about a side of length 2*b* is**

$$\frac{4}{3}Ma^2.$$

Thin uniform ring of mass *M* and radius *a*, about an axis through its centre and perpendicular to the plane of the ring.

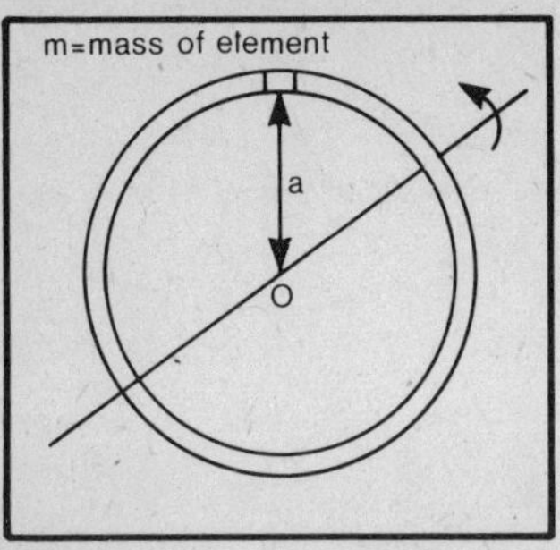

Figure 203.

The ring may be divided up into elements of mass m, all at distance a from the centre.

Hence

$$I = \sum ma^2 = a^2 \sum m = Ma^2,$$

since $\sum m = M$.

We use this moment of inertia of a uniform ring about an axis through its centre perpendicular to its plane, to calculate the moments of inertia of other uniform bodies which may be divided up into **ring-like elements**.

Uniform circular lamina of radius *a* and mass *M*, about an axis through its centre, perpendicular to its plane.

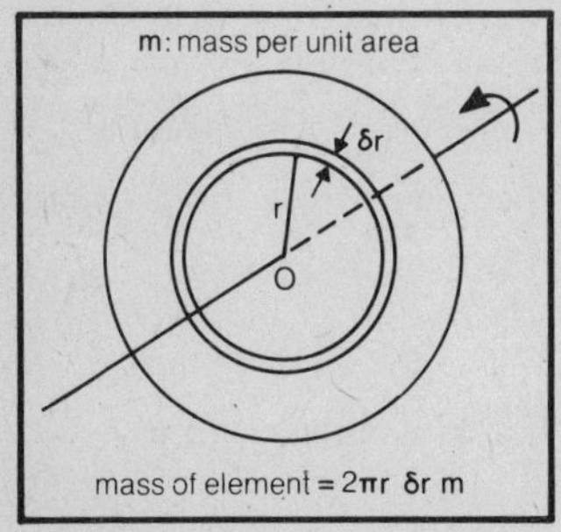

Figure 204.

Let the mass of the disc per unit area be m. We divide the lamina into concentric rings of radius r and thickness δr. Hence the moment of inertia of an element about the given axis is $2\pi r\ \delta r\ mr^2$. Therefore

$$I = \int_0^a 2\pi r^3 m\ dr = m\pi \frac{a^4}{2}$$

$$= \frac{Ma^2}{2}, \text{ since } M = \pi a^2 m.$$

Uniform hollow cylinder of mass *M* and radius *a*, about its axis.

We divide this solid into ring-like elements.

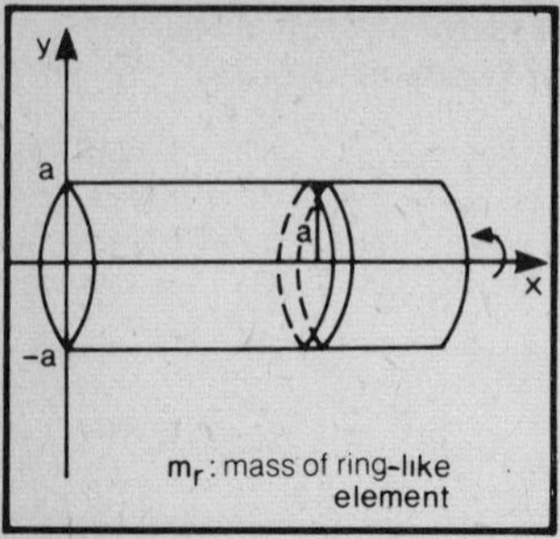

Figure 205.

Let the masses of the ring-like elements each be m_r. The moments of inertia of each of the elements about the given axis is $m_r a^2$. Hence,

$$I = \sum m_r a^2 = a^2 \sum m_r$$
$$= \boldsymbol{Ma^2} \text{ since } M = \sum m_r$$

Uniform hollow sphere of mass *M* and radius *a*, about a diameter.

We divide this solid into ring-like elements. We take their thicknesses as $a\,\delta\theta$, see figure 206. (δx as the thickness gives an unacceptable error of approximation, as noted when discussing the location of the centre of gravity of a hollow sphere). The radius of the ring-like element is $a \sin \theta$. Its moment of inertia about the given axis is

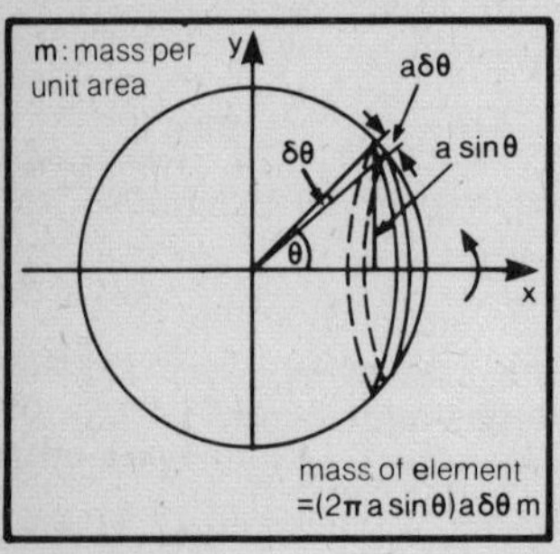

Figure 206.

$$m 2\pi a \sin\theta \, a\, \delta\theta \, (a \sin\theta)^2$$

where m is the mass per unit area.

Therefore, summing the moments of inertia of the elements,

$$I = 2ma^4\pi \int_0^\pi \sin^3\theta \, d\theta = 2ma^4\pi \int_0^\pi \sin\theta \, (1 - \cos^2\theta) \, d\theta$$

$$= 2ma^4\pi \left[-\cos\theta + \frac{1}{3}\cos^3\theta \right]_0^\pi = 2ma^4\pi \left[1 - \frac{1}{3} + 1 - \frac{1}{3} \right]$$

$$= \frac{8}{3} ma^4\pi = \frac{\boldsymbol{2Ma^2}}{\boldsymbol{3}}, \quad \text{since} \quad M = 4\pi a^2 m$$

Uniform solid sphere of radius *a* and mass *M*, about a diameter.

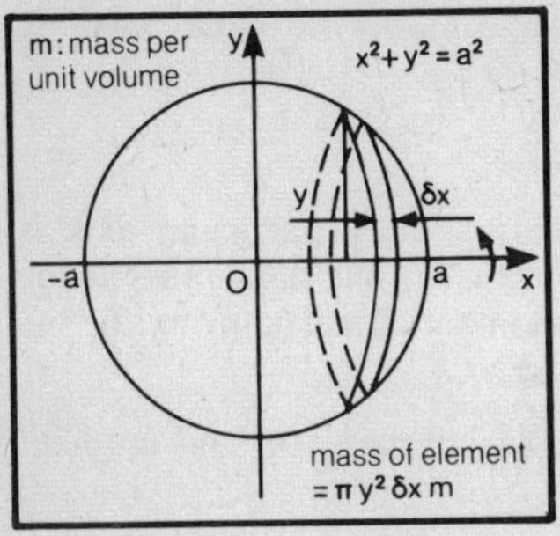

Figure 207.

We divide the solid into thin disc-like elements as shown in figure 207. The mass of an element is $m\pi y^2\ \delta x$ where m is the constant mass per unit volume.

The moment of inertia of an element is that of a uniform circular lamina of radius y, about an axis through its centre perpendicular to its plane. i.e.,

$$m\pi y^2\ \delta x\ \frac{y^2}{2}.$$

$$\text{Hence } I = m\frac{\pi}{a}\int_{-a}^{a} y^4\, dx = \frac{m\pi}{2}\int_{-a}^{a}(a^2-x^2)^2\, dx$$

$$= m\frac{\pi}{2}\left[a^4x - 2a^2\frac{x^3}{3} + \frac{x^5}{5}\right]_{-a}^{a} = \frac{8\pi}{15}ma^5 = \frac{2Ma^2}{5}$$

since $M = \frac{4}{3}\pi a^4 m$.

The moment of inertia of a compound body about an axis is the sum of the moments of inertia of the constituent bodies about the axis.

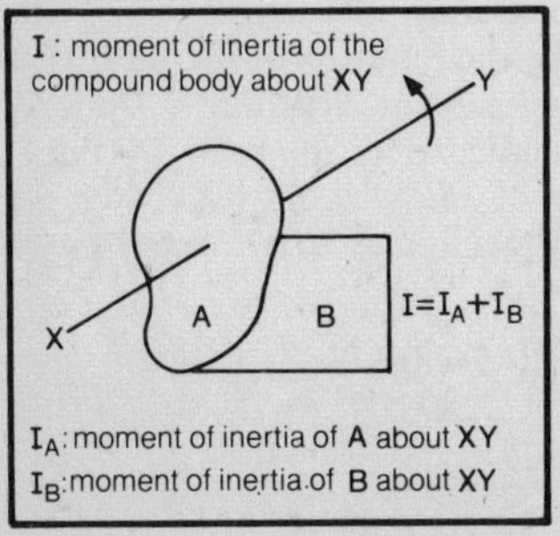

Figure 208.

This result is clear from the consideration of a compound body made up of body A and body B (see figure 208) about any axis.

The moment of inertia of the body about the given axis is the summation of all the quantities mr^2, with the usual notation, i.e., $I = \sum mr^2$ where the summation is over the whole body.

We may, without loss of generality, split the summation into two separate summations, one over A and one over B.

Thus, $\sum_{AB} mr^2 = \sum_{A} mr^2 + \sum_{B} mr^2$. Hence, $\boldsymbol{I} = \boldsymbol{I}_A + \boldsymbol{I}_B$.

This may be extended to cover any number of bodies forming the composite body.

Radius of gyration

The **radius of gyration** of a body about a particular axis, is the distance from the axis that a particle of the same mass as the body could be placed, so that its moment of inertia would be the same as that of the body about that axis.

If k is the radius of gyration of a body of mass, M, and moment of inertia I then

$$\boldsymbol{I = Mk^2} \quad \text{and} \quad k = \sqrt{\frac{I}{M}}.$$

Parallel and perpendicular axes theorem

There are many possible axes about which we may require the moment of inertia of a body. The theorem of parallel axes and the theorem of perpendicular axes are two theorems which provide connections between the moments of inertia of the same body about certain different axes, and so help to avoid repeated and complex integrations.

The theorem of parallel axes states that if the moment of inertia of a body of mass M about an axis through its centre of mass is Mk^2, **its moment of inertia about a parallel axis distant d from the first axis is $M(k^2 + d^2)$.**

Let I_G be the moment of inertia of a body about an axis through the centre of mass G and let I_A be the moment of inertia of the body about a parallel axis through a point A distant d from G, such that GA lies in a plane perpendicular to the axes. Taking P as a typical constituent particle of mass m in the body, its moment of inertia about the axis through A is ml^2, referring to figure 209.

From the cosine rule, $l^2 = r^2 + d^2 - 2rd\cos\theta$.

Summing for all such particles P,

$$\begin{aligned} I_A &= \sum ml^2 = \sum m(r^2 + d^2 - 2rd\cos\theta) \\ &= \sum mr^2 + \sum md^2 - 2d\sum mr\cos\theta. \end{aligned}$$

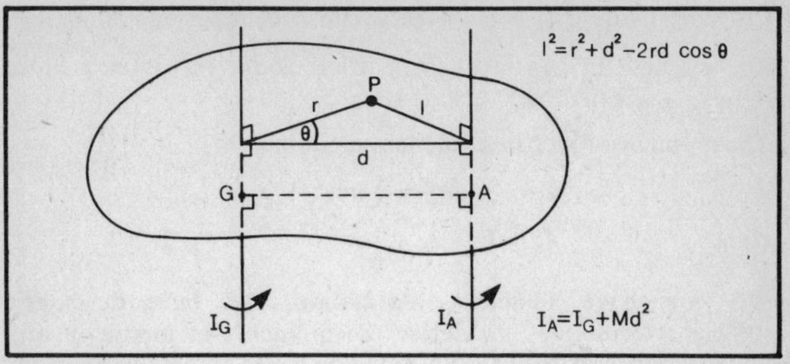

Figure 209.

Since $r\cos\theta$ is the distance from G, in the direction GA of P, then

$$\frac{\sum mr\cos\theta}{\sum m}$$

would give the coordinate of the centre of mass of the body in the direction GA. Because G is the centre of mass, then this coordinate must be zero. Therefore $\sum mr\cos\theta = 0$.

Hence $I_A = \sum mr^2 + \sum md^2 = I_G + \boldsymbol{Md^2} = \boldsymbol{M(k^2 + d^2)}$.

The perpendicular axes theorem, unlike the theorem of parallel axes, which applies to all rigid bodies, **may only be applied to laminae.**

The theorem of perpendicular axes states that if the moments of inertia of a plane body about two perpendicular axes in its plane are I_x and I_y, then the moment of inertia, I_z, of the body about an axis perpendicular to the plane of the body, and passing through the point of intersection of the first two axes, is given by $I_z = I_x + I_y$.

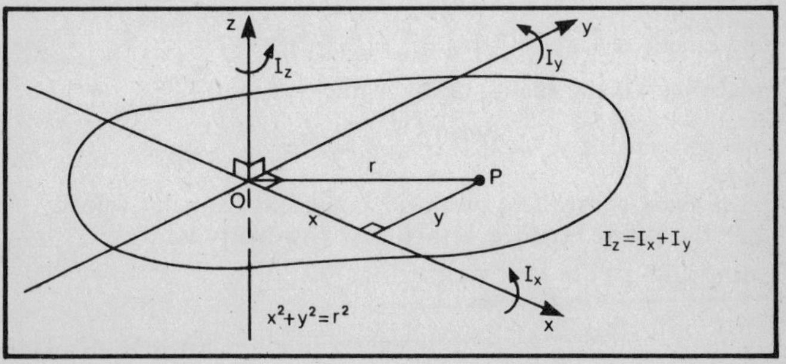

Figure 210.

Consider figure 210. P is a typical constituent particle of mass m.

The moment of inertia of the particle about OZ is mr^2. Since $r^2 = x^2 + y^2$, then $mr^2 = mx^2 + my^2$.

The moment of inertia of the lamina about OZ,

$$I_z = \sum mr^2 = \sum my^2 + \sum mx^2$$

Hence $$I_z = I_x + I_y.$$

We now have sufficient information, and have developed sufficient techniques, to derive the moments of inertia of any bodies about any axes that we may be asked to deal with at this level. We shall now consider further properties of a rigid body rotating about a fixed axis.

Kinetic energy of rotation

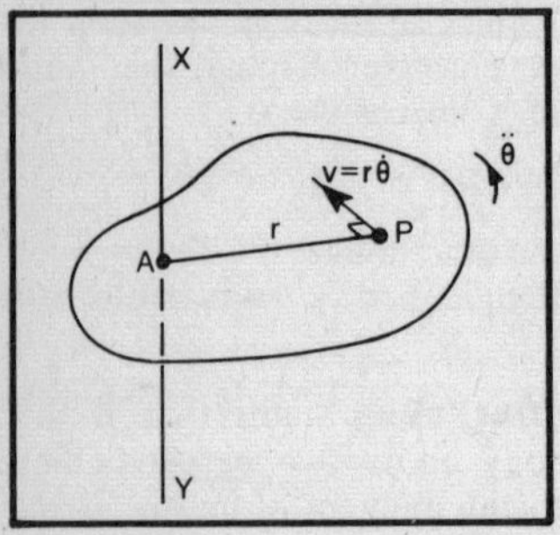

Figure 211.

Consider a rigid body rotating with angular velocity about a fixed axis XY. Figure 211 shows a cross section of the body perpendicular to the axis of rotation, passing through A.

P is a typical constituent particle of mass m and distance r from A.

The linear speed, v, of P is $r\dot{\theta}$.

The kinetic energy of P is $\frac{1}{2}mv^2 = \frac{1}{2}m(r\dot{\theta})^2$.

Hence the kinetic energy of the whole body is

$$\sum \tfrac{1}{2}m(r\dot{\theta})^2 = \tfrac{1}{2}\sum mr^2\dot{\theta}^2.$$

Since every constituent particle has the same angular velocity, $\dot{\theta}$, then the **total kinetic energy of the body** is,

$\frac{1}{2}\dot{\theta}^2 \sum mr^2 = \frac{1}{2}I\dot{\theta}^2$, $(I = \sum mr^2)$.

Again, we see parallels with linear motion: the **kinetic energy** of a body of mass M, moving with **linear** velocity V, is $\frac{1}{2}\boldsymbol{MV}^2$; the

kinetic energy of a rigid body rotating about a fixed axis, with moment of inertia I about that axis, and **angular** velocity $\dot{\theta}$, is $\frac{1}{2}I\dot{\theta}^2$.

Angular momentum of a rotating rigid body

The **linear momentum** of a particle of mass m, moving with linear velocity v, is mv. If that particle is distance r from a fixed axis, then its **moment of momentum** about that axis is rmv.

The **moment of momentum** of a particle is often called its **angular momentum** and always referred to a particular axis.

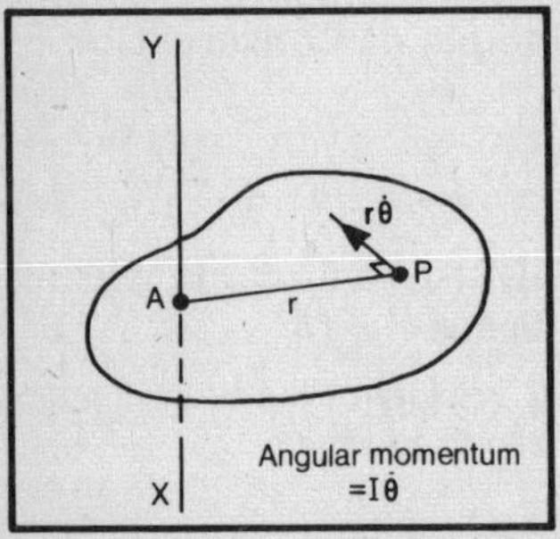

Consider a rigid body rotating about a fixed axis XY.

Figure 212 shows a cross-section of the body perpendicular to the axis. Let P be a typical constituent particle of mass m in that plane.

The **angular momentum** of P about XY is

$$rmv = rm(r\dot{\theta}) = mr^2\dot{\theta}.$$

Figure 212.

Hence, the **total angular momentum of the body about XY** is

$$\sum mr^2\dot{\theta} = \dot{\theta}\sum mr^2 \text{ (since } \dot{\theta} \text{ is common to all the particles)}$$

$= I\dot{\theta}$, where I is the moment of inertia of the body about XY.

The **linear momentum** of a body of mass M, moving with velocity v is Mv. The angular momentum of a body rotating about an axis is $I\dot{\theta}$, where I is its moment of inertia about that axis, and $\dot{\theta}$ is its angular velocity.

The **unit** of angular momentum is 1 kg m^2 s^{-1}. If $H = I\dot{\theta}$ is the angular momentum of a body about a particular axis, then

$$\frac{dH}{dt} = I\ddot{\theta}$$

and, as we saw at the beginning of this chapter, $C = I\ddot{\theta}$, where C is the sum of the moments of the external forces acting on the body.

Thus $$\frac{\mathrm{d}H}{\mathrm{d}t} = C.$$

Hence, **the rate of change of angular momentum of a body rotating about a fixed axis is equal to the resultant torque acting on the body.** (Again we see a parallel with linear motion, where the **rate of change of linear momentum of a particle is equal to the resultant force acting on the particle.**)

As with linear motion, there is an important corollary: **if the sum of the moments of the external forces acting on the body is zero, then the angular momentum remains constant.** This is called the **Principle of Conservation of Angular Momentum.**

Impulse of a torque

Consider a rigid body rotating freely about a fixed axis. With the usual notation, its equation of motion is $C = I\ddot{\theta}$.

Suppose that in time t the angular velocity of the body changes from ω_1 to ω_2. Integrating with respect to time:

$$\int_0^t C\,\mathrm{d}t = \int_0^t I\ddot{\theta}\,\mathrm{d}t = \int_{\omega_1}^{\omega_2} I\,\mathrm{d}\omega.$$

Hence, $\int_0^t C\,\mathrm{d}t = I\omega_2 - I\omega_1$ where $\int_0^t C\,\mathrm{d}t$ is called the **impulse of the torque C**, and $I\omega_2 - I\omega_1$ is the change in angular momentum it produces.

Clearly the impulsive torque is equal to the moment of the impulsive force applied to the body about the axis of rotation. Therefore, '**impulse torque = moment of impulse = increase in angular momentum.**'

(Compare with 'impulse of force = increase in linear momentum').

When C is **constant**, $Ct = I\omega_2 - I\omega_1$.

Conservation of mechanical energy applied to rotating rigid bodies

The **Principle of Conservation of Mechanical Energy** states that the total mechanical energy of a system remains constant, if no external force other than gravity does work, and no sudden changes in motion take place. (i.e., a **conservative system of forces**).

If the axis about which a rigid body is rotating is **smooth**, then it offers no resistance to the rotation and no work is done at the axis. Hence when a body is rotating **freely** about an axis, i.e., the axis is smooth, and no external force other than gravity is doing work, the total mechanical energy of that body remains constant, provided no sudden changes in motion occur.

Work done by a couple

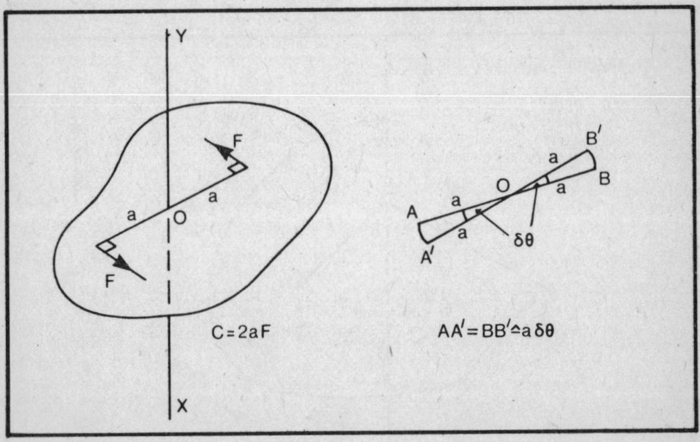

Figure 213.

Consider a body rotating about axis XY under the action of a resultant torque C. The couple may be considered to comprise two equal unlike parallel forces of magnitude F and distance $2a$ apart, as shown in figure 213.

When the body rotates through a small angle $\delta\theta$, the point of application of each force moves through a distance $a\,\delta\theta$. The work done by each force is approximately $Fa\,\delta\theta$ ($a\,\delta\theta$ is approximately a straight line).

If the work done by the couple in the small angular displacement $\delta\theta$ is δw, then $\delta w = 2Fa\,\delta\theta$. Since $C = 2aF$, then $\delta w = C\,\delta\theta$.

Summing these elements of work to find the work, W, done by the couple as the body rotates through an angle α, then

$$\int_0^w dw = \int_0^\alpha C\, d\theta. \quad \text{Therefore,} \quad \boldsymbol{W = \int_0^\alpha C\, d\theta}.$$

If C is **constant**, then $\boldsymbol{W = C\alpha = 2aF\alpha}$.

Compound pendulum

A body free to swing about a smooth horizontal axis, is called a compound pendulum. Let such a body be of mass m and let G be its centre of mass. Figure 214 shows a vertical section of the body through the centre of mass.

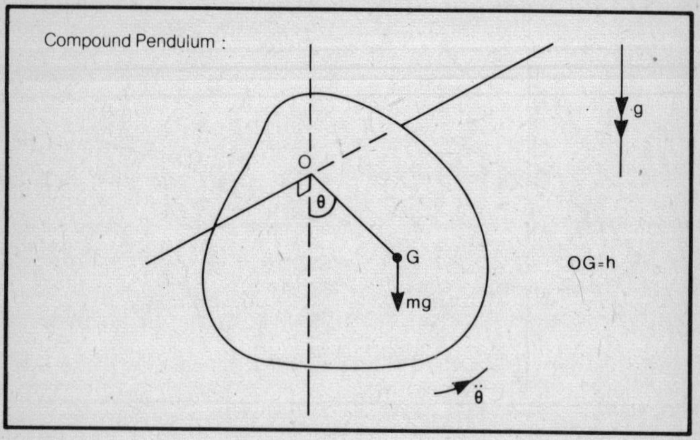

Figure 214.

From '$C = I\ddot{\theta}$,' taking moments about the horizontal axis, since the weight of the body is the only force with a moment about this axis, $\boldsymbol{-mgh \sin\theta = I\ddot{\theta}}$, where I is the moment of inertia of the body, the negative sign being necessary since the torque exerted on the body opposes the increase of θ.

If θ is sufficiently small, then $\boldsymbol{\sin\theta \simeq \theta}$ and the equation of motion is approximately

$$\boldsymbol{I\ddot{\theta} = -mgh\theta, \quad \ddot{\theta} = -\frac{mgh}{I}\theta.}$$

This is an **equation of angular simple harmonic motion** about the vertical, through O, with period

$$T = 2\pi\sqrt{\frac{I}{mgh}}.$$

The period of small oscillations of a simple pendulum is

$$2\pi\sqrt{\frac{l}{g}},$$

and we are often asked to find the **length of the equivalent simple pendulum** for a compound pendulum. What is required is the length, l, of a simple pendulum which would have the same period of motion.

Comparing the two expressions for the period T, we see that the length, l, of an **equivalent simple pendulum** is given by

$$l = \frac{I}{mh}.$$

Force exerted by the axis

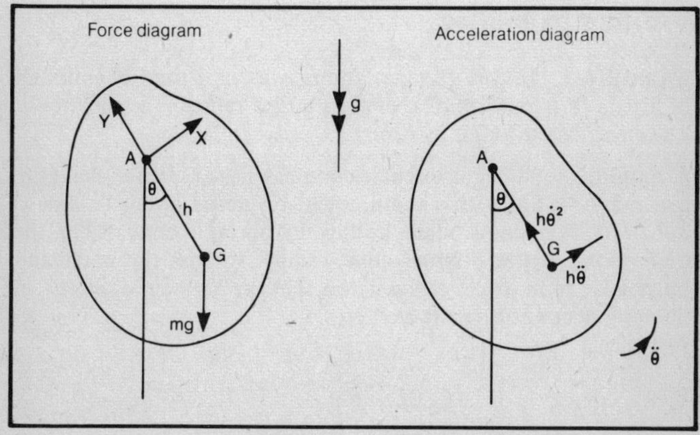

Figure 215.

We saw in Chapter 17 that the acceleration of the centre of mass of a system of particles is the same as that of a single particle of the same mass as that of the total mass of the system acted upon by the resultant of the force acting on the system. Clearly, this result applies to a rigid body which consists of a large number of particles.

Consider a body of mass m, rotating about a smooth horizontal axis. Figure 215 shows the cross-section, perpendicular to the axis, through the centre of mass, G, of the body. The forces acting on the body are its weight and the reaction at the axis. We resolve the reaction into components parallel and perpendicular to AG, as shown.

The radial and transverse components of acceleration of G are $h\dot{\theta}^2$ and $h\ddot{\theta}$ respectively. These components of acceleration are those which would be produced by the resultant of the forces acting on a particle of mass m placed at G.

Hence, applying Newton's second law radially and transversely,

$$Y - mg\cos\theta = mh\dot{\theta}^2 \quad \text{and} \quad X - mg\sin\theta = mh\ddot{\theta}.$$

We may therefore determine the reaction of the axis if we know $\dot{\theta}$ and $\ddot{\theta}$. $\dot{\theta}$ is usually best found by applying the Conservation of Mechanical Energy. $\ddot{\theta}$ may then be obtained by differentiating $\dot{\theta}$, or alternatively by using '$C = I\ddot{\theta}$.'

Worked examples

Example 1 In this question you may assume that the moment of inertia of a uniform disc of mass m and radius r, is $\frac{1}{2}mr^2$ about a normal axis through its centre.

A uniform, solid, right circular cone has base radius a, height h, and mass M. Show that the moment of inertia about its axis is $3Ma^2/10$. The cone is made hollow, by boring a cone of half the dimensions of the original cone axially, so that the inner and outer base radii are $\frac{1}{2}a$ and a. Show that the moment of inertia of the hollow cone about its axis is

$$\frac{93\,M'a^2}{280},$$

where M' is the mass of the hollow cone.

If $a = 1$ m and $M' = 500$ kg, calculate the kinetic energy of the hollow cone when it is rotating about its axis at a rate of 20 revolutions per minute. Give your answer in joules, correct to 2 significant figures.

(S.U.J.B.)

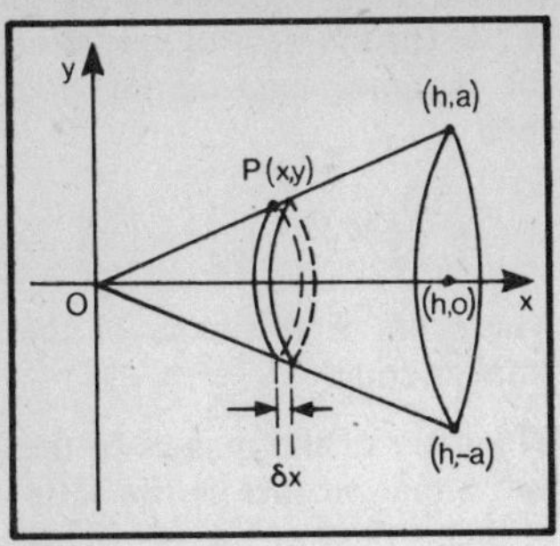

Figure 216.

Let ρ be the mass per unit volume of the cone.

If the cone is divided into elements by cuts parallel to its base, then each element is approximately a disc. The mass of the element is $\pi y^2\,\delta x\,\rho$.

Hence, the moment of inertia of the element about Ox is $\frac{1}{2}(\pi y^2\,\delta x\,\rho)y^2$.

Summing the moments of inertia of the elements over the whole cone,

$$I = \int_0^h \tfrac{1}{2}\pi y^4\,\rho\,dx.$$

From similar triangles, $\dfrac{y}{x} = \dfrac{a}{h}$.

$$\text{Therefore, } I = \tfrac{1}{2}\pi\rho\frac{a^4}{h^4}\int_0^h x^4\,dx = \frac{\pi\rho a^4 h}{10}$$

$$= \frac{3Ma^2}{10},$$

since $M = \frac{1}{3}\pi a^3 h\rho$.

The moment of inertia of the hollow cone about Ox, is the summation of the moments of inertia of all its constituent particles.

The summation over the hollow cone will be the same as the summation over the large solid cone, minus the summation over the smaller solid cone, which was removed.

Hence, the moment of inertia of the hollow cone about Ox is the moment of inertia of the solid cone of radius a about Ox, minus the moment of inertia about the same axis of the solid cone of radius $a/2$, which was removed.

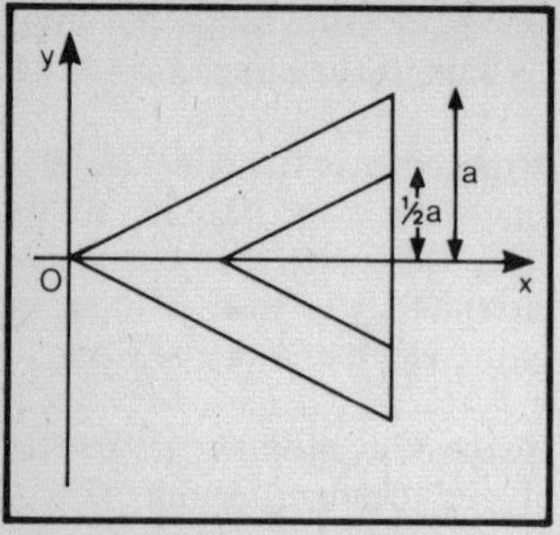

Figure 217.

If I' is the moment of inertia of the hollow cone about Ox, then

$$I' = \frac{3}{10}Ma^2 - \frac{3}{10}M_1\left(\frac{a}{2}\right)^2$$

where M_1 is the mass of the smaller cone.

The ratio of the masses of the two similar cones is the ratio of their volumes, i.e., $(1:2)^3 = 1:8$.

Hence, $M_1 = \dfrac{M}{8}$ and $M' = \dfrac{7M}{8}$.

Thus, $I' = \dfrac{3}{10}Ma^2 - \dfrac{3}{10}\dfrac{M}{8}\dfrac{a^2}{4} = \dfrac{93}{320}Ma^2$ and substituting

$$M = \frac{8}{7}M', \quad \boldsymbol{I' = \frac{93}{280}M'a^2}.$$

The kinetic energy of the hollow cone rotating about its axis is $\frac{1}{2}I\dot{\theta}^2$ where $\dot{\theta}$ is its angular velocity.

We know that its angular velocity is 20 revolution per minute,

i.e., $20 \times \dfrac{2\pi}{60}$ radians per second.

Therefore, the kinetic energy is

$$\frac{1}{2}\left(\frac{93}{280}\right) . 500 . \frac{4}{9}\pi^2 = \mathbf{360\ J}$$

correct to 2 significant figures.

Example 2 Show by integration, that the moment of inertia of a thin uniform square plate $ABCD$, of mass m, and side $2a$, about the edge AB, is $\frac{4}{3}ma^2$. Deduce that the moment of inertia of the plate about an axis through its centre perpendicular to the plate, is $\frac{2}{3}ma^2$.

The plate is rotating freely about the vertical axis perpendicular to the plate, through its centre, with angular velocity ω, when a corner picks up a stationary particle of mass $\frac{2}{3}m$ which adheres to the plate. Find the subsequent angular velocity of the plate and particle, the impulse which acted on the particle and the loss of kinetic energy due to the impact.

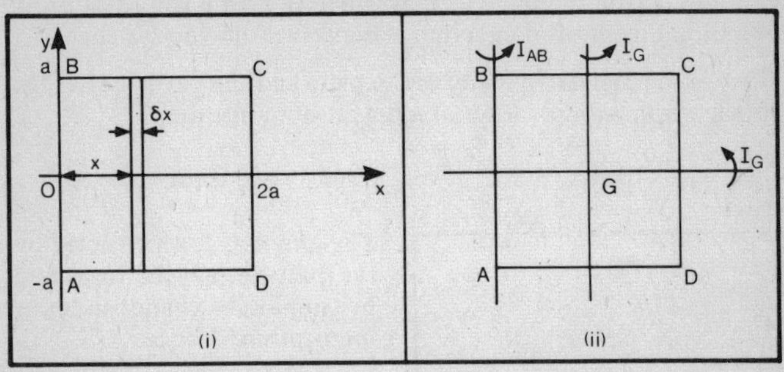

Figure 218.

Referring to figure 218(i), and taking ρ to be the constant mass per unit area, the mass of an element is $2a\,\delta x\,\rho$. Its moment of inertia about Oy is $x^2 2a\,\delta x\,\rho$, since its constituent particles are all distant x from Oy.

Summing for the whole body,

$$I_{AB} = \int_0^{2a} 2a\rho x^2\,dx = 2a\rho\left[\frac{x^3}{3}\right]_0^{2a}$$

$$= \frac{16}{3}a^4\rho = \frac{4}{3}\boldsymbol{ma^2}, \text{ since } m = 4a^2\rho.$$

By the **parallel axes theorem**, if I_G is the moment of inertia about an axis parallel to AB, passing through G, the centre of the square, then $I_{AB} = I_G + ma^2$.

Therefore, $I_G = \frac{4}{3}ma^2 - ma^2 = \frac{ma^2}{3}$, (see figure 218(ii)).

By symmetry, the moment of inertia about an axis perpendicular to this axis, but still in the plane of the square, and passing through G, is also I_G. Hence, by the perpendicular axes theorem, **the moment of inertia of the square about an axis passing through G perpendicular to the plane of the lamina** is

$$I_G + I_G = \frac{2}{3}\boldsymbol{ma^2}.$$

When the plate is rotating freely about this vertical axis, it picks up a stationary particle of mass $\frac{2}{3}m$ which adheres to the plate. When the particle is picked up, it will exert an equal and opposite impulse on the plate to that which the plate will exert on it. Hence, moment of momentum is conserved for the system.

If ω_1 is the angular velocity of the plate and the particle after the impact, from conservation of moment of momentum,

$$\frac{2}{3}ma^2\omega = \frac{2}{3}ma^2\omega_1 + \sqrt{2}\,a\frac{2}{3}m\sqrt{2}\,a\omega_1. \text{ Therefore, } \omega_1 = \frac{\omega}{3}.$$

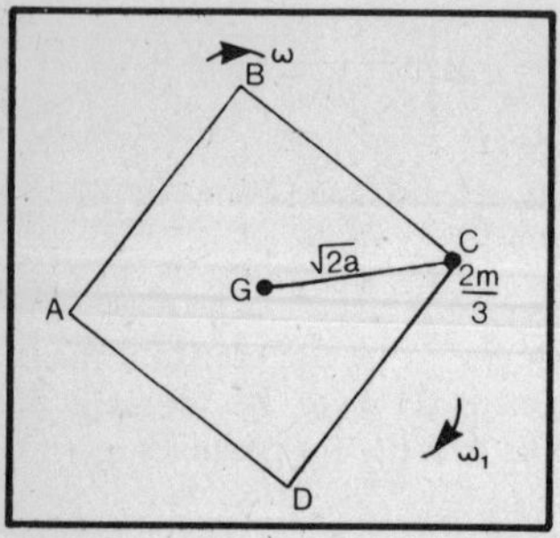

Figure 219.

The impulse, J, which acted on the particle may be measured by 'impulse = change in linear momentum.' Hence,

$$J = \frac{2}{3}m\frac{\omega}{3}\sqrt{2}\,a - 0$$

$$= \frac{\mathbf{2\sqrt{2}}}{\mathbf{9}}\,\boldsymbol{am\omega}.$$

Before impact the only kinetic energy of the system was that of the lamina, i.e.,

$$\frac{1}{2}\left(\frac{2}{3}ma^2\right)\omega^2.$$

After impact, the kinetic energy of the system is,

$$\frac{1}{2}\left(\frac{2}{3}ma^2 + \frac{2}{3}m(\sqrt{2}\,a)^2\right)\omega_1{}^2.$$

Hence, the **loss of kinetic energy** is,

$$\frac{ma^2}{3}\omega^2 - \frac{ma^2}{3}\left(\frac{\omega}{3}\right)^2 - \frac{2}{3}ma^2\left(\frac{\omega}{3}\right)^2 = \frac{\mathbf{2}}{\mathbf{9}}\boldsymbol{ma^2\omega^2}.$$

Example 3 Show that the moment of inertia of a uniform triangular lamina ABC of mass m, with base $AB = a$, and height h, about its base, is

$$\frac{mh^2}{6}.$$

The lamina is free to rotate about a fixed smooth horizontal axis along its base AB. When the lamina hangs vertically at rest, a horizontal impulse J is applied at C in the direction perpendicular to the plane ABC. Find the minimum value of J for the lamina to make a complete rotation.

(A.E.B.)

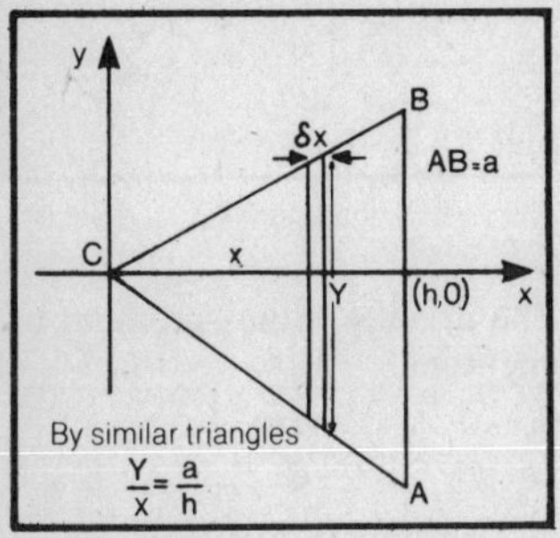

Figure 220.

Let ρ be the constant mass per unit area. The mass of the rod-like element, shown in figure 220, is $Y\,\delta x\,\rho$ and its moment of inertia about Oy is $(Y\,\delta x\,\rho)x^2$.

Hence, the moment of inertia of the lamina about Oy is

$$I_y = \int_0^h Y\rho x^2\,dx.$$

Since $Y = \dfrac{ax}{h}$ and $m = \tfrac{1}{2}ah\rho$,

then $$I_y = \rho\int_0^h \frac{ax^3}{h}\,dx = \frac{a\rho}{h}\frac{h^4}{4} = \frac{mh^2}{2}.$$

We are asked to find the moment of inertia about AB, I_{AB}. Let the moment of inertia about an axis parallel to AB through the centre of gravity be I_G. Hence, from the parallel axes theorem,

$$I_y = I_G + m\left(\frac{2}{3}h\right)^2 \quad \text{and} \quad I_{AB} = I_G + m\left(\frac{h}{3}\right)^2$$

$$I_{AB} = m\frac{h^2}{2} - \frac{4}{9}mh^2 + \frac{mh^2}{9} = \frac{mh^2}{6}.$$

(To find the initial angular velocity after the impulse, J, we could either use 'impulse = change in linear momentum' or 'impulse of torque = change in angular momentum.' The former would involve the instantaneous impulsive horizontal reaction X at the axis and an examination of the motion of the centre of mass, i.e.,

$$J - X = \frac{m2h\omega}{3}.$$

Since we are not asked to find this impulsive reaction, we shall use moment of momentum about the axis of rotation to eliminate X.)

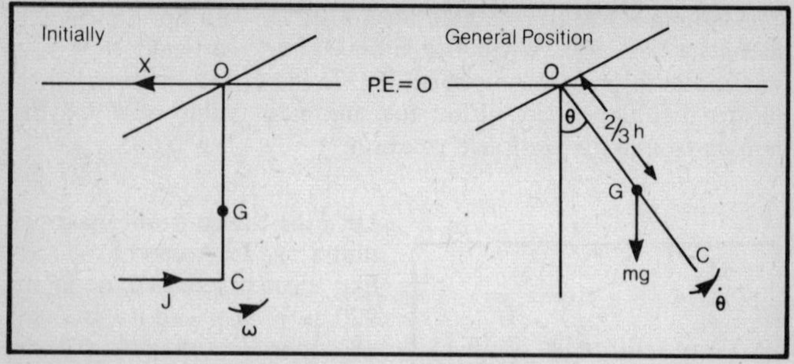

Figure 221.

If ω is the angular velocity after the impulse, from 'moment of impulse = increase in angular momentum,'

$$hJ = I_{AB}\omega - 0 = \frac{mh^2\omega}{6}, \quad \boldsymbol{J = \frac{mh\omega}{6}}.$$

In the **initial** position, taking the zero potential energy line as the horizontal through O, as shown,

then the P.E. $= -mg\frac{2}{3}h$ and K.E. $= \frac{1}{2}\left(\frac{mh^2}{6}\right)\omega^2$.

In the general position, where OC makes an angle θ with the downward vertical,

$$\text{P.E.} = -mg\frac{2}{3}h\cos\theta, \quad \text{and} \quad \text{K.E.} = \frac{1}{2}\left(\frac{1}{6}mh^2\right)\dot{\theta}^2.$$

After the impulsive force, the only external forces acting on the system, are the weight of the body and the reaction at the hinge. The reaction at the hinge does no work since its point of application does not move.

Hence, we are in a conservative system of forces and may use Conservation of Mechanical Energy. Thus,

$$\boldsymbol{-mg\frac{2}{3}h + \frac{1}{2}\frac{mh^2}{6}\omega^2 = -mg\frac{2}{3}h\cos\theta + \frac{1}{2}\frac{mh^2}{6}\dot{\theta}^2.}$$

If the lamina makes a complete revolution, then when $\theta = \pi$, its angular velocity must be greater than zero, i.e.,

$$\dot{\theta}^2 = \frac{12}{mh^2}\left(\frac{2}{3}mgh(-1-1) + \frac{mh^2}{12}\omega^2\right) > 0$$

($\dot{\theta}$ will not be negative).

Hence, $\omega^2 > \dfrac{16g}{h}$, $\omega > 4\sqrt{\dfrac{g}{h}}$ and $J = \dfrac{mh\omega}{6} > \dfrac{2}{3}m\sqrt{gh}$.

Example 4 Figure 222 shows a thin uniform rod OA of length $2a$ and mass m, free to rotate about a fixed horizontal axis through O. The rod is released from rest when it is horizontal. Find its angular velocity when it makes an angle of θ with the downward vertical through O. (The moment of inertia of a rod of length $2a$ and mass m about an axis perpendicular to the rod through its centre is $ma^2/3$.)

Find the magnitudes of the components parallel and perpendicular to the rod of the reaction at the axis. Find θ such that the horizontal component of the reaction of the axis is maximum. Find the radial and transverse components of the reaction for this value of θ.

The moment of inertia of a rod of length $2a$ and mass m, about an axis perpendicular to the rod through an end, is, by the parallel axis theorem,

$$\frac{1}{3}ma^2 + ma^2 = \frac{4}{3}ma^2.$$

We shall apply Conservation of Mechanical Energy to the system (the reaction at the pivot does no work).

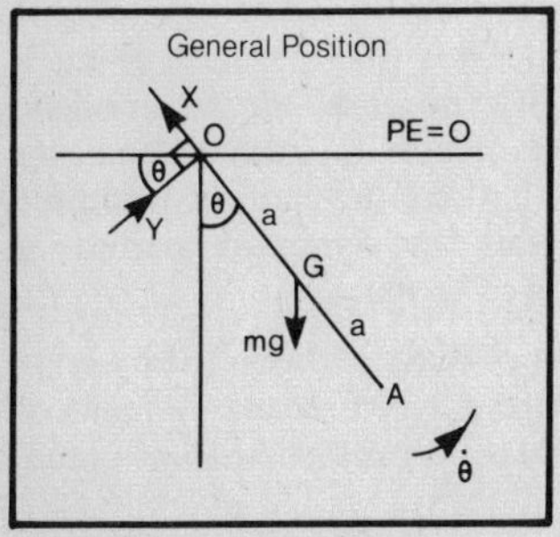

Figure 222.

Initially, P.E. = 0, K.E. = 0.

In the general position,

P.E. = $-mga\cos\theta$ and

K.E. $= \dfrac{1}{2}\left(\dfrac{4}{3}ma^2\right)\dot{\theta}^2$.

Hence, $\dot{\theta} = \sqrt{\dfrac{3g}{2a}\cos\theta}$.

We have seen that the acceleration of the centre of mass of a system, is that which would be produced by the resultant force of the system acting on a particle of the same mass as that of the total mass of the system, placed at the centre of mass.

Hence, parallel to the rod, $X - mg\cos\theta = m\dot{\theta}^2 a$,

and perpendicular to the rod, $Y - mg\sin\theta = ma\ddot{\theta}$.

Since $\dot{\theta}^2 = \frac{3g}{2a}\cos\theta$, then $2\dot{\theta}\,\ddot{\theta} = -\frac{3g}{2a}\sin\theta\,\dot{\theta}$

$$\text{and } \ddot{\theta} = -\frac{3g}{4a}\sin\theta$$

Hence, $\boldsymbol{X} = mg\cos\theta + \frac{3}{2}mg\cos\theta = \frac{5}{2}\boldsymbol{mg\cos\theta}$,

and $\boldsymbol{Y} = mg\sin\theta - \frac{3}{4}gm\sin\theta = \boldsymbol{\frac{mg}{4}\sin\theta}$.

The horizontal component of the reaction is

$$X\sin\theta - Y\cos\theta = \frac{5}{2}mg\cos\theta\sin\theta - \frac{mg}{4}\sin\theta\cos\theta$$

$$= \frac{9}{8}mg\sin\theta\cos\theta.$$

This will be a maximum when $\sin 2\theta$ is a maximum, i.e., when $\theta = 45°$.

In this position, $Y = \boldsymbol{\frac{mg}{4\sqrt{2}}}$ and $X = \boldsymbol{\frac{5mg}{2\sqrt{2}}}$.

Example 5 A rigid body of mass m is free to rotate about a fixed horizontal axis, L. Its radius of gyration about this axis is k, and h is the distance of its centre of gravity from the axis of rotation. Show that for small oscillations the motion is simple harmonic and find its period. Find the minimum period of oscillation for the body if h is allowed to vary.

Figure 224 shows a uniform lamina of mass M bounded by concentric circles of radii a and b where $a > b$. Find its moment of inertia about an axis through its centre perpendicular to its plane.

The lamina is free to rotate in its own plane about a fixed horizontal axis through a point on its surface. Show that the axis which gives the least period of oscillation is at a distance from the centre of the body, of,

$$\sqrt{\frac{a^2 + b^2}{2}}.$$

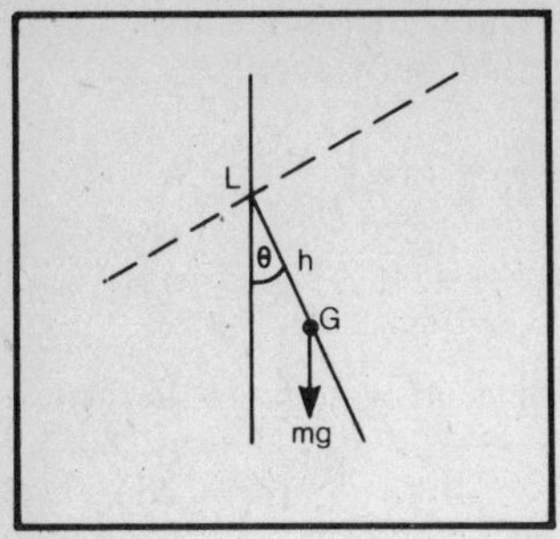

Figure 223.

Figure 223 shows a section of the body, perpendicular to the axis, L, passing through G.

If k is the radius of gyration of the body about an axis parallel to L through G, then its moment of inertia about L is $m(k^2 + h^2)$ where m is the mass of the body.

Taking moments about L, and applying '$C = I\ddot{\theta}$.'

$$-mgh \sin \theta = m(k^2 + h^2)\ddot{\theta}.$$

For small oscillations, $\sin \theta \simeq \theta$, therefore,

$$\ddot{\theta} = -\frac{gh}{(k^2 + h^2)}\theta,$$

which is an equation of motion for **angular simple harmonic motion** of period

$$T = 2\pi\sqrt{\frac{k^2 + h^2}{gh}}.$$

The minimum value of T occurs when $\sqrt{\dfrac{k^2 + h^2}{h}}$ is a minimum.

Let $P = \dfrac{k^2 + h^2}{h}$. Then $\dfrac{dP}{dh} = -\dfrac{k^2}{h^2} + 1$.

$\dfrac{dP}{dh} = 0$ when $k^2 = h^2$, $k = h$.

Since $\dfrac{d^2P}{dh^2} = \dfrac{2k^3}{h^3} > 0$ when $k = h$,

then $k = h$ gives a **minimum** value of P.

Hence, the **period of oscillation is minimum when $h = k$** and is $\sqrt{2h}$.

Let the constant mass per unit area of the body be ρ. Taking ring-like elements distance r from the centre O and of thickness δr, the moment of inertia of an element about XY is $2\pi r\ \delta r\ \rho r^2$.

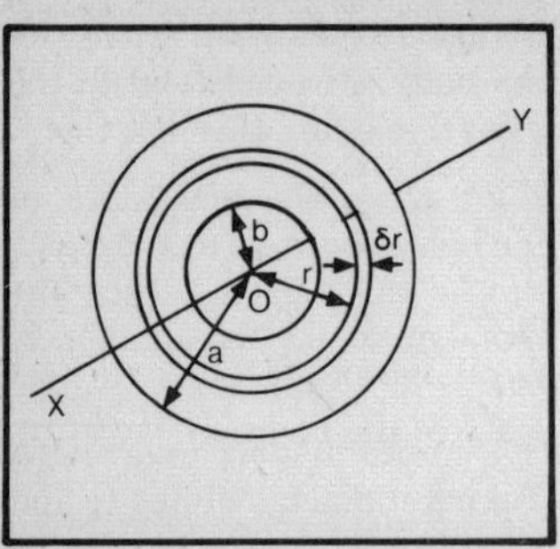

Figure 224.

Hence, the moment of inertia of the body about XY is.

$$I_{XY} = 2\pi\rho \int_b^a r^3 \, dr$$

$$= \frac{2\pi\rho}{4}(a^4 - b^4).$$

Since $M = \pi(a^2 - b^2)\rho$ then

$$\boldsymbol{I_{XY} = \frac{M}{2}(a^2 + b^2)}.$$

We have shown in the first part of the question, that the period of oscillation will be least when the distance between the centre of gravity and the point of suspension, is equal to the radius of gyration, k', of the body about an axis through its centre of gravity, parallel to the axis of rotation.

The axis of rotation is perpendicular to the body since it rotates in its own plane. Hence, the radius of gyration, k', is given by

$$Mk'^2 = \frac{M}{2}(a^2 + b^2).$$

Thus, $\boldsymbol{k' = \sqrt{\frac{(a^2 + b^2)}{2}}}$.

This is the distance from the centre of the body of the point of suspension for **minimum period of oscillation**.

Example 6 A wheel has a horizontal cylindrical axle of radius a. The system of the wheel and axle has mass M and radius of gyration k about its axle. The axis is smooth. A light thin inextensible string is wound around the axle, with one point attached to a point on the axle, and the other attached to a particle of mass m. The string lies in a vertical plane with part of the string hanging vertically. The system is released from rest. Find the acceleration of the particle.

When the particle has descended distance $8a$ it falls off and the wheel is brought to rest in n revolutions by a constant braking couple of magnitude C. Find C.

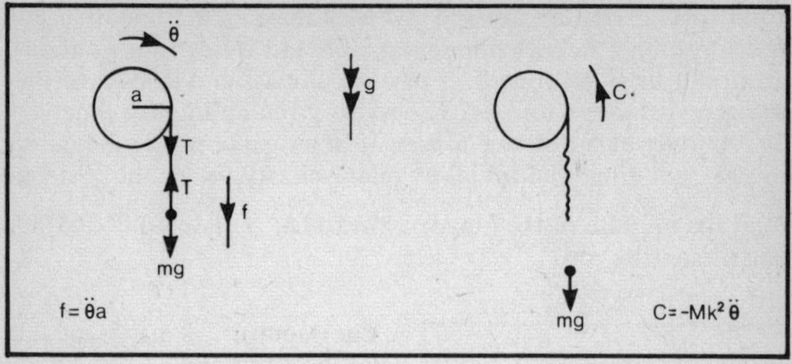

Figure 225.

The moment of inertia, I, of the wheel and axle is Mk^2. The equation of motion for the wheel is '$C = I\ddot{\theta}$,' hence, $aT = Mk^2\ddot{\theta}$.

Since the string cannot slip on the axle, then $f = \ddot{\theta}a$. The equation of motion for the particle, from Newton's second law, is $mg - T = mf$.

Therefore, the **acceleration of the particle, *f*, is**

$$\frac{a^2mg}{a^2m + Mk^2}$$

When the particle has descended distance $8a$ it will have velocity, v, given by '$v^2 = u^2 + 2as$.'

Therefore, $$v^2 = \frac{16a^3mg}{a^2m + Mk^2},$$

and the wheel will have angular velocity ω

$$= \frac{1}{a}\sqrt{\frac{16a^3mg}{a^2m + Mk^2}}.$$

The particle falls off and the wheel is brought to rest by constant retarding couple C in n revolutions. Its angular acceleration $\ddot{\theta}$ is given by $C = -Mk^2\ddot{\theta}$, and applying '${\omega_2}^2 = {\omega_1}^2 + 2\alpha\theta$,'

$$0 = \frac{1}{a^2}\left(\frac{16a^3mg}{a^2m + Mk^2}\right) - \frac{2C2\pi n}{Mk^2}$$

Thus, $$C = \frac{4mMk^2ag}{\pi n(Mk^2 + ma^2)}.$$

Example 7 A uniform rod AB is of mass m and length $2a$. It lies at rest on a smooth horizontal table and is free to turn about a smooth fixed pivot at A. A particle of mass $m/3$ is fixed to the rod at a distance c from A. The rod is given an impulse J acting horizontally at a distance d from A in a direction perpendicular to the rod. Find the initial angular velocity, ω, of the system.

Find the impulse of the hinge on the rod at A and impulse of the particle on the rod.

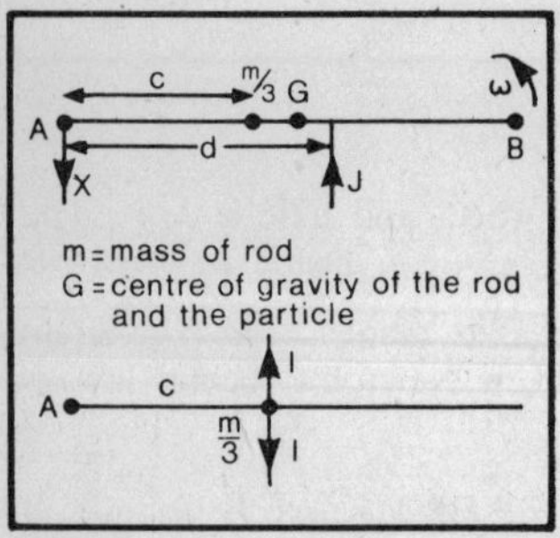

Figure 226.

The moment of inertia of the uniform rod about A is

$$\frac{4}{3}ma^2.$$

The moment of inertia of the rod and particle about A is

$$\left(\frac{4}{3}ma^2 + \frac{mc^2}{3}\right).$$

Since 'moment of impulse = change in angular momentum,'

then $$dJ = \left(\frac{4ma^2}{3} + \frac{mc^2}{3}\right)\omega.$$

Hence, $$\omega = \frac{3dJ}{m(4a^2 + c^2)}.$$

The impulsive reaction, X, on the hinge at A will be perpendicular to the rod. The centre of gravity, G, of the rod and the particle is given by

$$\frac{4}{3}mAG = \frac{cm}{3} + ma, \quad \text{i.e.,} \quad AG = \frac{(c + 3a)}{4}.$$

The linear motion of the system corresponds to that produced by the resultant external forces acting on the total mass at the centre of gravity. Applying 'Impulse = change in momentum,'

then $$J - X = \frac{4}{3}m\left(\frac{c + 3a}{4}\right)\omega.$$

Thus, $X = J - \frac{m}{3}(c + 3a)\omega,$

where $\omega = \frac{3dJ}{m(4a^2 + c^2)}.$

The rod and the particle will impart equal and opposite internal impulsive reactions, I, on each other. Using 'moment of impulse = change in angular momentum' for the particle,

$$cI = \frac{mc^2}{3}\omega.$$

Thus, $I = \frac{mc}{3}\omega,$

where $\omega = \frac{3dJ}{m(4a^2 + c^2)}.$

Example 8 A uniform circular lamina of mass m and radius a is rotating freely in a vertical plane with angular velocity Ω about a fixed horizontal axis through its centre O. P is a point on the circumference of the lamina. When OP is horizontal, and P is moving upwards, a particle of mass m adheres to the point P. The particle was previously at rest. If the lamina just comes to rest with P vertically above O, find Ω.
Find the horizontal and vertical components of the reaction of the axis on the lamina immediately after the particle is picked up.

The moment of inertia of the uniform circular lamina of mass m and radius a about the axis perpendicular to its plane through its centre is $ma^2/2$.

When the particle adheres to P, the lamina and the particle will exert **equal and opposite impulses** on each other. Hence, **moment of momentum will be conserved**.

Refer to figure 227.

The moment of inertia, I, of the lamina and the particle together, about the fixed axis, is

$$\frac{ma^2}{2} + ma^2 = \frac{3}{2}ma^2.$$

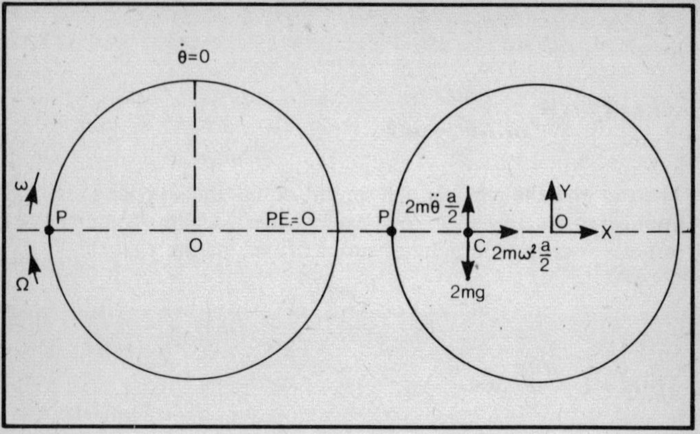

Figure 227.

Let ω be the angular velocity immediately after impact, then,

$$\frac{ma^2}{2}\Omega = \frac{3}{2}ma^2\omega.$$

Thus, $$\omega = \frac{\boldsymbol{\Omega}}{\mathbf{3}}.$$

Since the reaction at the axis does no work after the particle has adhered to the lamina, we are in a conservative system of forces. Hence, the **mechanical energy is conserved**.

Initially, P.E. = 0, K.E. = $\frac{1}{2}\left(\frac{3}{2}ma^2\omega^2\right)$.

In the position where P is vertically above O, and the lamina is at rest, P.E. = mga, K.E. = 0.

Hence,

$$\frac{3}{4}ma^2\omega^2 = mga, \quad \omega^2 = \frac{4g}{3a}$$

and $$\Omega^2 = \frac{12g}{a}, \quad \boldsymbol{\Omega} = \mathbf{2}\sqrt{\frac{\mathbf{3g}}{\mathbf{a}}}.$$

The **centre of gravity** of the system, C, is such that,

$$2mgOC = amg, \quad OC = \frac{a}{2}.$$

The angular acceleration, α, of the system immediately after impact is given by '$C = I\ddot{\theta}$' so that

$$-amg = \frac{3}{2}ma^2\alpha \quad \text{and} \quad \alpha = -\frac{2g}{3a}.$$

The acceleration of the centre of mass is the same as if the total mass were concentrated there, and the resultant force acted there. Thus,

horizontally, $X = 2m\left(\omega^2 \frac{a}{2}\right)$,

and **vertically,** $Y - 2mg = 2m\left(\alpha \frac{a}{2}\right)$.

Therefore, the horizontal component of reaction at the instant immediately after the particle is picked up

is $$\boldsymbol{X = \frac{4}{3}mg}$$

and the vertical component is

$$\boldsymbol{Y = \frac{4}{3}mg.}$$

(If we were asked to find, for more than one position of the lamina, the reaction at the axis, then we should write the **general** equations of motion and substitute particular values into them.)

Key terms

The **moment of inertia**, I, of a body about a fixed axis is a measure of its reluctance to accept change in its rotational motion, in a similar way to a body's mass being a measure of its resistance to change in its linear motion. $\boldsymbol{I = \sum mr^2}$, with the usual notation.

The moments of inertia of some standard bodies are given at the end of this chapter.

The radius of gyration, k, of a body about an axis is the distance from the axis that a particle of the same mass, m, as that of the body, would have to be placed so that it had the same moment of inertia, I, as the body, i.e., $I = mk^2$.

The parallel axis theorem states that if the moment of inertia of a body of mass m about an axis through its **centre of mass** is mk^2 then its moment of inertia about a **parallel** axis distant d from the first axis is $m(k^2 + d^2)$.

The perpendicular axes theorem states that if the moment of inertia of a **lamina** about two perpendicular axes **in the plane of the lamina** are I_a and I_b then the moment of inertia about the axis **perpendicular** to the plane of the lamina, **passing through the point of intersection** of the first two axes is $I_a + I_b$.

When a torque C acts on a body of moment of inertia I producing an angular acceleration $\ddot{\theta}$ then $\boldsymbol{C = I\ddot{\theta}}$.

The **kinetic energy of a rotating body** of moment of inertia I and moving with angular velocity $\dot{\theta}$ is $\frac{1}{2}I\dot{\theta}^2$.

The **angular momentum**, or **moment of momentum** of a body is $\boldsymbol{I\dot{\theta}}$.

The **rate of change of angular momentum** of a body rotating about a fixed axis is equal to the **resultant torque acting on the body**, i.e., $C = I\ddot{\theta}$.

The principle of conservation of angular momentum states that if there is **no resultant torque** acting on a body then the **total moment of momentum remains constant.**

The **impulse of a torque**, C, is given by

$$\int_0^t C\,dt$$

and is equal to the **moment of the resultant impulse** and is measured by the **increase in angular momentum it produces**, i.e.,

'**moment of impulse = impulse of torque = increase in angular momentum.**'

The work, W, done by a couple, C, is given by

$$W = \int_0^\alpha C\,d\theta.$$

A compound pendulum is a rigid body free to swing about a smooth fixed axis. For **small oscillations** the motion is approximately **angular simple harmonic**.

The length of the equivalent simple pendulum is the length of the simple pendulum which would have the **same period** of motion.

To find the **force exerted by the axis on a rotating rigid body**, we use the fact that **the acceleration of the centre of mass of a system, is the same as that of a single particle whose mass is the total mass of the system acted on by the resultant of the forces acting on the system**. We write down the **equations of motion** for the particle at the centre of mass in the **radial** and **transverse** directions.

Moments of inertia of some standard bodies

Body, mass m	**Axis**	I
Rod, length $2a$	Perpendicular to rod, through centre.	$\frac{ma^2}{3}$
Rod	Parallel to rod, distant d from it.	md^2
Rectangular lamina, length $2a$, width $2b$	Parallel to sides of length $2a$, passing through centre.	$\frac{mb^2}{3}$
Ring of radius a	Perpendicular to plane of ring, through centre.	ma^2
Disc of radius a	Perpendicular to disc, through centre.	$\frac{ma^2}{2}$
Solid sphere, radius a	A diameter	$\frac{2}{5}ma^2$
Hollow sphere, radius a	A diameter	$\frac{2}{3}ma^2$
Solid cylinder, radius a	The axis	$\frac{ma^2}{2}$
Hollow cylinder, radius a	The axis	ma^2
Solid cone, right circular, radius a, height h	The axis	$\frac{3}{10}ma^2$

Index